AF327485

Workshop on

DIQUARKS III

Workshop on

DIQUARKS III

Villa Gualino, Torino, Italy
28 – 30 October 1996

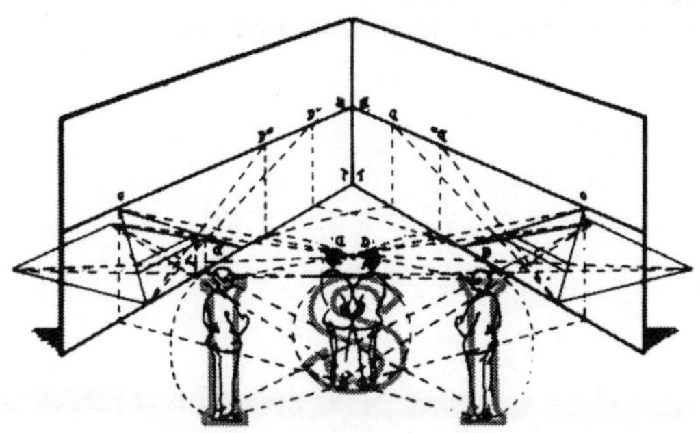

Edited by **ISI:**

Mauro Anselmino
Enrico Predazzi
University of Torino

Published by

World Scientific Publishing Co. Pte. Ltd.

P O Box 128, Farrer Road, Singapore 912805

USA office: Suite 1B, 1060 Main Street, River Edge, NJ 07661

UK office: 57 Shelton Street, Covent Garden, London WC2H 9HE

British Library Cataloguing-in-Publication Data
A catalogue record for this book is available from the British Library.

DIQUARK III

Copyright © 1998 by World Scientific Publishing Co. Pte. Ltd.

ISBN 981-02-3635-2

This book is printed on acid-free paper.

Printed in Singapore by Uto-Print

CONTENTS

Preface   vii

Historical and Sentimental Remarks on Quarks and Diquarks   1
    D. Lichtenberg

Exclusive Final States in Two Gamma Collisions at LEP   7
    T. Barillari

The Nucleon Electromagnetic Form Factors in the Time-like
Region: New Results from the Fenice Experiment   17
    C. Bini

Nucleon and Δ in a Covariant Quark-Diquark Model   27
    V. Keiner

Baryon Structure in a Covariant Diquark-Quark Model   36
    G. Hellstern, R. Baürle, U. Zückert, R. Alkofer
    and H. Reinhardt

Exclusive Photo- and Electro-production of Mesons in the
GEV Region   46
    G. Folberth, R. Rossmann and W. Schweiger

The $\pi\gamma$ Transition Form Factor and Its Impact on
Charmonium Decays into Two Pions   56
    P. Kroll

Production of Meson Pairs Involving $L \neq 0$ Mesons
in Photon-Photon Collisions   70
    L. Houra-Yaou, P. Kessler and J. Parisi

Covariant Formulation of Light Front Dynamics and
Relativistic Systems   81
    J.-F. Mathiot

Hadron Structure and Odderon Exchange   90
    M. Rueter

vi

Cosmic Diquarks   99
S. Fredriksson

The Search for Chargino-Neutralino Pairs with the
ATLAS Detector at the LHC   109
S. Muanza

Physics at LISS   116
L. C. Bland

The Study of the Elementary Photo- and Electro-production
of Kaons at Jefferson Lab   128
M. Iodice, E. Cisbani, S. Frullani, F. Garibaldi,
G. M. Urciuoli, R. De Leo, R. Perrino and M. Sotona

The Physics at ELFE   141
J. M. Laget

Diquark Model of Exotic Mesons   146
D. B. Lichtenberg, R. Roncaglia and E. Predazzi

Diquark Model of Dibaryons   156
D. B. Lichtenberg, R. Roncaglia and E. Predazzi

Which Scalar Meson is the Glue-State?   161
M. Boglione

Study of Multiquarks System in a Chiral Quark Model   169
M. Genovese, J.-M. Richard, S. Pepin and Fl. Stancu

Heavy Baryons in a Quark-Diquark-Picture   179
D. Ebert, R. N. Faustov and V. O. Galkin

Higher Twist Effects in DIS: An Outline of Diquark Contributions   188
M. Anselmino and F. Caruso

The Proton Spin Structure in a Light-Cone
Quark-Spectator-Diquark Model   198
B.-Q. Ma

Quark Distribution Functions in a Diquark Spectator Model 208
    P. J. Mulders and J. Rodrigues

Spin-dependent Fragmentation Functions for Baryons in a
Diquark Model 218
    A. D. Adamov and G. R. Goldstein

Fast Antibaryon Production in pp Collisions and
String Fusion 228
    N. Armesto, E. G. Ferreiro, C. Pajares and
    Yu. M. Shabelski

Non-Factorizable Contributions to the Large Rapidity
Gap at HERA 238
    R. Fiore, L. L. Jenkovszky, F. Paccanoni and
    E. Predazzi

Medium Effects in the Production of Hadronic Resonances on
Nuclei (Mass Shift and Collision Broadening) 246
    L. A. Kondratyuk and Ye. S. Golubeva

PREFACE

The Third Workshop on Diquarks was held in Torino, Villa Gualino, on October 28-30, 1997, eight years after the first and four years after the second.

Although, initially, we had some reluctance to organize a third Workshop on this subject given how much our knowledge of QCD has progressed in the last few years, it was, finally, judged that the notion of diquarks is still sufficiently useful and vital and that many short cuts induced by diquarks can still be very efficient in order to simulate non perturbative effect in the intermediate energy region (from a few GeV to some 10-20 GeV) where a whole new generation of machines are planned and will, hopefully, become operative in the near future (upgraded CEBAF, ELFE, LISS, etc.). *A posteriori*, we believe that it was the correct decision.

The Meeting was made possible by the joint sponsorship of the Istituto Nazionale di Fisica Nucleare (INFN), of the University of Torino and of the Institute for Scientific Interchanges (ISI) that have provided the financial support and the technical support for both holding the Workshop and publishing the Proceedings. To these Institutions go our warmest thanks.

We thank all the participants and in particular, those who have submitted written contributions for the Proceedings. These contributions are presented in the same order in which they were presented at the Meeting.

M. Anselmino
E. Predazzi

HISTORICAL AND SENTIMENTAL REMARKS ON QUARKS AND DIQUARKS

DON LICHTENBERG

*Physics Department, Indiana University, Bloomington,
IN 47405, USA*

A little of the lighter side of quark and diquark history is presented. Also, some brief remarks are made about the present and future of diquarks.

1   The Two Cultures

The twentieth century English novelist C.P. Snow called attention to the two cultures in modern life: the culture of those who are scientists and of those who are not. He implied that each culture is largely ignorant of the other. Although I agree that non-scientists are often woefully illiterate in matters of science, I like to think that scientists are often well versed in the arts.

In this talk I should like to contribute to enhancing the literary and historical knowledge of my audience, thereby helping to bridge the gap between the two cultures.

2   Quarks and Diquarks

The idea of fractionally-charged hadronic constituents was supposedly invented independently in 1964 by Murray Gell-Mann, who named them "quarks," and by George Zweig, who named them "aces." Zweig did not publish his paper at the time. We all know that Gell-Mann's name, quarks, prevailed. (It was not even close.)

In Gell-Mann's original paper introducing quarks,[1] entitled, "A schematic model of baryons and mesons," there is the sentence (I have omitted a few irrelevant words):

"We...refer to the members...of the triplet as 'quarks'[6])"

Reference 6 reads in part as follows:

"James Joyce, Finnegan's Wake..."

Let me point out a minor mistake in Gell-Mann's footnote. Joyce's novel actually has the title *Finnegans Wake* without the apostrophe. Gell-Mann will have to take consolation from the fact that he was right gramatically.

The quote from *Finnegans Wake* is

"Three quarks for Muster Mark!"

What can that mean? I looked up the word "quark" in the unabridged Oxford English dictionary and found the verb "quark," an obsolete word meaning to "croak." The noun is listed as "quarking," as in the phrase, "The quarking of the frogs." To my knowledge, Joyce was the first to use "quark" as an English noun.

However, there is a German word, "quark," which is a kind of cheese. At the present time, quark is even sold in some cities in the USA, and I have an empty quark carton here in front of me. Because quark is a modest, mild sort of cheese, some Germans use the word "quark" to mean a trifle or something worthless. Joyce was fond of making puns in more than one language, and I guess that he knew the German, as well as the English meaning of the word.

Gell-Mann also was the first to mention a diquark. A footnote to his 1964 quark paper reads:

"*There is the alternative possibility that the quarks are unstable under decay into baryon plus anti-di-quark or anti-baryon plus quadri-quark. In any case some particle of fractional charge would have to be absolutely stable."

The quarks have the whimsical names *up, down, strange, charmed, bottom*, and *top*. It is easy to find thousands of literary references to one of these at a time. For example, consider *bottom*. Shakespeare said in *Midsummer Night's Dream*:

"...it shall be called Bottom's Dream, because it hath no bottom"

I find this a particularly good quote because in Europe there is no bottom quark, but rather a *beauty* quark.

It is a little more difficult to find literary references in which a single sentence contains the names of both quarks belonging to a single family. I have found several examples, and shall mention three of them here, one for each of the three families.

W.R. Mandale, a nineteeth century English writer, wrote

"Up and down the City Road
In and out the Eagle
That's the way the money goes—
Pop goes the weasel!"

For the second family, I have a quote from Henry James, an American novelist of the late nineteeth and early twentieth century. He wrote in *Portrait of a Lady*:

"...everything she now saw had all the charm of strangeness"

And, for the third family, I have a quote from the American composer Cole Porter, who died a little over 30 years ago. In the musical *Anything Goes*, there are the lines

"But if baby
I'm the bottom
You're the top"

I did not try to find a quote with top and beauty.

3 A short review

I am pleased to have had a part in writing a review of diquarks that appeared in 1993,[2] the same year that the Superconducting Super Collider was canceled. The demise of the SSC signaled a shift of the center of gravity of research in experimental high energy physics from America to Europe. Our diquark review had four European authors and only one American, signalling a comparable shift of research in theoretical high energy physics to Europe.

Useful collections of papers on diquarks can be found in the proceedings of the first Torino diquark meeting in 1988[3] and the second in 1992.[4] Sverker Fredriksson points out that, so far, these conferences have appeared in Olympic years. For what it's worth, these are also years of election of American presidents. Also in 1992 there was a conference in Bad Honnef, Germany, on related subjects, and the proceedings have been published.[5]

As far as I know, the first people to use diquarks in a model of baryons were Ida and Kobayashi in 1966.[6] The following year Tassie and I[7] independently used a diquark-quark model for baryons, calling it a "boson-fermion" model. In our paper we pointed out that the low-mass mesons could still each be made of quark and antiquark, but mesons of higher mass might be made of a diquark and antidiquark. We did not consider such (exotic) mesons because at that time there was no evidence for them. (The situation hasn't changed all that much to the present.)

In the same year that Ida and Kobayashi introduced diquarks, Miyazawa[8] introduced an algebraic scheme of supersymmetry, which in effect postulated a broken supersymmetry between a diquark and an antiquark. The words "supersymmetry" and "diquark" do not appear in Miyazawa's paper, but he still should be given credit for the idea. The situation is similar to that of the old OZI (Okubo, Zweig, Iizuka) rule, which is thought of today in terms of quarks. But Okubo's version was algebraic, with no mention of quarks. This fact is not surprising, in view of the fact that Okubo's paper appeared in 1963, the year before quarks were invented. As far as I know, Miyazawa was the first

4

to introduce the idea of supersymmetry into particle physics. It is now called "hadron supersymmetry" to distinguish it from the more fundamental supersymmetry of elementary particles proposed some years later. We know that broken hadron supersymmetry exists, but do not yet know about elementary particle supersymmetry.

In the first few years after the early papers on diquarks, the subject did not receive much attention. However, in the mid 70s, the subject picked up. In 1975 Eguchi[9] proposed a quark-diquark structure for some baryons in a string model, and the following year Johnson and Thorn[10] made a similar proposal in a bag model. Rosenzweig[11] again brought up the idea of diquark-antidiquark mesons and extended the idea to diquarks containing charmed quarks.

Also in the mid 70s there appeared an important paper on diquarks by Pavkovic.[12] To my knowledge, this was the first paper to use diquarks in the parton model and make applications to high-energy scattering, hadron form factors, and scaling. In his acknowledgements, Pavkovic attributes the idea to Bjorken.

In the 1980s, lots of papers appeared on diquarks, and work in this field is still going on today. Using the concept of diquarks seems to help us understand lepton-nucleon and hadron-hadron collisions, electron-positron annihilation, some problems in nuclear physics, some exclusive processes, and hadron supersymmetry. I apologize to the authors of the many important papers on these subjects for not giving references. Just look at our review.[2]

4 Present and Future

Clearly, in the few minutes left I cannot summarize the present state of our understanding of diquarks. I just want to give a personal qualitative answer to a still controversial question: How big are diquarks? I also want to give a qualitative answer the related question: How heavy are diquarks? For simplicity, I consider only diquarks made of light quarks.

It is useful first to consider the comparable problem for quarks. We know that at least two different concepts of quarks are useful: current quarks and constituent quarks. I like to think of a constituent quark as a current quark dressed with a sea of gluons and quark-antiquark pairs. This sea adds size and mass to the quark. If one strikes a quark gently with a probe, the quark will move adiabatically, dragging along the sea with its extra size and inertia. Such a probe can measure the large mass and extended size of a constituent quark. (Some theoretical input is needed.) On the other hand, if one strikes a quark with an energetic probe, giving the quark a large momentum transfer, the quark will move quickly, leaving most of its sea behind. Such a probe can measure the small mass and pointlike size of a current quark.

The situation is similar for diquarks. The size and mass of a diquark depend on how you look at it. However, there is the additional complication that if you probe a diquark too violently, it will break up, so that the very concept of a diquark is most useful at energies and momentum transfers which are not too high.

I hope that the these remarks will lessen, if not end, the controversy about the mass and size of diquarks.

At present we use diquarks as input to calculations to model some of the effects of nonperturbative QCD. I do not know what the future will bring, but I hope that this situation will change. I hope that some time in the future, physicists will be able to calculate with non-perturbative QCD (or with some successor theory), and two-quark correlations (diquarks) will be the output rather than the input. But I do not foresee such a situation developing very soon, and so I predict with some confidence that the use of diquarks as input to model calculations will be fruitful for a considerable time to come.

Acknowledgments

I should like to thank the organizers of the conference, Mauro Anselmino and Enrico Predazzi, for preparing what promises to be a stimulating meeting, as well as for inviting me to participate. Also, I should like to thank those members of the audience whose remarks in the discussion period following my talk caused me to add a few short items to the written version of this history.

References

1. M. Gell-Mann, Phys. Lett. **8**, 214 (1964).
2. M. Anselmino, E. Predazzi, S. Ekelin, S. Fredriksson, and D.B. Lichtenberg, Rev. Mod. Phys. **65**, 1199 (1993).
3. *Workshop on Diquarks*, ed. by M. Anselmino and E. Predazzi, World Scientific, Singapore (1989).
4. *Workshop on Diquarks II*, ed. by M. Anselmino and E. Predazzi, World Scientific, Singapore (1994).
5. *Quark Cluster Dynamics*, ed. by K. Goeke, P. Kroll, and H.-R. Petry, Springer-Verlag, Berlin (1993).
6. M. Ida and R. Kobayashi, Prog. Theor. Phys. **36**, 846 (1966).
7. D.B. Lichtenberg and L.J. Tassie, Phys. Rev. **155**, 1601 (1967).
8. H. Miyazawa, Prog. Theor. Phys. **36**, 1266 (1966).
9. T. Eguchi, *Phys. Lett.* B **59**, 457 (1975).

10. K. Johnson and C.B. Thorn, *Phys. Rev.* D **15**, 1934 (1976).
11. C. Rosenzweig, *Phys. Rev. Lett.* **36**, 697 (1976).
12. M.I. Pavkovic, *Phys. Rev.* D **13**, 2128 (1976).

EXCLUSIVE FINAL STATES IN TWO GAMMA COLLISIONS AT LEP

T. BARILLARI

OPAL Collaboration - University of Bologna, Italy

A variety of interesting physics can be studied in the exclusive final states in two gamma collisions at LEP. Preliminary results are now available, in OPAL, for $\gamma\gamma \to p\bar{p}$ events at LEP1 and for LEP running at a center-of-mass energy of 161 GeV. For $\gamma\gamma \to \Lambda\bar{\Lambda}$ more statistic is needed; it seems that the appropriate statistics can be obtained with the data taking at LEP2. The studies can then be extended to the other baryons, Ξ^-, Σ^0, etc. L3 has studied the formation of resonances in two-photon interactions at LEP1. The radiative width of many states: $\eta^{'}$, a_2 , $f_2^{'}$, η_c and χ_{c2} have been measured with good accuracy. The helicity status of the a_2 has been determined.

1 Introduction

LEP, the Large Electron Positron collider operating at CERN (Geneva), has been designed to provide collision points at four large general purpose detectors: ALEPH, DELPHI, L3 and OPAL. In the LEP1 phase (1989-1995) it has been delivered an integrated Luminosity of approximately 140 pb^{-1} per experiment at center-of-mass energies close to 91 GeV. In the LEP2 phase, it is expected to reach center-of-mass energies up to 190 GeV and deliver a total integrated Luminosity of approximately 500 pb^{-1}. The main purpose of the LEP1 program was to study the Z^0 boson. The aim of LEP2 is to measure and study the W^+W^- pair production and to search for new particles such as the Higgs boson and supersymmetric particles.

A by-product of colliding electrons and positrons is two-photon physics: each electron or positron can emit a virtual space-like photon and the interaction between the two photons can be observed. In the annihilation process (fig. 1a), the whole event may be visible in the detectors and typically 20 charged particles are recorded. In two-photon processes instead a large part of the incident energy is taken by the scattered electrons which often escape undetected inside the beam pipe. The final state X (fig. 1b) has a low mass and the decay particles are few and with low momentum. The cross-section for $e^+e^- \to e^+e^-X$, where the system X is produced in the scattering of two virtual photons $\gamma^*\gamma^* \to X$, grows like $(ln\ s/m^2_{electron})^2$ whereas the one-photon annihilation cross-section decreases like s^{-1}. Thus the $e^+e^- \to e^+e^-X$ processes become increasingly important at higher energy colliders (fig. 2). The relevant kinematic variables for the two-photon processes are defined in fig. 3. Since the photon propaga-

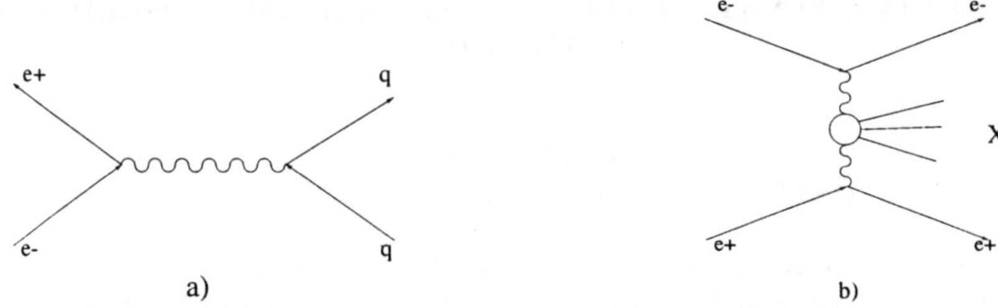

Figure 1: a) One photon annihilation $e^+e^- \to q\bar{q}$. b) Two photon exchange $e^+e^- \to e^+e^-X$

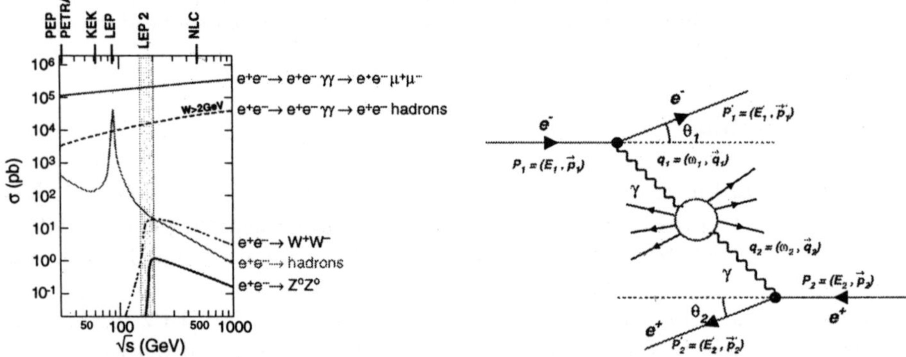

Figure 2: Cross sections as function of $\sqrt{s}$ for various processes measured in e^+e^- colliders

Figure 3: Kinematic of the two-photon process.

tors naturally peak at $Q^2=0$, the majority of the events involve almost real photons. Two-photon events can be classified according to the detection of the scattered electrons. If no electron is detected one has the 0-tag (untagged) events which form the major part of the data. When one (two) of the scattered electrons is (are) detected one has the single (double) tag events.

A typical detector built for LEP is the OPAL detector,[1] which is a multipurpose apparatus. It has been designed to provide measurements of charged particles and electromagnetic energy over nearly the full solid angle. It consists of a system of central tracking chambers inside a solenoid which provides a uniform magnetic field of 0.435 T. The solenoid coil is surrounded by a time-of-flight counter array, an electromagnetic calorimeter made of lead glass, an hadron calorimeter made of iron and wire chambers and muon detectors. Electromagnetic detectors at small polar angles are used to measure the integrated luminosity (L) and to identify scattered electrons.

In general all LEP detectors have a good solid angle coverage and they can cover a large Q^2 range with various tagging systems; for example in L3 at 170 GeV one can use the Very Small Angle Tagger (VSAT, $0.2 \leq Q^2 \leq 0.8\ GeV^2$), the Luminosity monitor (LUMI, $0.2 \leq Q^2 \leq 0.8\ GeV^2$), and the endcap Electromagnetic Calorimeter (ECAL, $Q^2 \geq 40\ GeV^2$).

In $e^+e^- \rightarrow e^+e^- X$ untagged events the observed system X has a small transverse momentum. The low multiplicity and the low momentum characterizing the $\gamma\gamma$ events are a challenge for the trigger of the LEP experiments, since they were mainly designed for high-energy events.

2 Exclusive processes

At LEP it is possible to study $\gamma\gamma$ exclusive processes, such as:

- $\gamma\gamma \rightarrow \pi^+\pi^-$

- $\gamma\gamma \rightarrow K^+K^-$

- $\gamma\gamma \rightarrow B\overline{B}$ where $B = \Lambda$, p, Ξ^-

- $\gamma\gamma \rightarrow 4\pi + n\pi^0$

- $\gamma\gamma \rightarrow VV$ where $V = \rho$, ω, J/Ψ

- $\gamma\gamma \rightarrow R$ where R = Resonances $= \eta^{'}$, f_2, a_2, $f_2^{'}$, η_c, χ_{c_2}...

For high masses of the $\gamma\gamma$ system these channels are a test of QCD models.[2,3]

2.1 Untagged $\gamma\gamma \rightarrow p\overline{p}$ events

It is interesting to study the baryon antibaryon production in the reaction $\gamma\gamma \rightarrow B\overline{B}$. Measurements of the production cross-section of such events can be compared with analytic calculations,[4] for a sufficiently high invariant mass of the two photon system, $W_{\gamma\gamma}$.

The reaction $\gamma\gamma \rightarrow p\overline{p}$ has been studied at various experiments.[5] The best measurement have been performed at CLEO,[6] a general purpose detector [10] located at the e^+e^- storage ring CESR [11] operating at $\sqrt{s} \sim 10.58$ GeV. The two-photon production cross-section for $\gamma\gamma \rightarrow p\overline{p}$, measured by the CLEO experiment, was consistent with the prediction of the quark-diquark model[4−7−8].

An interesting range between 2.0 GeV $< W_{\gamma\gamma} <$ 3.5 GeV can be explored at

LEP in the OPAL experiment.

In light of the studies made with the PC Monte Carlo,[9] a criteria to select exclusive untagged $\gamma\gamma \to p\bar{p}$ events, has been developed. The cuts have been optimised to select the interesting events, while rejecting background events such as beam wall interactions and cosmics rays.

It has been mainly required to observe two good tracks with opposite charge. To eliminate part of the large number of low mass $\gamma\gamma \to e^+e^-$ and $\gamma\gamma \to \mu^+\mu^-$ events, it has been required that the dE/dx probability of each track to be a proton or an antiproton had to be greater than 0.1 %. The estimate of the $\gamma\gamma \to p\bar{p}$ events seen with these loose selection criteria is of about 200 such events for the LEP1 data with a luminosity of 96 pb^{-1}.

The number of events selected by the exclusive untagged $\gamma\gamma \to p\bar{p}$ selection are given in table 1 together with all the other experimental results on this subject. In the table it is also reported the estimated number of $\gamma\gamma \to p\bar{p}$

Table 1: List of experiments that have studied the reaction $\gamma\gamma \to p\bar{p}$. The number of $\gamma\gamma \to p\bar{p}$ events estimated for LEP1 data and expected at LEP2 in the OPAL detector is also given.

Experiments e+e-	$\sqrt{s}$ (GeV)	L (pb^{-1})	$W_{\gamma\gamma}$ (GeV)	$p\bar{p}$ Events
Tasso (DESY) 1982	15 - 18.3	19.685	2.0 - 2.6	8
Tasso (DESY) 1983	17	74	2.0 - 3.1	72
Jade (DESY) 1986	17.4 - 21.9	59.3 + 24.2	2.0 - 2.6	41
TPC/2γ (SLAC) 1987	14.5	75	2.0 - 2.8	50
Argus (DESY) 1989	4.5 - 5.3	234	2.6 - 3.	60
CLEO (CESR) 1993	5.29	1310	2.0 - 3.25	484
OPAL (CERN)	LEP1	96.3	2.0 - 3.5	~200 preliminary
OPAL (CERN)	LEP 161	10	2.0 - 3.5	~20 preliminary
OPAL (CERN)	LEP 200	500	2.0 - 3.5	~1000 expected

events expected in OPAL, at LEP2, with a total integrated luminosity of 500 pb^{-1}. From this table we can see that in OPAL, at LEP1, with a factor of 10 less luminosity it is possible to see almost half of the $\gamma\gamma \to p\bar{p}$ events seen by CLEO.

2.2 *Untagged $\gamma\gamma \to \Lambda\bar{\Lambda}$ events*

The study of two-photon production of baryons can be extended also to $\gamma\gamma \to \Lambda\bar{\Lambda}$ events. In the $\gamma\gamma \to \Lambda\bar{\Lambda}$ processes, the decay products $p\pi^-\bar{p}\pi^+$ usually have low transverse momentum since the available energy in the Λ decay is small, and the $\gamma\gamma \to \Lambda\bar{\Lambda}$ cross-section is strongly peaked near the $\Lambda\bar{\Lambda}$ threshold. CLEO[6] has extended the analysis of $\gamma\gamma \to p\bar{p}$ events to include $\gamma\gamma \to \Lambda\bar{\Lambda}$ events. The CLEO analysis is based on 3400 pb^{-1} of data. They have searched for exclusive untagged $\gamma\gamma \to \Lambda\bar{\Lambda}$ events. After applying selection criteria to minimize the background,[7] they saw a clear enhancement of the $\Lambda\bar{\Lambda}$ signal by plotting $m_{p\pi^-}$ versus $m_{\bar{p}\pi^+}$, fig. 4. The small amount of background seen in data is consistent with non-resonant $\gamma\gamma \to p\pi^-\bar{p}\pi^+$ production. Using the

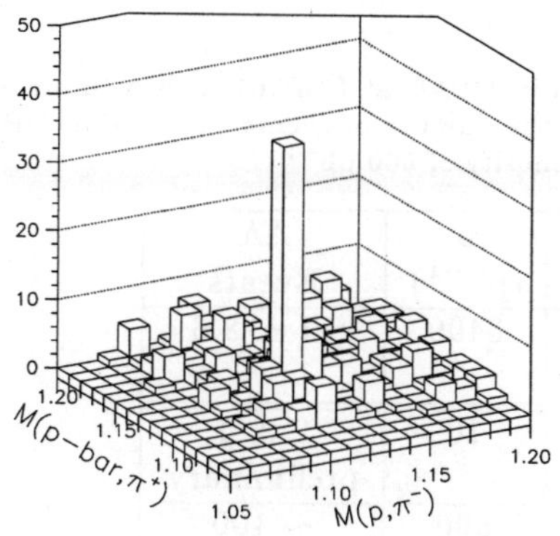

Figure 4: $m_{p\pi^-}$ versus $m_{\bar{p}\pi^+}$, data

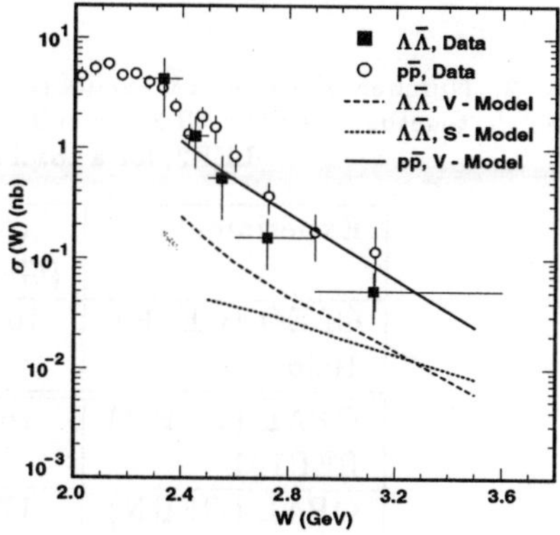

Figure 5: CLEO results for $\sigma_{\gamma\gamma\to\Lambda\bar{\Lambda}}(W_{\gamma\gamma}), \sigma_{\gamma\gamma\to p\bar{p}}(W_{\gamma\gamma})$ for $|\cos\theta^*| < 0.6$. Vertical error-bars include systematic uncertainties.

Monte Carlo signal combined with a linear background, they have measured 51.0 ± 8.6 events in data. Without separating Λ from Σ^0 they have measured the inclusive cross-section of $e^+e^- \to e^+e^-(\Lambda/\Sigma^0)(\bar{\Lambda}/\bar{\Sigma^0})$ events to be 2.0 ± 0.5 (stat) ± 0.5 (syst) pb for $|\cos\theta^*| < 0.6$. They have also identified the Σ^0 and estimated the relative contribution of the $e^+e^- \to e^+e^-(\Sigma^0\bar{\Sigma^0})$

and $e^+e^- \rightarrow e^+e^-(\Lambda/\Sigma^0)(\overline{\Lambda}/\overline{\Sigma^0})$, and extracted the exclusive $e^+e^- \rightarrow \Lambda\overline{\Lambda}$ cross-section to be 1.6 ± 0.6 (stat) ± 0.4 (syst) pb for $|\cos\theta^*| < 0.6$. The measurement of the cross-section for $e^+e^- \rightarrow e^+e^-(\Lambda/\Sigma^0)(\overline{\Lambda}/\overline{\Sigma^0})$ has been done as a function of $m_{\Lambda\overline{\Lambda}}$ and using the $\gamma\gamma \rightarrow (\Sigma^0\overline{\Sigma^0}/\Lambda\overline{\Sigma^0})$ contamination.

As seen in (fig. 5), the $\gamma\gamma \rightarrow \Lambda\overline{\Lambda}$ cross-section measured by CLEO, appears to be larger than the quark-diquark model prediction. Indeed, the cross-section appears to be similar to the $\gamma\gamma \rightarrow p\overline{p}$ cross-section for a value of $W_{\gamma\gamma}$ above the threshold, although the $\gamma\gamma \rightarrow \Lambda\overline{\Lambda}$ cross-section appears to have steeper $W_{\gamma\gamma}$ dependence. The large CLEO $\gamma\gamma \rightarrow \Lambda\overline{\Lambda}$ cross-section is an unexpected result.

Other two-photon production channels that CLEO is interested in measuring, are: $\gamma\gamma \rightarrow \Sigma^+\Sigma^- \rightarrow p\pi^0\overline{p}\pi^0$ and $\gamma\gamma \rightarrow \Xi^-\overline{\Xi^-}$.

OPAL has observed exclusive untagged $\gamma\gamma \rightarrow \Lambda\overline{\Lambda}$ events. The number of events approximately selected in OPAL at LEP161 is given in Table 2 together with the CLEO results and the number of $\gamma\gamma \rightarrow \Lambda\overline{\Lambda}$ events expected in OPAL at LEP2. It is also possible to look for exclusive untagged $\gamma\gamma \rightarrow \Xi^-\overline{\Xi^-}$ events and compare the results with the quark-diquark model predictions.

Table 2: Number of $\gamma\gamma \rightarrow \Lambda\overline{\Lambda}$ events selected in OPAL at LEP161 with a luminosity of 10 pb^{-1}, together with the CLEO results and the number of events expected in OPAL at LEP2, for a total luminosity of 500 pb^{-1}

Experiments	$\sqrt{s}$ (GeV)	L (pb^{-1})	$\Lambda\overline{\Lambda}$ Events
CLEO (CESR) 1996	10.6	3400	51 ± 8.6
OPAL (CERN) LEP161	161	10	2 ± 1 preliminary
OPAL (CERN) LEP200	175	500	~ 100 expected

2.3 Untagged $\gamma\gamma \rightarrow 4\pi$ events

Measurements of exclusive meson finale states and resonance formation by two-photon interactions also yield important information about the structure of hadronic states. In OPAL a large quantity of $e^+e^- \rightarrow e^+e^-\gamma\gamma \rightarrow e^+e^-4\pi$ events have been measured. The Monte Carlo used to study these $\gamma\gamma \rightarrow 4\pi$ events is the same[9] used for exclusive untagged $\gamma\gamma \rightarrow p\overline{p}$ events. For the $\gamma\gamma \rightarrow 4\pi$ events the cuts have been optimised to select the interesting events and to reject background events. The criteria applied to select low energy

$\gamma\gamma \rightarrow 4\pi$ events is similar to that of $\gamma\gamma \rightarrow p\bar{p}$: four good charged tracks are required with the dE/dx signature of a pion and a total charge equal to zero. The distribution of the total transverse momentum of the event, $\mid \Sigma\vec{p_T} \mid^2$, is shown in fig. 6 a) and c) for data taken $\sqrt{s} \simeq 91$ GeV and $\sqrt{s} = 161$ GeV. In fig. 6 b) and d) the mass spectrum for the $\gamma\gamma \rightarrow 4\pi$ events selected with a $\mid \Sigma\vec{p_T} \mid$ cutoff of 500 MeV is shown.

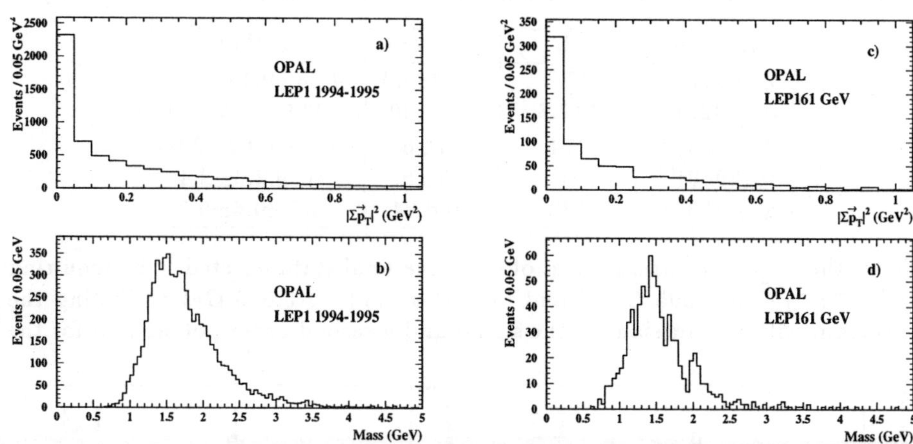

Figure 6: a) and c) Distribution of the total transverse momentum, $\mid \Sigma\vec{p_T} \mid^2$, of the 4 π events at LEP1, 1994-1995 data, and LEP161 GeV. b) and d) Mass spectrum for all the $\gamma\gamma \rightarrow 4\pi$ events at LEP1, 1994-1995 data, and LEP161.

2.4 Exclusive states in L3

The formation of resonances in two-photon interactions has been studied with the L3 detector[13] at LEP1. The cross-section for two-photon resonance production is given by:[14]

$$\sigma(e^+e^- \rightarrow e^+e^-R) = \int \sigma(\gamma\gamma \rightarrow R)dL_{\gamma\gamma}(W^2_{\gamma\gamma}, q_1^2, q_2^2), \tag{1}$$

with the Breit-Wigner cross-section:

$$\sigma(\gamma\gamma \rightarrow R) = 8\pi\frac{(2J_R + 1)}{M_R}\Gamma^R_{\gamma\gamma} \frac{M_R\Gamma_R}{(W^2_{\gamma\gamma} - M^2_R)^2 + M^2_R\Gamma_R}. \tag{2}$$

Here $L_{\gamma\gamma}$ is the two photon luminosity function. M_R is the mass, Γ_R the total width and J_R the spin of the resonance R. The resonance R has to be a neutral meson with positive charge conjugation (C=+1) and angular momentum J=0 or 2, since spin 1 production by two quasi-real photon is highly suppressed according to the Yang-Landau theorem. The resonances studied in L3 are listed in Table 3 together with the number of events observed for the given integrated luminosity L. The η' is detected in its decay $\eta'(958) \to \gamma\rho^0$, $\rho^0 \to \pi^+\pi^-$. Fig. 7

Table 3: List of resonances studied in L3. All the unpublished results are preliminary

Resonance	L (pb^{-1})	N. event	$\Gamma_{\gamma\gamma}$(keV)
$\eta'(958)$	97.7	1213±40	4.38±0.14±0.51
$a_2(1320)$	117.7	317±20	1.03±0.006±0.10
$f_2'(1525)^{15}$	114	31±6	0.13±0.03±0.03
$\eta_c(2980)^{16}$	30	17±5	8. ±2.3±2.4
$\chi_{c2}(3555)$	142	10.6±4	1.5±0.6±0.3

shows the $\pi^+\pi^-\gamma$ invariant mass of exclusive final state selected [12] by requiring $|\vec{p_T}(\pi^+\pi^-)|^2 > 0.005\ GeV^2$ and $|\vec{p_T}(\pi^+\pi^-\gamma)|^2 < 0.015\ GeV^2$. Fitting the spectrum with a Gaussian for the signal and a second order polynomial for the

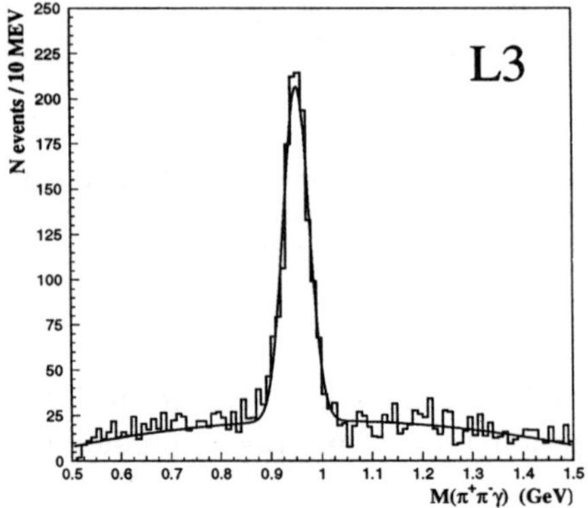

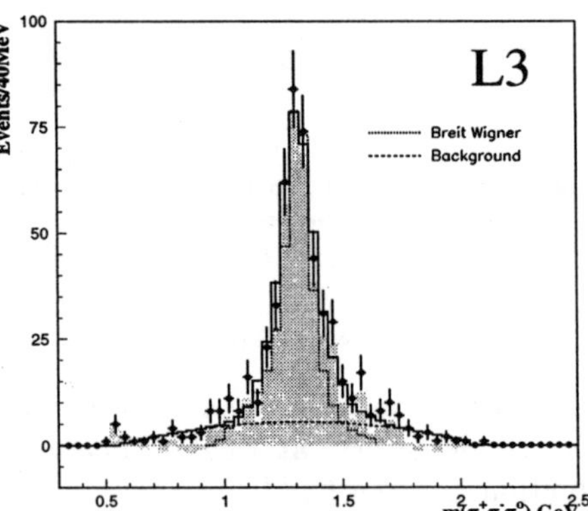

Figure 7: The $\pi^+\pi^-\gamma$ invariant mass spectrum. Notice the η' peak

Figure 8: The $\pi^+\pi^-\pi^0$ mass spectrum of a a_2 sample. The solid line is the result of the fit, the shaded histogram is the spectrum after subtraction of the background. The dotted line is the Breit-Wigner distribution for the a_2.

background, 1213 ± 40 events are found. For a total luminosity of 97.7 pb^{-1} and a selection efficiency, evaluated by Monte Carlo simulation, of 2.52 %. The two photon width of the $\eta^{'}$ is $\Gamma_{\gamma\gamma}(\eta^{'}) = 4.38 \pm 0.14$ (stat) ± 0.51 (syst) keV. L3 has also measured the a_2 radiative width and its helicity status. Events in the $\pi^+\pi^-\pi^0$ channel have been selected by requiring two oppositely charged tracks and two photons. The p_T threshold for charged tracks is 60 MeV. The π^0 signal region is defined as $100 < m(\gamma\gamma) < 160$ MeV in the mass spectrum of the two photon. In addition the transverse momentum, $| \vec{p}_T(\pi^+\pi^-\pi^0) |^2$ is required to be smaller than 0.0015 GeV^2. The $a_2(1320)$ sample is selected by requiring $m(\pi^{\pm}\pi^0) < m_\rho + 250$ MeV as the a_2 decays dominantly through $\rho\pi$. In fig. 8 the mass spectrum of the $a_2(1320)$ sample is fitted to a Breit-Wigner function; the mass and the width obtained are M $= 1319.5 \pm 2.8$ (stat) ± 13 (syst) MeV, $\Gamma = 117 \pm 5$ (stat) ± 14 (syst) MeV. The radiative width, obtained by normalization to Monte Carlo expectation, is $\Gamma_{\gamma\gamma}(a_2) = 1.03 \pm 0.06$ (stat) ± 0.10 (syst) keV. L3 has studied the χ_{c2} through its decay to $\gamma J/\Psi$, with the subsequent decay of the J/Ψ to e^+e^- or $\mu^+\mu^-$. Little background is expected from othe χ_c states because χ_{c1} formation is suppressed for quasi-real photon and Br$(\chi_{c0} \to \gamma l^+l^-) \sim 10^{-4}$. In fig. 9 is shown the distribution of the mass difference $m(l^+l^-\gamma)$ - $m(l^+l^-)$, with $l =$ e or μ for the selected events. For χ_{c2} decaying to $\gamma J/\Psi$ the mass difference should peak at 0.459

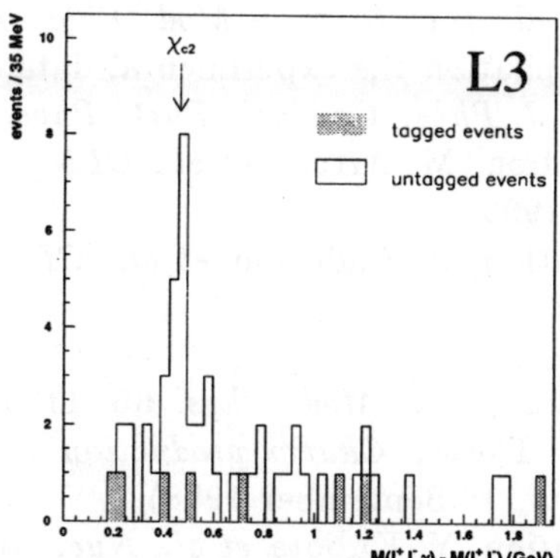

Figure 9: The mass difference $M(l^+l^-\gamma) - M(l^+l^-)$ spectrum for the selected events. The arrow shows the region where the χ_{c2} events are expected.

GeV. After subtraction of the background computed from the right hand side of the distribution, an enhancement of 10.6 ± 4.0 events remains. With an integrated luminosity of 142 pb^{-1} and a selection efficiency of 7.6 % the two-photon width is calculated to be $\Gamma_{\gamma\gamma}(\chi_{c2}) = 1.5 \pm 0.6$ (stat) ± 0.3 (syst) keV, a value within the range of known experimental values and in agreement with theoretical calculations.

Acknowledgments

I would like to thank: M. Anselmino for giving me the opportunity to present this work; M. N. Focacci Kienzle for the helpful advices and discussions through all this work, and providing me with the L3 plots. I am also grateful to the OPAL Collaboration, in particular to G. Giacomelli, G. Hanson, D. Charlton, B. Schmitt, S. Braibant, P. Giacomelli, R. McPherson and R. Nisius for their help in all the stages of the analysis .

References

1. OPAL Collaboration, K. Ahmet *et al.*, *Nucl. Instrum. Methods* A **305**, 275 (1991)
2. S.J. Brodsky *et al.*, *Proceedings of the Workshop on LEP200 Physics*, Aachen, 222 (1986).
3. A. Bawa *et al.*, *Proceedings of the Workshop on Physics at LEP2*, CERN 96-01, 291 (1996).
4. M. Anselmino *et al.*, *Int. Jour. of Mod. Phys.* A **4**, 5213 (1989).
5. For a recent compilation the experimental data, see:
   D. Morgan *et al.*,*J. Phys. G: Nucl. Part. Phys.* **20**, A1-A147 (1994).
6. CLEO Collaboration, M. Artuso *et al.*, *CLNS Preprint 93/1245* CLEO 93-10A August, 1993.
7. CLEO Collaboration, S. Anderson *et al.*, *CLNS 96/1448* CLEO 96-19 January 17, 1997.
8. See for example:
   M. Anselmino *et al.*, *Rev. Mod. Phys.* **65**, 1199 (1993).
9. F. Linde, Ph.D. Thesis, *Charm production in Two-Photon Collisions* (Leiden University, 15 September 1988).
10. CLEO Collaboration, Y. Kubota *et al.*, *Nucl. Instrum. Methods* A **320**, 66 (1992).
11. D. Rubin *Proceedings of the 1995 Particle Accelerator Conference* **1**, 481 (1995).
12. G. Ambrosi, L3 Collaboration in ICHEP96 (Warsaw) (1996).
13. L3 Collaboration, B. Adeva *et al.*, *Nucl. Instrum. Methods* A **289**, 35 (1990)
14. V.M. Budnev *et al.*, *Physics Reports* **15**, 181 (1975).
15. L3 Collaboration, *Phys. Lett.* B **363**, 118 (1995)
16. L3 Collaboration, *Phys. Lett.* B **318**, 575 (1993)

THE NUCLEON ELECTROMAGNETIC FORM FACTORS IN THE TIME-LIKE REGION: NEW RESULTS FROM THE FENICE EXPERIMENT

C.Bini[a]

Universitá "La Sapienza" and INFN Roma - Italy

The electromagnetic form factors of the neutron in the time-like region have been measured for the first time in the FENICE experiment at the e^+e^- storage ring ADONE. The results show that in the q^2 range between the nucleon-antinucleon threshold and ~ 6 GeV^2, the neutron form factor is larger than the proton one and $|G_E^n| \ll |G_M^n|$. Furthermore neutron, proton and multihadronic data closer to the threshold seem to indicate a possible non-trivial behaviour of the F.F. and of the hadronic cross-section around the $N\overline{N}$ threshold. Finally hints for a consistent imaginary part of the F.F. come from an analysis of the J/ψ decays. All these indications call for a more accurate experimental investigation. The possibilities of doing a new e^+e^- experiment using existing facilities are discussed.

1 Introduction

The electromagnetic form factors of the nucleons have been extensively measured in the space-like region through elastic scattering experiments[1]. In the time-like region they are accessible only for momentum transfer q^2 in excess of the nucleon-antinucleon threshold, namely $4M_N^2 = 3.53$ GeV^2, M_N being the nucleon mass, in e^+e^- annihilation experiments by looking at $N\overline{N}$ final states[2], or in $p\overline{p} \to e^+e^-$ experiments [3]. In both cases q^2 is simply given by the square of the center of mass energy.

The FENICE experiment has performed the first measurement of the neutron electromagnetic form factors in the time-like region measuring the cross-section for the process $e^+e^- \to n\overline{n}$ at the ADONE e^+e^- storage ring at the Frascati INFN laboratory in the q^2 range $3.6 \div 5.9$ GeV^2 [4]. Furthermore a new measurement of the proton form factor in the same energy range [5] and a systematic study of the e^+e^- multihadronic cross-section around the $N\overline{N}$ threshold [6] have also been done. Finally the J/ψ decays in nucleon-antinucleon have been measured [7].

---

[a]representing the FENICE collaboration A.Antonelli, R.Baldini, P.Benasi, M.Bertani, M.E.Biagini, V.Bidoli, C.Bini, T.Bressani, R.Calabrese, R.Cardarelli, R.Carlin, C.Casari, L.Cugusi, P.Dalpiaz, G.DeZorzi, A.Feliciello, M.L.Ferrer, P.Ferretti Dalpiaz, P.Gauzzi, P.Gianotti, E.Luppi, S.Marcello A.Masoni, R.Messi, M.Morandin, L.Paoluzi, E.Pasqualucci, G.Pauli, N.Perlotto, A.Perrone, F.Petrucci, M.Posocco, M.Preger, G.Puddu, M.Reale, L.Santi, R.Santonico, P.Sartori, M.Savrie, S.Serci, M.Spinetti, S.Tessaro, C.Voci, F.Zuin.

In this talk the results of the FENICE experiment are presented and discussed and some possible experimental perspectives are also outlined.

2 The results of the FENICE experiment

The FENICE experiment is based on a non-magnetic apparatus[8] designed for the detection of anti-neutrons through their annihilation in matter. It consists of layers of streamer tubes as tracking devices and scintillators intervealed with iron slabs that act as a distributed target for $\bar{n}$ annihilations. The scintillators allow to use the time of flight technique for background rejection and signal evaluation and are used for the trigger. The overall angular coverage is $\sim 76\%$ of 4π and all the detector is surrounded by an active RPC veto to reject cosmic rays at trigger level.

2.1 The neutron form factors

Due to the low efficiency for neutrons in the expected energy range, the selection of $n\bar{n}$ final states is done using the antineutron annihilation only. At every center of mass energy a sample of candidate events is obtained after a selection procedure that is partly authomatic and partly based on visual scanning. At the end the $1/\beta$ spectrum is done, β being the $\bar{n}$ velocity measured by the time of flight method, and the signal is evaluated as a peak over a smooth cosmic ray annihilations background that turns out to be the main backgrund source for this channel at this energies. In Tab.1 the results of this analysis including the $e^+e^- \to n\bar{n}$ cross-section evaluation are reported. At $E_{c.m.} = 1.9\ GeV$ that is at the center of mass energy closest to the $N\overline{N}$ threshold, we have found no signal, hence only an upper limit on the cross-section has been derived.

Table 1: $e^+e^- \to n\bar{n}$ cross-section measurement. At every center of mass energy, the integrated luminosity L_{int}, the overall efficiency ϵ_{tot} and the final number of events N_{evts} are shown toegheter with the cross-section σ. The uncertainties are statistical and include also the background subtraction.

$E_{c.m.}(MeV)$	$L_{int}(nb^{-1})$	ϵ_{tot}	N_{evts}	$\sigma(e^+e^- \to n\bar{n})$
1900	34	0.17	<6.5 (90% C.L.)	<1.1 (90% C.L.)
1920	80	0.14	17.8 ± 7.4	1.59 ± 0.66
2000	92	0.17	17.0 ± 6.6	1.09 ± 0.42
2100	100	0.25	22.1 ± 6.2	0.93 ± 0.26
2440	57	0.27	10.0 ± 4.5	0.65 ± 0.29

In order to extract the absolute value of the magnetic and eletric form factors $|G^n_M|$ and $|G^n_E|$ from the cross-section $\sigma(e^+e^- \to n\overline{n})$ the following relation has to be used:

$$\sigma(e^+e^- \to n\overline{n}) = \frac{4\pi\alpha^2\beta}{3q^2}[|G^n_M(q^2)|^2 + \frac{2M_N^2}{q^2}|G^n_E(q^2)|^2] \tag{1}$$

The different contributions from $|G^n_M|$ and $|G^n_E|$ can be obtained looking at the angular distribution of the neutron-antineutron pairs that is expected to be

$$\frac{d\sigma}{d\Omega} = \frac{\alpha^2\beta}{4q^2}[(1 + \cos^2\theta)|G^n_M(q^2)|^2 + \frac{4M_N^2}{q^2}\sin^2\theta|G^n_E(q^2)|^2] \tag{2}$$

with θ being the polar angle in the center of mass system. It can be seen from (2) that if $|G^n_M| \sim |G^n_E|$ the distribution gets almost isotropic when q^2 is close to the nucleon-antinucleon threshold. On the other hand a $(1 + \cos^2\theta)$ distribution is expected if the magnetic contribution is larger than the electric one (that is what happens for $q^2 \gg 4M_N^2$).

Fig.1(a) shows the neutron form factors as a function of q^2. The form factor values have been extracted by (1) assuming $|G_E| \ll |G_M|$, the latter being justified by an analysis of the angular distribution (see Fig.2)(a).

2.2 The proton form factors

In the case of the $p\overline{p}$ final states the selection is done using both particles. Events with two colinear tracks are looked for, one of them giving an annihilation star inside the detector, the other one having a range that depends on the center of mass energy. The main background sources are multihadrons interacting in the detector and beam-gas or beam-pipe interaction events. Both categories are rejected using a strict request on the colinearity of the two tracks in both views, and by using the time of flight. Fig.1(b) shows the results compared with previous worldwide data. The usual hypothesys $|G^p_E| \sim |G^p_M|$ has been done to extract the form factors from (1). The assumption is checked by looking at the θ distribution (see Fig.2(b)) for the candidate events at center of mass energies larger than $2\,GeV$ (for lower energies events the measurement of θ being affected by too large uncertainties).

2.3 The multihadronic cross-section

A measurement of the total multihadronic cross-section in the center of mass energy range $1.82 \div 2.44\,GeV$ has been done by looking at events with at least three tracks, one of them charged, pointing to the interaction point. In fig.3

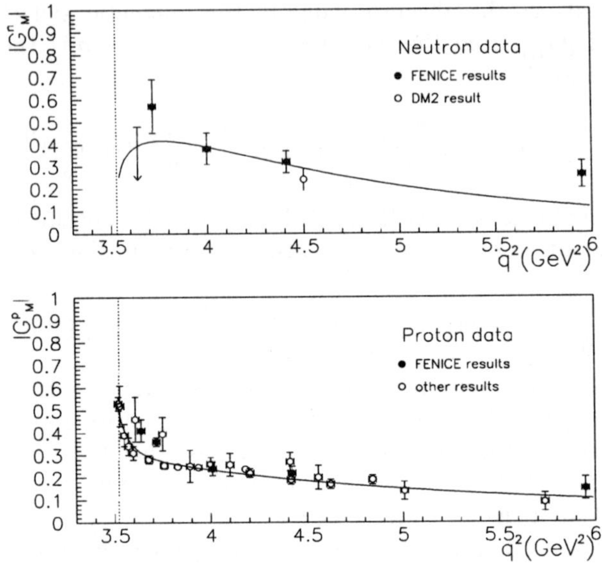

Figure 1: (a) Neutron magnetic form factor as a function of q^2. FENICE datand an indirect evaluation from DM2 (open square). Proton magnetic form factor as a function of q^2. FENICE data (full squares) compared with worldwide data in the low q^2 region (open squares). The two lines have been drawn to drive the eye.

the FENICE data are compared with those of the old ADONE experiment $\gamma\gamma2$[9], that constitute the most complete data set at these energies and with other experiments.

We have concentrated our measurements mainly at energies around the $N\overline{N}$ threshold, to look for eventual structures. In the next section we'll discuss this point.

2.4 The J/ψ decays in nucleon-antinucleon

The branching ratios of $J/\psi \rightarrow n\overline{n}$ and $p\overline{p}$ have also been measured. The experimental method is essentially the same outlined in the previous sections for both channels, the searched final states being the same. The normalization is done with respect to the number of multihadrons. We give here the results:

$$B.R.(J/\psi \rightarrow n\overline{n}) = (1.8 \pm 0.5) \times 10^{-3}$$

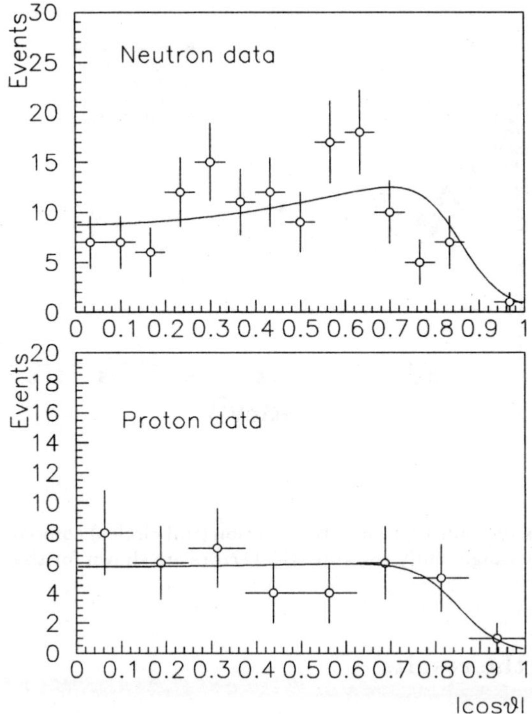

Figure 2: Angular distribution for $n\overline{n}$ events at all energies (upper plot) and for $p\overline{p}$ events at $E_{c.m.} \geq 2000 MeV$ (lower plot).

$$B.R.(J/\psi \to p\overline{p}) = (2.03 \pm 0.34) \times 10^{-3}$$

For what concern the neutron-antineutron channel it is the second measurement after the only one done at DORIS[11] and it improves the uncertainty by about a factor of two. For the case of the proton, it is in agreement with PDG[12] world average.

A simple parametrization of the $J/\psi \to N\overline{N}$ decay amplitudes in term of e.m. and strong contributions, allows to connect the ratio between the two branching ratios with the relative phase between the strong and the main e.m. amplitudes[13]. A $\sim 90^{\circ}$ phase between the two amplitudes better account for our results[14]. In the following we'll see some implications of this statement.

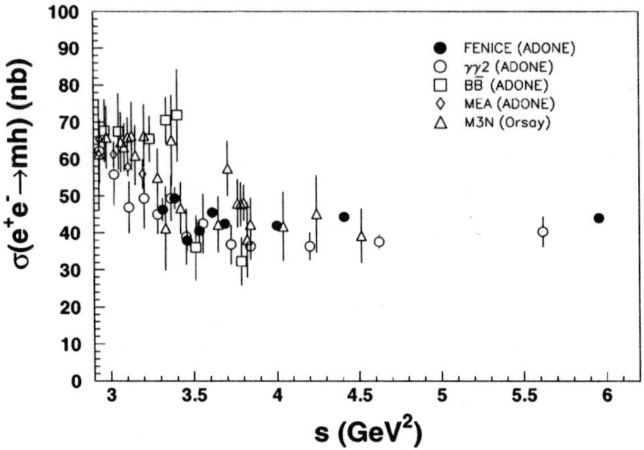

Figure 3: FENICE data on multihadronic cross-section (full circles) compared with previous data in the same energy range. Only the statistical errors are shown for the FENICE points.

3 Discussion of the results

The main result of the FENICE experiment is that in the time-like region in the q^2 range $3.6 \div 6\, GeV^2$ the neutron form factor is larger than the proton one by a factor between 1 and 2. This result is in disagreement with PQCD based predictions [15]. and also with simple models based on analiticity [16]. Different dispersion relation approaches [17,18] have also been recently tried, giving some inconsistencies with the FENICE data. Finally the data can be well accounted for in VMD based models [19].

The second important indication is that while for the proton the hypothesys $|G_E^p| \sim |G_M^p|$ seems to be well verified, for the neutron $|G_E^n| \ll |G_M^n|$ like in the space-like region at the corresponding q^2. We stress that since at threshold, by definition $|G_E^{p,n}| = |G_M^{p,n}|$ holds, the neutron behaviour has to be non-trivial in the region just above threshold.

A non-trivial behaviour around the nucleon-antinucleon threshold can be observed in the proton data and in the multihadronic cross-section data. So in Ref.[6] we have considered the hypothesys of the existence of a narrow resonance just below threshold strongly coupled to the $N\overline{N}$ final states and also interfering with the multihadronic amplitude. A new analysis is in progress that includes also new unpublished data from the DM2 collaboration, where a similar effect can be found [10].

4 Experimental perspectives

4.1 A new asymmetric e^+e^- experiment

To improve the statistical accuracy of the results presented here, a new experiment will be proposed [20] based on a different experimental approach.

e^+e^- collisions with a c.m. energy around 2 GeV can be obtained by $\sim 100~MeV$ electrons coming from a LINAC against a $\sim 10~GeV$ positron beam circulating in a storage ring. In this configuration, the produced final states, almost isotropic in the c.m. reference system, are all boosted in the forward direction in the laboratory system. The particles will come out in a very narrow cone of the order of 5° wide. A magnet put along the flight path will allow an analysis of the charged particles, while the neutral products (and especially the neutrons and anti-neutrons) will go in the forward direction and can be detected with an hadronic calorimeter that should also be able to measure the time of flight.

For what concern the $n\overline{n}$ final state the detector will have to measure a neutron and an antinuetron both with energies of the order of 5 GeV. This is a clear advantage respect to the standard e^+e^- configuration where neutron and anti-neutron have to be detected with kinetic energies down to few tens of MeV. The n and $\overline{n}$ have to be disentangled with respect to other neutral particles, essentially photons coming from $e^+e^- \rightarrow \gamma\gamma$, $e^+e^-\gamma$ processes or from π^0 decays.

A first evaluation [20] shows that by using the TTF LINAC and the Petra storage ring at DESY, a luminosity of $5 \cdot 10^{30}~cm^{-2}s^{-1}$ can be achieved with minor modifications on existing facilities.

4.2 A possible measurement of the phase of the strong J/ψ decay amplitude respect to the proton f.f.

The analysis of the decays of the J/ψ in mesons and in baryons, indicates [14,21,22] that the strong and the e.m. amplitudes involved in the decay are almost orthogonal each other. The implications of this statement are not clear at the moment. A phase can be due either to the e.m. amplitude (that is to the electromagnetic form factors of mesons and baryons) or to the strong amplitude. The first possibility is in contrast with very simple arguments

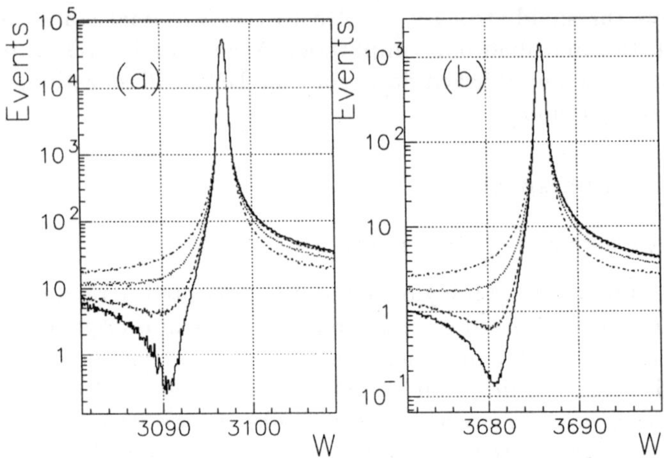

Figure 4: Expected interference pattern for the J/ψ (a) and the $\psi^{'}$ (b) for different values of the relative phase $\Delta\phi$ between the strong and the electromagnetic contribution; $\Delta\phi = 0^{o}$ (solid line), $\Delta\phi = 30^{o}$ (dashed line), $\Delta\phi = 60^{o}$ (dotted line) and $\Delta\phi = 90^{o}$ (dashed-dotted line). All curves are convoluted with the beam energy spread.

based on analyticity like dispersion relation approaches [23], the second cannot be easily accomodated in a coherent perturbative approach to the J/ψ decay mechanism.

A new direct measurement of the relative phase between strong and e.m. decay amplitudes, can be obtained by looking at the interference pattern between the $p\bar{p} \to J/\psi \to e^{+}e^{-}$ and the $p\bar{p} \to e^{+}e^{-}$ processes in $p\bar{p}$ interactions around the J/ψ mass. Fig.4 shows the expected interference pattern for four different values of the phase difference. The same study can be done at the $\psi^{'}$.

The Fermilab experiment E835 [24] can do the measurement and will do it in the following months. The vertical scale in Fig.4 is in number of events that can be collected by E835 integrating a luminosity of 1 pb^{-1} for each c.m. energy value.

References

1. The most recent data in the space-like region can be found in:
   P.E.Boosted et al. *Phys. Rev. Lett.* 68 (1992) 3841;
   S.Rock et al. *Phys. Rev. Lett.* 49 (1993) 1139;

A.Lung et al. *Phys. Rev. Lett.* 70 (1993) 718;
2. M.Castellano et al., *Nuovo Cimento* A14 (1973) 1;
   B.Delcourt et al., *Phys. Lett.* B86 (1979) 395;
   D.Bisello et al., *Nucl. Phys.* B224 (1983) 379;
3. G.Bassompierre et al., *Nuovo Cimento* A73 (1983) 347;
   G.Bardin et al., *Phys. Lett.* B255 (1991) 154;
   E760 coll. T.A. Armstrong et al. *Phys. Rev. Lett.* 70 (1993) 12;
4. Preliminary results are reported in: FENICE Coll. A.Antonelli et al.
   Phys. Lett. B313 (1993) 283;
5. FENICE Coll. A.Antonelli et al. *Phys. Lett.* B334 (1994) 431;
6. FENICE Coll. A.Antonelli et al. *Phys. Lett.* B365 (1996) 427;
7. FENICE Coll. A.Antonelli et al. *Phys. Lett.* B301 (1993) 317;
8. FENICE Coll. A.Antonelli et al. *Nucl. Instr. and Meth.* A337, (1993)
   34;
9. $\gamma\gamma2$ Coll. C.Bacci et al. *Phys. Lett.* B86 (1979) 234;
10. A reanalysis of the DM2 data on the multihadronic cross-section is in
    progress, M.Schioppa private comunication;
11. BONANZA Coll. H.J. Besch et al. *Phys. Lett.* B78 (1978) 374;
12. Review of Particle Properties *Phys. Rev.* D54 (1996) 1;
13. M.Claudson, S.L.Glashow, M.B.Wise *Phys. Rev.* 25 (1982) 13;
14. R.Baldini, C.Bini, E.Luppi submitted to *Phys. Lett.* B;
15. S.J.Brodsky, G.R.Farrar *Phys. Rev.* D11 (1975) 1309;
    V.L.Chernyak, I.R.Zhitnitski *Nucl. Phys.* B246 (1984) 52;
16. A.A. Logunov, N. Van Hieu,I.T. Todorov *Annals of Physics* 31 (1965)
    203;
17. H.W.Hammer, U.G.Meissner and D.Drechsel Note MKPH-T-96-05
    (1996);
    H.W.Hammer et al., *Phys. Lett.* B367 (1996) 323;
18. Z.Dziembowski and A.Szczurek *Phys. Lett.* B387 (1996) 875;
19. G.Hohler et al. *Nucl. Phys.* 114 (1976) 505;
    J.G.Korner and M.Kuroda, *Phys. Rev.* D16 (1977) 2165;
    S.Dubnicka, *Nuovo Cimento* A100 (1988) 1;
    M.M.Giannini, E.Santopinto and M.I.Krivoruchenko, Proceed. of the
    Intern. Conf. on Mesons and Nuclei at Intermediate Energies, Dubna
    (1994);
20. FENICE Coll. and P.Patteri, INFN Internal Note (1996);
21. L.Kopke and N.Wermes, *Phys. Rep.* 174 (1989) 67;
22. A.Seiden,H.F.W.Sadrozinski and H.E.Haber, *Phys. Rev.* D38,3 (1988)
    824;
23. R.Baldini et al., in preparation;

24. The E835 experiment is the upgrade of the E760 experiment: E760 coll. T.A. Armstrong et al. *Phys. Rev.* D47 (1993) 772;

Nucleon and Δ in a covariant quark-diquark model

Volker Keiner

Institut für Theoretische Kernphysik,
Universität Bonn, Nussallee 14-16, D-53115 Bonn, FRG

Abstract

We develop a formally covariant quark-diquark model of the nucleon. The nucleon is treated as a bound state of a constituent quark and a diquark interacting via a quark exchange. We include both scalar and axial-vector diquarks. The underlying Bethe-Salpeter equation is transformed into a pair of coupled Salpeter equations. The space- and timelike electromagnetic form factors of the nucleon are calculated in the Mandelstam formalism for squared momentum transfers up to 3 $(\mathrm{GeV}/\mathrm{c})^2$ above threshold. The model is applied also to the $\Delta(3/2)$.

1 Introduction

Many attempts have been made to describe the nucleon as a bound state of a quark and a diquark. The most fundamental approach uses the Nambu–Jona-Lasinio model[1,2,3] to derive static properties of the nucleon and ground state baryons. Most diquark models favour the existence of two kinds of diquarks, scalar and axial-vector ones. To take into account the truly three quark structure and the Pauli-principle a quark exchange between quark and diquark is assumed to dominate the short range interaction[3]. We adopt these ideas in a constituent quark model based on the Bethe-Salpeter equation. Following Salpeter[4] we assume an instantaneous interaction and obtain a Salpeter-type equation. As the only interaction we consider an (instantaneous) quark exchange between the quark and the diquark. Involving only scalar and axial-vector diquark channels we deduce a pair of coupled integral equations similar to the framework of Ref.[5]. Within a basis of positive parity amplitudes with spin-1/2 we obtain a bound state solution for this equation by making use of the Ritz variational principle. Then, electromagnetic transition currents are calculated using the Mandelstam formalism[6]. For details of the calculation see two recent papers[7,8]. We compute the electromagnetic form factors of the nucleon in both the space- and timelike regions. Extending the model to the $\Delta(3/2)$ resonance, the $N - \Delta$ transition form factors and the ratios $E2/M1$ and $C2/M1$ are calculated.

2   The model

In momentum space the *Bethe-Salpeter amplitude* for a bound state of a quark (index 2) and a scalar or axial-vector diquark (index 1) with total four momentum $P = p_1 + p_2$ and relative momentum p fulfills the *Bethe-Salpeter equation*:

$$\chi_P(p)_{(\mu)} = \Delta_1(p_1)_{(\mu)}^{(\nu)} S_2(p_2) \int \frac{d^4 p'}{(2\pi)^4} \left(-iK(P,p,p')\,\chi_P(p')\right)_{(\nu)} . \qquad (1)$$

Δ_1 and S_2 are the propagators of the constituent diquark and the quark, respectively. For the v-diquark we choose $\Delta_{\mu\nu} = -ig_{\mu\nu}/(m_1^2 - p_1^2 + i\varepsilon)$. Following Salpeter[4] we neglect the time (i.e. energy) dependence of the interaction kernel by assuming $K(P,p,p') = V(\vec{p},\vec{p}')$ (*instantaneous interaction*) in the rest frame of the bound state. Then, one can easily perform the p^0 integration. Defining in the rest frame of the bound state the *Salpeter amplitude*

$$\Phi(\vec{p}) := \left(\int \frac{dp^0}{2\pi} \chi_P(p^0, \vec{p}) \right)_{P=(M,\vec{0})} , \qquad (2)$$

one gets from Eq. (1) with standard techniques:

$$\Phi(\vec{p}) = \frac{1}{2\omega_1} \left(\frac{\Lambda_2^+(-\vec{p})\gamma^0}{M - \omega_1 - \omega_2} + \frac{\Lambda_2^-(-\vec{p})\gamma^0}{M + \omega_1 + \omega_2} \right) \int \frac{d^3 p'}{(2\pi)^3} V(\vec{p},\vec{p}')\,\Phi(\vec{p}') , \qquad (3)$$

with $\Lambda_2^{\pm}(-\vec{p}) = \frac{\omega_2 \pm H_2(-\vec{p})}{2\omega_2}$ the usual Dirac projectors and $\omega_i = \sqrt{\vec{p_i}^{\,2} + m_i^2}$. If we define

$$\Psi(\vec{p}) \;:=\; \gamma^0\,\Phi(\vec{p}) , \quad W(\vec{p},\vec{p}') := V(\vec{p},\vec{p}')\,\gamma^0 , \qquad (4)$$

we can rewrite Eq. (3) in a Schrödinger-type equation:

$$\begin{aligned}
(\mathcal{H}\,\Psi)(\vec{p}) \;&=\; M\,\Psi(\vec{p}) \\
&=\; \frac{\omega_1 + \omega_2}{\omega_2} H_2(\vec{p})\,\Psi(\vec{p}) + \frac{1}{2\omega_1} \int \frac{d^3 p'}{(2\pi)^3} W(\vec{p},\vec{p}')\,\Psi(\vec{p}') \\
&=:\; (\mathcal{T} + \mathcal{V})\,\Psi(\vec{p}) .
\end{aligned} \qquad (5)$$

Of special importance in the Bethe-Salpeter approach is the correct normalization of the amplitudes. This will guarantee the correct behaviour of the form factors at $q^2 = 0$. In the instantaneous approximation the usual *normalization condition* for the BS-amplitudes leads to

$$\int \frac{d^3 p}{(2\pi)^3} (2\omega_1)\,\overline{\Phi}(\vec{p})\,\gamma^0\,\Phi(\vec{p}) = 2\,M , \qquad (6)$$

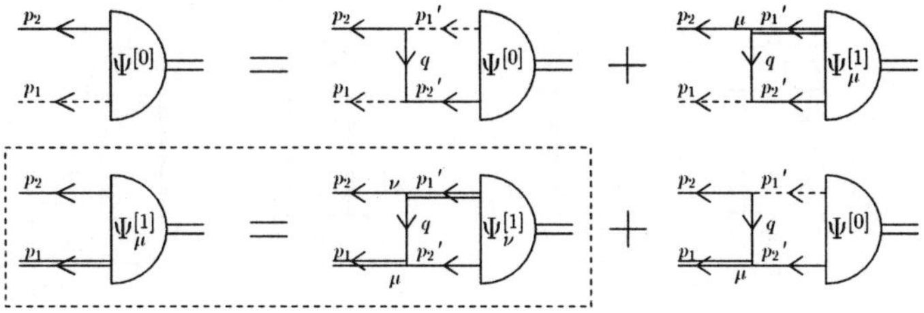

Figure 1: The quark-diquark Salpeter equations. The framed part is the equation for the $\Delta(3/2)$ (see Sec. 4).

with $\overline{\Phi}$ the adjoint Salpeter amplitude, fulfilling $\overline{\Phi}(\vec{p}) = \Phi^{\dagger}(\vec{p})\gamma^0$. This leads to the definition of a *scalar product* for $\Phi(\vec{p})$:

$$\langle \Phi_1 \,|\, \Phi_2 \rangle = \int \frac{d^3 p}{(2\pi)^3} \, (2\omega_1) \, \overline{\Phi}_1(\vec{p}) \, \gamma^0 \, \Phi_2(\vec{p}) \, . \tag{7}$$

For $M \in \mathcal{R}^+$, the Hamiltonian $\mathcal{H}$ (Eq. (5)) has to be hermitian (and positive definite) with respect to the above scalar product. In the instantaneous approximation the interaction kernel is of the form

$$W(\vec{p}, \vec{p}') \quad \sim \quad -g^2 \, \frac{1}{\omega_q^2} \, \left(-\vec{\gamma}(\vec{p} + \vec{p}') + m_q \right) \gamma^0 \, , \tag{8}$$

with ω_q the energy of the exchanged quark and g the dimensionless quark-diquark coupling parameter. To obtain a stable solution of the Salpeter equation (5) we have to introduce a form factor of the diquark. It is chosen to be of the form $\sim \exp(-\lambda^2 k^2)$, with λ parametrizing the extension of the diquark. Transitions between scalar and v-diquarks lead to a system of coupled Salpeter equations shown graphically in Fig. 1.

3 Nucleon form factors

The electromagnetic current is the sum of the diquark- and quark currents, see Fig. 2. In the *Mandelstam formalism* one finds e.g. for the quark current:

$$\langle P's' \,|\, j_\mu^q \,|\, Ps \rangle \quad = \quad e_q \int \frac{d^4 p}{(2\pi)^4} \overline{\Gamma}_{P'}(p') S_2^F(p_2') \gamma_\mu S_2^F(p_2) \Gamma_P(p) \Delta_1^F(p_1), \tag{9}$$

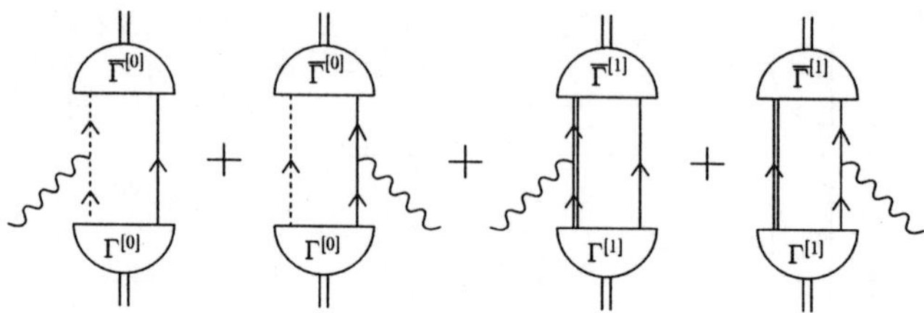

Figure 2: The electromagnetic current is the sum of the diquark currents and the quark currents.

	m_q	$m_S = m_V$	g^N	g^Δ	λ	κ_V	κ_{SV}
Set A	440 MeV	800 MeV	17.76	8.50	0.30 fm	1.1	-
Set B	440 MeV	800 MeV	17.76	8.50	0.30 fm	-0.07	2.4

Table 1: The parameters of the model.

with the vertex function $\Gamma_{P=(M,\vec{0})}(p) = \Gamma(\vec{p}) = -i \int \frac{d^3 p'}{(2\pi)^3} V(\vec{p},\vec{p}')\Phi(\vec{p}')$ in the instantaneous approximation. The form factors are obtained from:

$$J_\mu := e\,\overline{u}_{s'}(P') \left(\gamma_\mu\, F_1^N(q^2) + \frac{i\sigma_{\mu\nu} q^\nu}{2M}\, \kappa_N\, F_2^N(q^2) \right) u_s(P) , \qquad (10)$$

with $q = P' - P$. The parameters used are given in Tab. 1. The scalar and v-diquark parameters are chosen to be equal. g^N is the quark-diquark coupling parameter for the nucleon (cf. Eq. (8)) and g^Δ that for the Δ (see Sec. 4). κ_V is the anomalous magnetic moment of the v-diquark introduced via

$$j_{\mu;ba}^{V-V} \sim -(p+p')_\mu g_{ba} + (1+\kappa_V)(p_b - p_b')g_{\mu a} + (1+\kappa_V)(p_a' - p_a)g_{\mu b} . \quad (11)$$

We compare the results of the parameter Set A with those of Set B where we allow also for photon-induced scalar–v-diquark transitions according to

$$j_\mu^{S-V} \sim \frac{(1+\kappa_{SV})}{M_N} \varepsilon_{\mu\nu\rho\lambda}\, \epsilon^\nu\, p'^\rho_2\, p_2^\lambda . \qquad (12)$$

Fig. 3 shows the Sachs form factors of the nucleon in the spacelike region. Note that κ_V and κ_{SV} only affect the spin-flip currents, i.e. the magnetic form factors. We find a very good description of all form factors up to $-q^2 = 3$ GeV^2. The resulting static properties are listed in Tab. 2. The inclusion of scalar–v-diquark transitions is a possibility to describe also the neutron magnetic

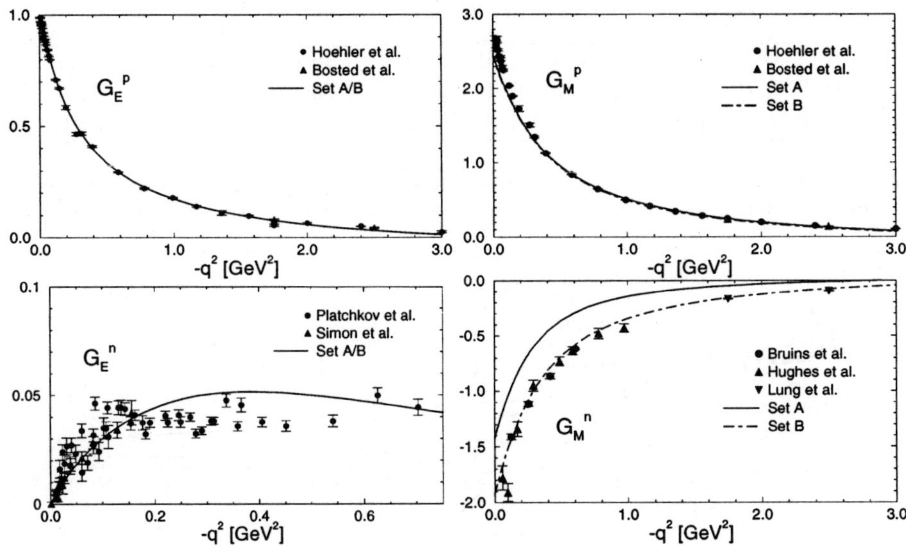

Figure 3: The spacelike Sachs form factors of the nucleon. For the experimental data see the analysis of Ref. [9].

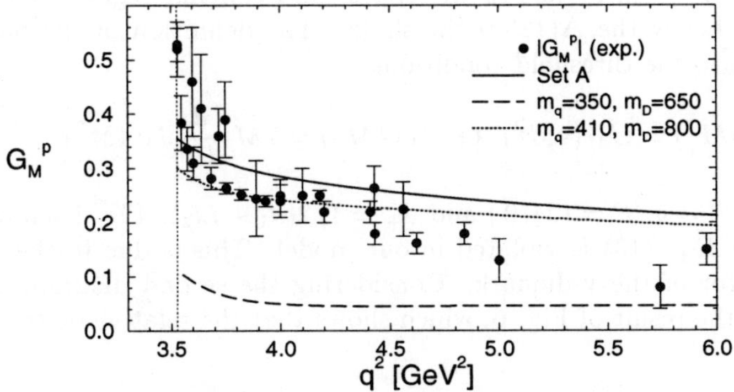

Figure 4: The timelike magnetic form factor of the proton. For the experimental data see Refs. [10,11].

	$\sqrt{\langle r^2\rangle^p_E}$	$\langle r^2\rangle^n_E$	$\sqrt{\langle r^2\rangle^p_M}$	$\sqrt{\langle r^2\rangle^n_M}$	μ_p/μ_N	μ_n/μ_N
Set A	0.79 fm	-0.110 fm^2	0.72 fm	0.86 fm	2.45	-1.42
Set B	0.79 fm	-0.110 fm^2	0.74 fm	0.75 fm	2.44	-1.91
exp.	0.847 fm	-0.113 fm^2	0.836 fm	0.889 fm	2.793	-1.913

Table 2: Static nucleon properties as they result from the threshold behaviour of the electromagnetic nucleon form factors. For the experimental data see the analysis of Ref. [9].

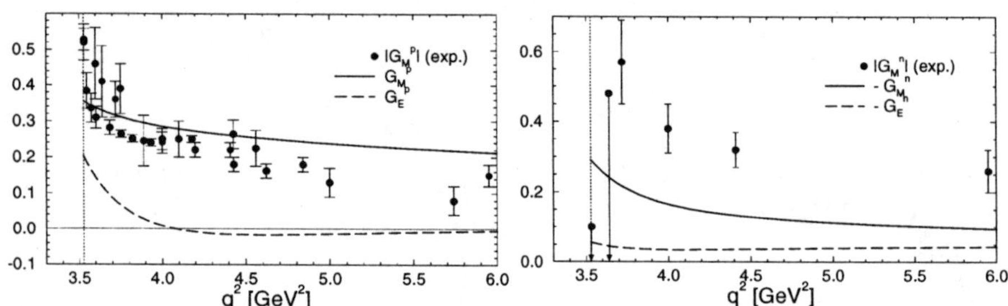

Figure 5: The timelike magnetic and electric form factors of the proton (left panel) and of the neutron (right panel). For the experimental data see Refs. [10,11].

form factor. Fig. 4 shows the proton magnetic form factor in the timelike region. We find that the threshold value is very mass dependent, see the dashed and dotted curves. The dotted curve corresponds to a best fit to the electromagnetic form factors in the space- and timelike region. The masses, however, lie below the $\Delta(1232)$ threshold. The definition of the Sachs form factors leads to the threshold conditions:

$$G_E(4M^2) = G_M(4M^2) \;\leftrightarrow\; J_+(4M^2) = 2M\,\frac{d}{dP'}J_0(4M^2)\,, \qquad (13)$$

with $P' = |\vec{P'}| = q$, $P = (M,\vec{0})$ and $J_+ = 1/2(J_1 + iJ_2)$. Fig. 5 shows $G^{n,p}_{E,M}$. We find that Eq. (13) is violated in our model. This is due to the choice of the propagator of the v-diquark. Considering the second diagram in Fig. 2 alone yields the result of Fig. 6, which shows that the relativistic treatment is correct.

4 $N - \Delta$ transition form factors

In our model, the $\Delta(3/2)$ appears as a bound state of a v-diquark and a quark, see the framed part in Fig. 1. To fix the Δ mass at 1232 MeV we introduce

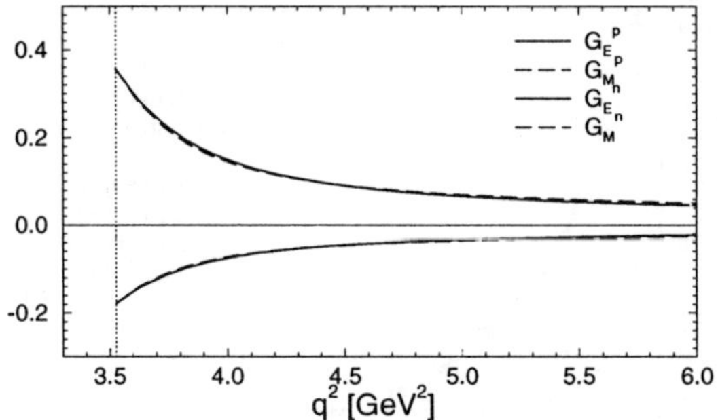

Figure 6: The timelike nucleon form factors as they result from the quark current with a scalar diquark as spectator *alone*. The upper two curves are the proton form factors, the negative ones are those of the neutron.

the quark-diquark coupling parameter g^Δ, see Tab. 1. The Δ amplitude is normalized according to $\langle \Psi_\Delta \,|\, \Psi_\Delta \rangle = 2M_\Delta$. The electromagnetic $N - \Delta$ transition is decomposed via [12]

$$e\, J_\mu(q^2) \;=\; e\,\sqrt{2/3}\,\overline{u}^\beta_{s'}(P')\, J_{\beta\mu}\, u_s(P) \,, \tag{14}$$

$$\text{with}\quad J_{\beta\mu} \;=\; G_M(q^2)\, G^M_{\beta\mu} + G_E(q^2)\, G^E_{\beta\mu} + G_C(q^2)\, G^C_{\beta\mu} \,, \tag{15}$$

with G_M, G_E, G_C the magnetic dipole, electric and Coulomb quadrupole form factors, respectively. In the rest frame of the incoming nucleon this leads to:

$$\left(\begin{array}{c} G_M(q^2) \\ G_E(q^2) \end{array} \right) \;=\; \sqrt{3/2}\, \frac{2\sqrt{2}}{g(q^2)} \left(\begin{array}{cc} \frac{\sqrt{3}}{4} & \frac{1}{12} \\ \frac{1}{4\sqrt{3}} & -\frac{1}{12} \end{array} \right) \left(\begin{array}{c} J_+(q^2) \\ J'_+(q^2) \end{array} \right) \tag{16}$$

$$G_C(q^2) \;=\; -\sqrt{3/2}\, \frac{4 M_\Delta M_N}{g(q^2)\, \sqrt{6\, Q^+ Q^-}}\, J_0(q^2) \,, \tag{17}$$

with $Q^\pm = (M_\Delta \pm M_N)^2 - q^2$, $g(q^2) = ((M_\Delta + M_N)/(2M_N))\sqrt{Q^-}$ and $J'_+ = \langle +3/2 \,|\, J_+ \,|\, +1/2 \rangle$. Fig. 7 shows the calculated form factors and the experimental G_M [13]. We find that G_M comes out a factor of 2 too low. Here, the inclusion of scalar to v-diquark transitions leads only to a small improvement. Note that G_C is negative, and $G_E(q_s^2) \approx G_C(q_s^2) \approx 0$ at the pseudothreshold $q_s^2 = (M_\Delta - M_N)^2$. Figs. 8 and 9 show the ratio $E2/M1$ and the absolute value $|C2/M1|$, respectively. With Set B we find the correct threshold value $E2/M1 = -3.4\,\%$ [14]. For $C2/M1$ we obtain a positive sign, with the absolute value describing the data well. Note that we follow consistently the definitions

34

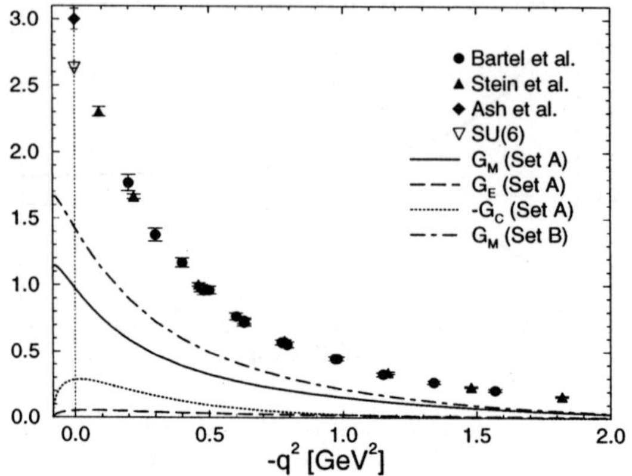

Figure 7: The N-Δ transition form factors.

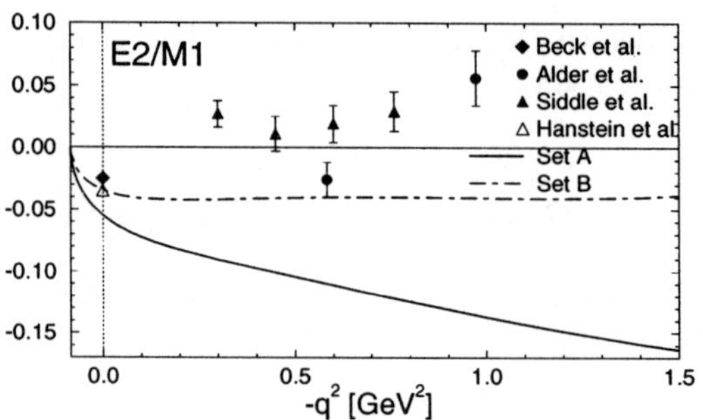

Figure 8: The ratio $E2/M1$.

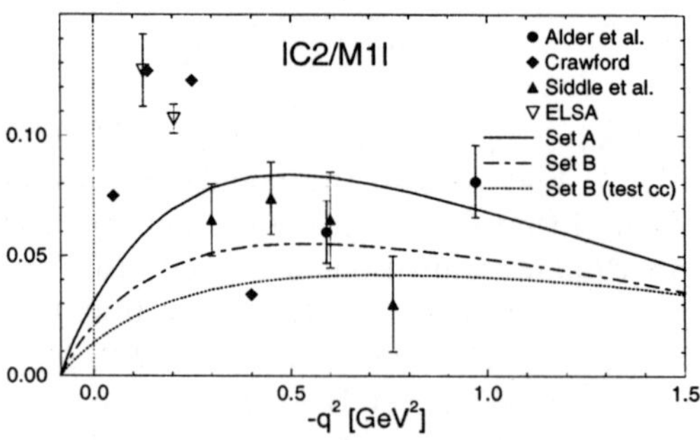

Figure 9: The ratio $|C2/M1|$.

of Ref. [12]. The two lines of Set B indicate the variance of the calculated ratio due to the fact that the $N - \Delta$ current is only partially conserved.

5 Summary

We developed a formally covariant quark-diquark model of the nucleon. The nucleon is assumed to be composed of a quark and a scalar/axial-vector diquark which interact via an instantaneous quark exchange. With a single set of parameters we are able to describe the nucleon electromagnetic form factors in the space- and timelike region for momentum transfers up 3 GeV^2 above threshold. In this picture, the $\Delta(3/2)$ is a bound state of a quark and a v-diquark. The results for the $N - \Delta$ transition agree only qualitatively with experiment, but we find the correct value for $E2/M1$.

Acknowledgments

I am indebted to H.R. Petry and B.C. Metsch. This work was supported by the Deutsche Forschungsgemeinschaft.

References

1. N. Ishii, W. Bentz, K. Yazaki, Phys. Lett. B **318**, 26 (1993)
2. S. Huang, J. Tjon, Phys. Rev. C **49**, 1702 (1994)
3. A. Buck, R. Alkofer, H. Reinhardt, Phys. Lett. B **286**, 29 (1992)
4. E.E. Salpeter, Phys. Rev. **87**, 328 (1952)
5. H. Meyer, Phys. Lett. B **337**, 37 (1994)
6. S. Mandelstam, Proc. Roy. Soc. **233**, 248 (1955)
7. V. Keiner, Z. Phys. A **354**, 87 (1996)
8. V. Keiner, Phys. Rev. C **54** (1996)
9. P. Mergell, U.-G. Meißner, D. Drechsel, Nucl. Phys. **A596**, 367 (1996)
10. H.W. Hammer, U.-G. Meißner, D. Drechsel, [hep-ph/9604294], to appear in Phys. Lett. (1996)
11. C. Bini, see the contribution in these proceedings
12. R.C.E. Devenish, T.S. Eisenschitz, J.G. Körner, Phys. Rev. D **14**, 3063 (1976)
13. V. Keiner, Bonn TK-96-28, [nucl-th/9610014] (1996)
14. O. Hanstein, D. Drechsel, L. Tiator, [nucl-th/9605008], submitted to Phys. Lett. B

Baryon Structure in a Covariant Diquark–Quark Model[*]

G. Hellstern,

R. Bäurle, U. Zückert, R. Alkofer, H. Reinhardt
*Institute for Theoretical Physics, Auf der Morgenstelle 14,
72076 Tübingen, Germany*

The baryon structure is investigated in a covariant diquark-quark model. In this approach baryons emerge as relativistic bound states of a constituent quark and a 0^+ or 1^+ diquark. After solving the Bethe-Salpeter Equation for the scalar diquark quark system in ladder approximation we couple various external currents to the constituents of the baryon to probe its internal structure. The quark and the diquarks are assumed to be confined which is implemented by suitable choices for the propagators. This leads to nontrivial vertex functions between the constituents and the external current. Baryonic matrix elements are then evaluated to extract observable formfactors.

1   Introduction

The main goal of our work is to understand the baryon structure up to a momentum transfer of $Q^2 \approx 1 - 2$ GeV. This momentum regime is particulary interesting because of the interplay between hadronic degrees of freedom on the one side and their intrinsic quark structure on the other side. To describe hadrons as bound states of quarks nonperturbative methods of QCD are unavoidable, whereas for large Q^2 the perturbative results of QCD should be met. So we are looking for an interpolating model to describe baryon structure at the intermediate energy region.

From the experimental side there is also an interest in this kind of calculations, because proposed and already started experiments (e.g. at COSY, ELSA, TJNAF) explore the baryon structure to a very high precision in this momentum regime.

This workshop is devoted to diquarks as a tool to parametrize complicated or unknown structures. We will exploit this philosophy in the following. Since a fully relativistic Faddeev equation which determines the nucleon as bound state of three quarks (bound by some kind of gluon interaction) is almost impossible to solve it is worthwhile to study the approximation where two quarks are bound to a diquark and interact with the third quark through quark exchange. Such a picture of a baryon has been derived by path integral techniques in the context of the Nambu–Jona-Lasinio model[1] and in the Global Color Model[2].

[*] Supported by COSY under contract 41315266.

2 Bethe-Salpeter Equation for Baryons

The basic assumption of our model is that a baryon is a bound state of a constituent quark and a scalar (0^+) or axialvector (1^+) diquark which are bound due to quark exchange [1]. The corresponding Bethe-Salpeter equation (BSE) for the nucleon is then given by [a],

$$
\Phi(P,p) = \int \frac{d^4k}{(2\pi)^4} g(-k-p)\gamma_5 S(-k-p)\gamma_5 S(\tfrac{1}{2}P+k)D(\tfrac{1}{2}P-k)\Phi(P,k)
$$

$$
+ \int \frac{d^4k}{(2\pi)^4} g(-k-p)\gamma^\mu S(-k-p)\gamma_5 S(\tfrac{1}{2}P+k)D^{\mu\nu}(\tfrac{1}{2}P-k)\Phi^\nu(P,k)
$$

$$
\Phi^\mu(P,p) = \int \frac{d^4k}{(2\pi)^4} g(-k-p)\gamma_5 S(-k-p)\gamma^\mu S(\tfrac{1}{2}P+k)D(\tfrac{1}{2}P-k)\Phi(P,k)
$$

$$
+ \int \frac{d^4k}{(2\pi)^4} g(-k-p)\gamma^\lambda S(-k-p)\gamma^\mu S(\tfrac{1}{2}P+k)D^{\lambda\rho}(\tfrac{1}{2}P-k)\Phi^\rho(P,k)
$$

$$\tag{1}$$

where $\Phi(P,p)$ and $\Phi(P,p)^\mu$ are the baryon (scalar and axial vector) vertex functions with amputated quark and diquark legs. The quark propagator is denoted by $S(q)$ and the diquark propagator by $D(q)$ and $D^{\mu\nu}(q)$ respectively. $g(q)$ describes a possible extension of the diquark quark vertices. A solution of the Bethe-Salpeter equation which assumes a static quark exchange can be found in ref. [3]. Although 1^+ diquarks are partially included in our calculation [4], we will restrict ourselves in the following to scalar diquarks so that only the first line of equation (1) survives.

Because $\Phi(P,p)$ describes a spin $\tfrac{1}{2}$ particle it is useful to write it as a product of an amplitude $\Psi(P,p)$, which depends on the total and the relative momentum of the nucleon, and a Dirac spinor $u(P,S)$: [5]

$$
\Phi(P,p) := \Psi(P,p)u(P,S). \tag{2}
$$

After inserting this ansatz in the Bethe-Salpeter Equation (1), we multiply with the adjoint spinor $\bar{u}(P,S)$ from the right hand side and sum over the nucleon spins. It is the immidiately seen, that only the projections

$$
\chi(P,p) := \Psi(P,p)\Lambda^+, \qquad \Lambda^+ = \frac{(-iP\cdot\gamma + M_B)}{2M_B} \tag{3}
$$

---

[a]We use a Euclidean space formulation with $\{\gamma_\mu,\gamma_\nu\} = 2\delta_{\mu\nu}$, $\gamma_\mu^\dagger = \gamma_\mu$, $p\cdot q = \sum_{i=1}^4 = p_i q_i$ and assume isospin symmetry.

38

to positive energy survive. The projected amplitude $\chi(P,p)$ is then by construction an eigenfunction of Λ^+,

$$\chi(P,p) = \Psi(P,p)\Lambda^+ = \chi(P,p)\Lambda^+. \tag{4}$$

This leads to the possibility to decompose further the Dirac structure of χ : The decomposition

$$\chi(P,p) = \tfrac{1}{2}(1 - i\tfrac{\gamma P}{M_B})S_1 + \tfrac{1}{2}(-i\gamma \cdot p_T - \tfrac{1}{2M_B}(\gamma \cdot p \gamma \cdot P - \gamma \cdot P \gamma \cdot p))S_2, \tag{5}$$

is the only one compatible with equation (4). $p_T^\mu = p^\mu + P^\mu \frac{Pp}{M_B^2}$ denotes the transversal relative momentum. The still unknown functions $S_1(P^2, pP, p^2)$ and $S_2(P^2, pP, p^2)$ are of course determined by the solution of the BSE.

In the rest frame of the bound state (where the actual calculations are done) we obtain

$$\chi_{\text{r.f.}}(P,p) = \begin{pmatrix} \mathbf{1}S_1(P,p) & 0 \\ \vec{\sigma}\vec{p}S_2(P,p) & 0 \end{pmatrix}. \tag{6}$$

It is now obvious that the first two columns of the 4×4 matrix correspond to spinors with positive energy, spin up ("large" component) and spin down ("small" component) respectively. In the calculation of baryonic matrix elements we will restrict ourselves, in a first approximation, therefore to S_1.

Finally we expand the scalar functions $S_j(P,p), j = 1, 2$ in terms of Gegenbauer Polynomials[6]

$$S_j(P,p) = \sum_{k=0}^{\infty} i^k f_k^j(\sqrt{p^2})C_k^1(\frac{p \cdot P}{\sqrt{P^2}\sqrt{p^2}}), \tag{7}$$

and solve the Bethe-Salpeter Equation in the rest frame of the nucleon either by diagonalization or by iteration. In this way, the bound state mass is determined by the eigenvalue of the integral equation, while the expansion coefficients f_k^j correspond to the eigenvectors.

2.1 Parametrization of confined constituents

Since neither quarks nor diquarks have ever been observed as asymptotic states in experiments, there is not much justification necessary to include this feature of QCD in a calculation of baryon properties.

We implement confinement by using propagators which have no poles. In case of the quark propagator this has to be understood as an effective

parametrization of the behavior obtained in Dyson-Schwinger studies of QCD [7].
The quark self-energies which appear in

$$S(p) = -i\gamma \cdot p\sigma_v(p) + \sigma_s(p) = [i\gamma \cdot pA(p) + B(P)]^{-1} \qquad (8)$$

are then choosen as

$$\sigma_v(p) = \frac{1 - \exp(-(1 + \frac{p^2}{m_q^2})}{p^2 + m_q^2} \quad \text{and} \quad \sigma_s(p) = m_q\sigma_v(p). \qquad (9)$$

Obviously the exponentials remove the mass poles however, thereby introducing an essential singularity at $p^2 = -\infty$.

In ref.[8] it has been shown that confined diquarks can be obtained when one uses an appropriate irreducible quark-quark kernel (beyond ladder approximation). We simulate this by using

$$D(p) = \frac{-iZ(p)}{p^2 + m_d^2} \qquad (10)$$

where

$$Z(p) = 1 - \exp(-(1 + \frac{p^2}{m_d^2})). \qquad (11)$$

Note that an elementary scalar diquark would correspond to $Z \equiv 1$. With such a prescription of confined constituents we especially get rid of the unphysical quark-diquark thresholds which plague baryon calculations within the pure hadronized NJL-model [9].

To estimate the effects of an intrinsic diquark structure we did the calculation with a pointlike quark-diquark coupling

$$g(k) = g_0 \qquad (12)$$

as well as with a diquark formfactor (see also ref. [10])

$$g(k) = g_0 \frac{\Lambda^2}{k^2 + \Lambda^2}, \qquad (13)$$

where Λ is of the order of $2m_q$ and which simulates the diquark structure in a crude way.

2.2 Numerical results

In Fig.1 the numerical results for the dominant expansion functions f_k^j without and with a diquark formfactor are shown. The parameters here and in the

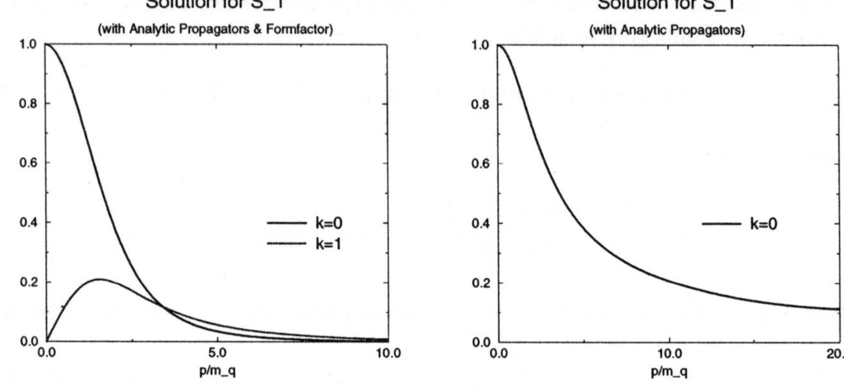

Figure 1: Numerical Solution of the Bethe-Salpeter equation. The results in the left figure are obtained with a diquark formfactor, in the right figure with a pointlike coupling.

following are fixed to $m_q = m_d = 0.5$ GeV and the coupling is chosen to obtain $M_B = 0.95$ GeV. In both figures the functions fall off rapidly for large relative momenta, but the asymptotic behavior is quite different: Assuming a diquark structure leads to a narrower amplitude in momentum space. This will also have an effect for the observables, where a possible diquark structure just enters via the nucleon vertex function. We observe that in both cases the expansion in terms of Gegenbauer Polynomials converge rather rapidly.

3 Hadronic matrix elements

3.1 Mandelstam's formalism

Matrix elements of operators in the Bethe-Salpeter approach are calculated according to Mandelstam's formalism[11]

$$\langle \hat{O} \rangle = \int \frac{d^4p}{(2\pi)^4} \int \frac{d^4p'}{(2\pi)^4} \bar{\psi}(P', p') \Gamma_{\hat{O}}(p', P'; p, P) \psi(P, p). \tag{14}$$

Note that P and p (P' and p') are the total and relative momenta of the diquark-quark states before and after the interaction; ψ is the normalized diquark-quark Bethe-Salpeter wave function related to the vertexfunction by multiplying with the quark and diquark propagator. The 5-point function $\Gamma_{\hat{O}}$ describes how the external current couples to all internal lines of the diquark-quark system.

In the following we work in a generalized impulse approximation, where we consider only the coupling to the quark and to the diquark,

$$\Gamma_{\hat{O}}(p', P'; p, P) = \Gamma_{\hat{O}}^{q}(p', P'; p, P) + \Gamma_{\hat{O}}^{d}(p', P'; p, P), \tag{15}$$

and neglect the coupling to the exchanged quark.

3.2 Vertexfunctions of the constituents

In case of electromagnetic (e.m.) formfactors, we still have to know the quark-photon and the diquark-photon vertexfunctions. Using gauge symmetry, which is manifest in the Ward-Takahashi identity,

$$q_{\mu} i \Gamma_{\mu}(p, q, p+q) = S^{-1}(p+q) - S^{-1}(p) \tag{16}$$

one observes the connection of the longitudinal part of the vertexfunction and the inverse propagator.

A vertexfunction which solves the above identity for the quark is the Ball-Chiu vertex [12]

$$\Gamma_{\mu}^{BC}(p, k) = \frac{A(p^2) + A(k^2)}{2} \gamma_{\mu} + \frac{(p+k)_{\mu}}{p^2 - k^2}\{(A(p^2) - A(k^2))\frac{\gamma p + \gamma k}{2}$$
$$-i(B(p^2) - B(k^2))\}, \tag{17}$$

and in case of the diquark

$$\Gamma_{\mu}^{d}(p, k) = (p + k)_{\mu}\frac{D^{-1}(p) - D^{-1}(k)}{p^2 - k^2}. \tag{18}$$

These vertexfunctions are totally determined by the self-energy functions of the corresponding propagators.

4   Electromagnetic nucleon formfactors

Within our approximations we decompose the matrix element into a quark and a diquark part

$$J_{\mu}(Q^2) = Q_q J_{\mu}^{q}(Q^2) + Q_d J_{\mu}^{d}(Q^2), \tag{19}$$

42

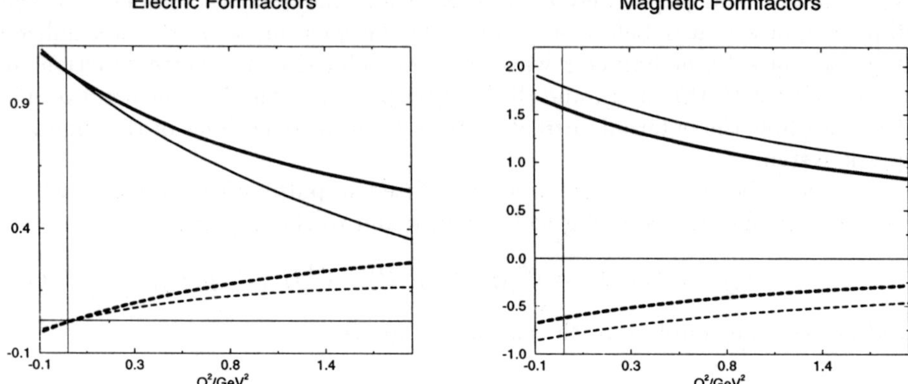

Figure 2: Electromagnetic Formfactors of the nucleon: The solid lines are the proton formfactors, the dotted lines the neutron formfactors; the thin curves are the results with a diquark formfactor, whereas the thick curves are the results assuming a pointlike quark-diquark coupling.

with the charge matrices in isospin space

$$Q_q = \tfrac{1}{2}(\tfrac{1}{3}\mathbf{1}_F + \boldsymbol{\tau}_z), \quad Q_d = \tfrac{1}{3}\mathbf{1}_F. \tag{20}$$

The quark part is given by

$$J_\mu^q(Q^2) = \int \frac{d^4k}{(2\pi)^4} \bar{\chi}(P_f, p_f) S_q(k_+) i\Gamma_\mu^{BC}(k_+, k_-) S_q(k_-) S_d(k_d) \chi(P_i, p_i), \tag{21}$$

whereas the diquark part is calculated with the loop diagram

$$J_\mu^d(Q^2) = \int \frac{d^4k}{(2\pi)^4} \bar{\chi}(P_f, p_f) S_d(k_+) i\Gamma_\mu^d(k_+, k_-) S_d(k_-) S_q(k_q) \chi(P_i, p_i). \tag{22}$$

From Lorentz invariance and discrete symmetries it is also known that the onshell e.m. nucleon current can be decomposed into

$$\begin{aligned} J_\mu(Q^2) &= \langle N(P_f, S_f)|j_\mu|N(P_i, S_i)\rangle \\ &= \bar{u}(P_f, S_f)[iM_B(F_e(Q^2) - F_m(Q^2))\frac{P_\mu}{P^2} + F_m(Q^2)\gamma_\mu]u(P_i, S_i). \end{aligned} \tag{23}$$

Note that in our formalism the Dirac spinors appearing on the right hand side are replaced by the Λ^+ projectors. The Lorentz invariant functions $F_e(Q^2)$

and $F_m(Q^2)$ denote the electric and magnetic formfactor, respectively, which depend on the photon momentum Q^2. When the left hand side of this equation is calculated with the above loop diagrams in the Breit frame, $F_e(Q^2)$ and $F_m(Q^2)$ can be extracted by taking appropriate traces.

In Fig.2 the numerical results for the e.m. formfactors are shown. We observe that an internal diquark structure (leading to a narrower vertex function in momentum space) seems to be necessary in our approach because otherwise the variation of the formfactors with Q^2 around 0 is only weak, the e.m. radii are then far too smalll. While the violation of charge conservation due to the impulse approximation can be compensated by choosing a suitable momentum distribution in the loop integrals, the absolute values of the magnetic moments are not large enough which means that 1^+ diquarks are needed for a phenomenologically satisfying description. Since our model is covariant, timelike formfactors are also accessible, which can be seen in the figure: The formfactors are continous at $Q^2 = 0$.

5 Pion-nucleon formfactor

When following Mandelstam's prescripton it is also possible to couple an external pion current to the nucleon. Because of parity there is no coupling to the (scalar) diquark, so there is just one diagram to calculate (in analogy to equation (21) but where the quark-photon vertex is replaced by the quark-pion vertex). The quark-pion vertex (= pion Bethe-Salpeter amplitude) in the chiral limit ($m_\pi = 0$) is determined by the scalar quark selfenergy

$$\Gamma_5^a(P^2 = 0, p^2) = \gamma_5 \frac{B(p^2)}{f_\pi} \tau^a, \tag{24}$$

as a consequence of chiral symmetry[13]. Furthermore the invariant parametrization of the pion-nucleon matrix element is given by

$$
\begin{aligned}
J_5^a(Q^2) &= \langle N(P_f, S_f)|j_5|N(P_i, S_i)\rangle \\
&= \bar{u}(P_f, S_f)[\gamma_5 g_{\pi NN}(Q^2)\tau^a]u(P_i, S_i).
\end{aligned}
\tag{25}
$$

In analogy to the electromagnetic formfactors, $g_{\pi NN}$ on the right hand side of this equation can be extracted when the left hand side is evaluated with the diagram where the pion couples to the quark and the diquark remains a spectator. Our numerical results for $g_{\pi NN}$ are displayed in Fig. 3. Again a diquark structure is unavoidable to get the expected variation of the formfactor with the pion momentum. The absolute value of $g_{\pi NN}$ at the pion mass shell ($Q^2 = 0$ in the chiral limit) is in remarkable agreement with the experimental value of about 14.

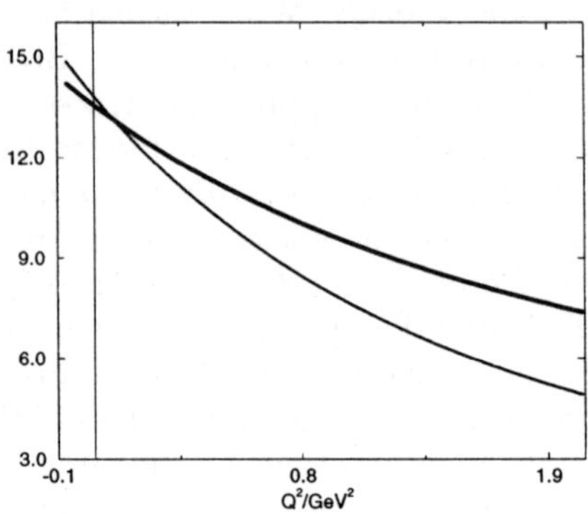

Figure 3: Numerical Solutions for the Pion-Nucleon coupling $g_{\pi NN}(Q^2)$. The thin curve denotes our results with the diquark formfactor, the thick curve our results without a formfactor.

6 Conclusion

In this talk we have presented a covariant bound state approach for the nucleon where quarks and diquarks interact through quark exchange. Using an effective parametrization of confinement we calculated within the needed approximations not only the nucleon mass and wave function but also electromagnetic and strong formfactors. It has been discussed that an internal diquark structure, entering through a diquark formfactor in the BSE is preferable to obtain a nucleon wave function which has a realistic width in momentum space. Furthermore the implementation of axialvector diquarks is necessary to improve the magnetic observables.

This work is just the first step towards a full satisfying baryon model of the proposed diquark-quark bound state approach. Nevertheless the results obtained so far are quite encouraging. To benefit from the advantages of our covariant and confining approach we will also apply this model to electromagnetic and strong NN^* transitions, virtual Compton scattering and Λ-production.

Acknowledgments

G.H. would like to thank Prof. Anselmino and Prof. Predazzi for the very stimulating and friendly athomosphere in Torino. He is also grateful to the Graduiertenkolleg ,,Hadronen und Kerne" in Tübingen for financial support.

References

1. H. Reinhardt, Phys. Lett. **B 244** (1990) 316.
2. R. T. Cahill, Aust. J. Phys. **42** (1989) 171.
3. A. Buck, R. Alkofer, H. Reinhardt, Phys. Lett. **B 286** (1992) 29.
4. In preparation.
5. H. Meyer, Phys. Lett. **B337** (1994) 37.
6. T. Nieuwenhuis, J. A. Tjon, to appear in Few Body Systems **21** (1996).
7. C. D. Roberts, A. G. Williams, Prog. Part. Nuc. Phys. **33** (1994).
8. A. Bender, C. D. Roberts, L. v. Smekal, Phys. Lett. **B 380** (1996) 7.
9. G. Hellstern, C. Weiß, Phys. Lett. **B 351** (1995) 64.
10. K. Kusaka, G. Piller, A. W. Thomas, A. G. Williams, hep-ph/9609277.
11. S. Mandelstam, Proc. Roy. Soc. **233** (1955) 248.
12. J. S. Ball, T.W. Chiu, Phys. Rev. **D 22** (1980) 2542.
13. R. Delbourgo, M. D. Scadron, J. Phys. **G 5** (1979) 1621.

EXCLUSIVE PHOTO- AND ELECTROPRODUCTION OF MESONS IN THE GEV REGION [a]

G. FOLBERTH, R. ROSSMANN, AND W. SCHWEIGER

Institute of Theoretical Physics, University of Graz, Universitätsplatz 5,
A-8010 Graz, Austria

We consider the reactions $\gamma p \to MB$, with MB being either $K^+ \Lambda$, $K^+ \Sigma^0$, or $\pi^+ n$, within a diquark model which is based on perturbative QCD. The model parameters and the quark-diquark distribution amplitudes of the baryons are taken from previous investigations of electromagnetic baryon form factors and Compton-scattering off protons. Reasonable agreement with the few existing photoproduction data at large momentum transfer is found for meson distribution amplitudes compatible with the asymptotic one ($\propto x(1 - x)$). We present also first results for hard electroproduction of the $K^+ \Lambda$ final state. Our predictions exhibit some characteristic features which could easily be tested in forthcoming electroproduction experiments.

1 Introduction

Our investigation of exclusive photo- and electroproduction of mesons is part of a systematic study of hard exclusive reactions [1,2,3] within a model which is based on perturbative QCD, in which baryons, however, are treated as quark-diquark systems. This model has already been applied to baryon form factors in the space-[1] and time-like region [2], real and virtual Compton scattering [3], two-photon annihilation into proton-antiproton [2] and the charmonium decay $\eta_c \to p\bar{p}$ [2]. A consistent description of these reactions has been achieved in the sense that the corresponding large momentum-transfer data ($p_\perp^2 \gtrsim 3$ GeV^2) are reproduced with the same set of model parameters. Further applications of this model include three-body J/Ψ decays [4] and the calculation of Landshoff contributions in elastic proton-proton scattering [5]. Like the usual hard-scattering approach (HSA) [6] the diquark-model relies on factorization of short- and long-distance dynamics; a hadronic amplitude is expressed as a convolution of a hard-scattering amplitude $\widehat{T}$, calculable within perturbative QCD, with distribution amplitudes (DAs) ϕ^H which contain the (non-perturbative) bound-state dynamics of the hadronic constituents. The introduction of diquarks does not only simplify computations, it is rather motivated by the requirement to extend the HSA from (asymptotically) large down to intermediate momentum transfers ($p_\perp^2 \gtrsim 3$ GeV^2). This is the momentum-transfer region where some experimental data already exist, but where still persisting non-perturbative

---

[a]Talk given by W. Schweiger

effects, in particular strong correlations in baryon wave functions, prevent the pure quark HSA to become fully operational. Diquarks may be considered as an effective way to cope with such effects. Photoproduction of mesons is one class of exclusive reactions where the pure-quark HSA obviously fails in reproducing the hard-scattering data[7] and where a modification of the perturbative QCD approach seems to be necessary in the few-GeV region.

The following section starts with an outline of the hard-scattering approach with diquarks and introduces the hadron DAs to be used in the sequel. The diquark model predictions for the various reaction channels are presented in Sec. 3 along with some general considerations on photo- and electroproduction reactions. Our conclusions are finally given in Sec. 4.

2 Hard Scattering with Diquarks

Within the hard-scattering approach a helicity amplitude $M_{\{\lambda\}}$ for the reaction $\gamma^{(*)}\,\mathrm{p} \to \mathrm{M\,B}$ is (to leading order in $1/p_\perp$) given by the convolution integral[6]

$$M_{\{\lambda\}}(\hat{s},\hat{t}) = \int_0^1 dx_1\,dy_1\,dz_1\,\phi^{\mathrm{M}\dagger}(z_1,\tilde{p}_\perp)\phi^{\mathrm{B}\dagger}(y_1,\tilde{p}_\perp)\widehat{T}_{\{\lambda\}}(x_1,y_1,z_1;\hat{s},\hat{t})\phi^{\mathrm{P}}(x_1,\tilde{p}_\perp)\,.$$

(1)

The distribution amplitudes ϕ^{H} are probability amplitudes for finding the valence Fock state in the hadron H with the constituents carrying certain fractions of the momentum of their parent hadron and being collinear up to a factorization scale $\tilde{p}_\perp$. In our model the valence Fock state of an ordinary baryon is assumed to consist of a quark (q) and a diquark (D). We fix our notation in such a way that the momentum fraction appearing in the argument of ϕ^{H} is carried by the quark – with the momentum fraction of the other constituent (either diquark or antiquark) it sums up to 1 (cf. Fig. 1). For our actual calculations the (logarithmic) $\tilde{p}_\perp$ dependence of the DAs is neglected throughout since it is of minor importance in the restricted energy range we are interested in. The hard scattering amplitude $\widehat{T}_{\{\lambda\}}$ is calculated perturbatively in collinear approximation and consists in our particular case of all possible tree diagrams contributing to the elementary scattering process $\gamma\mathrm{qD} \to \mathrm{q\bar{q}qD}$. A few examples of such diagrams are depicted in Fig. 2. The subscript $\{\lambda\}$ represents the set of possible photon, proton and Λ helicities. The Mandelstam variables s and t are written with a hat to indicate that they are defined for vanishing hadron masses. The calculation of the hard scattering amplitude involves an expansion in powers of $(m_{\mathrm{H}}/\hat{s})$ which is cut off after $\mathcal{O}(m_{\mathrm{H}}/\hat{s})$. Hadron masses, however, are fully taken into account in flux and phase-space factors.

The model, as applied in Refs. 1-3, comprises scalar (S) as well as axial-

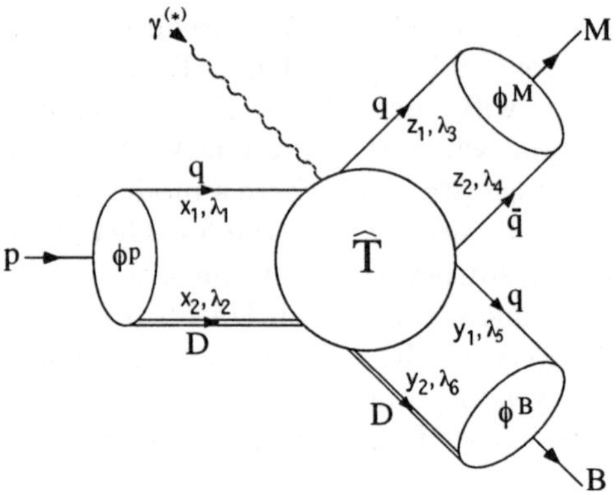

Figure 1: Graphical representation of the hard-scattering formula, Eq. (1), for $\gamma\,p \to M\,B$. x_i, y_i, and z_i denote longitudinal momentum fractions of the constituents, the λ_i's their helicities.

vector (V) diquarks. V diquarks are important, if one wants to describe spin observables which require the flip of baryonic helicities. The Feynman rules for electromagnetically interacting diquarks are just those of standard quantum electrodynamics. Feynman rules for strongly interacting diquarks are obtained by replacing the electric charge e by the strong coupling constant g_s times the Gell-Mann colour matrix t^a. The composite nature of diquarks is taken into account by multiplying each of the Feynman diagrams with diquark form factors $F_D^{(n+2)}(Q^2)$ which depend on the kind of the diquark ($D = S, V$), the number n of gauge bosons coupling to the diquark, and the square of the 4-momentum Q^2 transferred to the diquark. The form factors are parameterized by multipole functions with the power chosen in such a way that in the limit $p_\perp \to \infty$ the scaling behaviour of the pure quark HSA is recovered.

Having outlined how diquarks are treated perturbatively, a few words about the choice of the quark-diquark DAs (which incorporate the q-D bound-state dynamics) are still in order. In Refs. 1-3 a quark-diquark DA of the form ($c_1 = c_2 = 0$ for S diquarks, x ..... longitudinal momentum fraction carried by the quark)

$$\phi_D^B(x) \propto x(1-x)^3(1+c_1 x + c_2 x^2)\exp\left[-b^2\left(\frac{m_q^2}{x}+\frac{m_D^2}{(1-x)}\right)\right], \quad D = S, V,$$

$$(2)$$

in connection with an SU(6)-like spin-flavour wave function, turned out to be quite appropriate for octet baryons B. The origin of the DA, Eq. (2), is a nonrel-

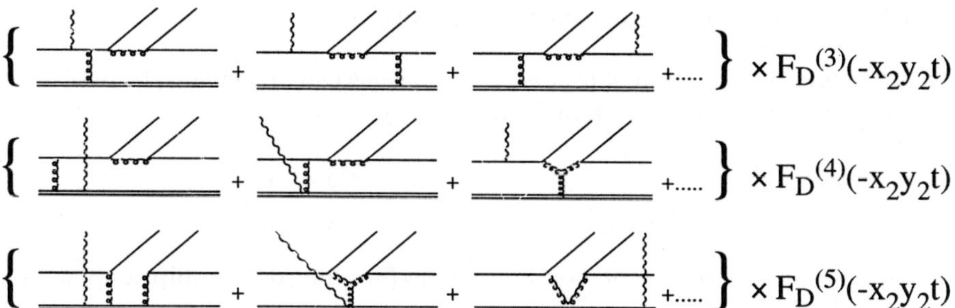

Figure 2: A few representative examples for three-, four-, and five-point contributions to $\gamma\,p \to M\,B$. As indicated, diagrams are first calculated in Born order for point-like diquarks and afterwards multiplied with appropriate vertex functions (diquark form factors).

ativistic harmonic-oscillator wave function [8]. Therefore the masses appearing in the exponential have to be considered as constituent masses (330 MeV for light quarks, 580 MeV for light diquarks, strange quarks are 150 MeV heavier than light quarks). The oscillator parameter $b^2 = 0.248$ GeV^{-2} is chosen in such a way that the full wave function gives rise to a value of 600 MeV for the mean intrinsic transverse momentum of a quark inside a nucleon.

For meson DAs various models can be found in the literature. In order to check the sensitivity of our calculation on the choice of the meson DAs we employ two qualitatively different forms. On the one hand, the asymptotic DA

$$\phi_{\text{asy}}(x) \propto x(1 - x)\,, \tag{3}$$

which solves the $\tilde{p}_\perp$ evolution equation for $\phi(x, \tilde{p}_\perp)$ in the limit $\tilde{p}_\perp \to \infty$, and, on the other hand, the DAs

$$\phi^{\pi}_{\text{CZ}}(x) \propto \phi_{\text{asy}}(2x - 1)^2\,, \quad \phi^{K}_{\text{CZ}}(x) \propto \phi_{\text{asy}}[0.08 + 0.6(2x - 1)^2 - 0.25(2x - 1)^3]\,, \tag{4}$$

which have been proposed in Ref. 9 on the basis of QCD sum rules. The "normalization" of the meson DAs is determined by the experimental decay constants for the weak π, $K \to \mu\nu_\mu$ decays. The analogous constants f_S and f_V for the q-D DAs of baryons are free parameters of the model. They are, in principle, determined by the probability to find the q-D state (D = S, V) in the baryon B and by the transverse-momentum dependence of the corresponding wave function.

For further details of the diquark model we refer to the publication of R. Jakob et al. [1]. The numerical values of the model parameters for the present study are also taken from this paper.

3 Photo- and Electroproduction Reactions

Exclusive photoproduction of pseudoscalar mesons can, in general, be described by four independent helicity amplitudes $M_{\lambda_M,\lambda_B,\lambda_\gamma,\lambda_p}$ [10]

$$N = M_{0,-\frac{1}{2},+1,+\frac{1}{2}}, \qquad S_1 = M_{0,-\frac{1}{2},+1,-\frac{1}{2}},$$
$$D = M_{0,+\frac{1}{2},+1,-\frac{1}{2}}, \qquad S_2 = M_{0,+\frac{1}{2},+1,+\frac{1}{2}}. \tag{5}$$

N, S_1, S_2, and D represent non-flip, single-flip, and double-flip amplitudes, respectively. Two additional amplitudes,

$$\tilde{N} = M_{0,+\frac{1}{2},0,+\frac{1}{2}}, \qquad S_3 = M_{0,-\frac{1}{2},0,+\frac{1}{2}}, \tag{6}$$

occur if electroproduction is considered. Electroproduction and photoproduction are related as far as in the one-photon approximation the electroproduction cross section can be expressed in terms of four response functions $d\sigma_U$, $d\sigma_L$, $d\sigma_T$, and $d\sigma_I$ for the production of a pseudoscalar meson by a virtual photon, $\gamma^* p \to M B$ [11]. The two helicity amplitudes in Eq. (6) correspond to the longitudinal polarization degree of the virtual photon. The two response functions $d\sigma_U$ and $d\sigma_L$ are just cross sections for (spin averaged) transversely and longitudinally polarized photons, respectively. The other two functions are transverse-transverse ($d\sigma_T$) and longitudinal-transverse ($d\sigma_I$) interference terms. In the limit of vanishing photon-virtuality ($Q^2 \to 0$) $d\sigma_U$ reduces to the ordinary photoproduction cross section, the ratio $d\sigma_T/d\sigma_U$ becomes the (negative) photon asymmetry Σ [10] and $d\sigma_L$ as well as $d\sigma_I$ become zero.

From the q-D flavour functions (D = S, V)

$$\chi_S^P = uS_{[u,d]}, \qquad \chi_V^P = [uV_{\{u,d\}} - \sqrt{2}dV_{\{u,u\}}]/\sqrt{3}, \tag{7}$$

$$\chi_S^n = dS_{[u,d]}, \qquad \chi_V^n = -[dV_{\{u,d\}} - \sqrt{2}uV_{\{d,d\}}]/\sqrt{3}, \tag{8}$$

$$\chi_S^{\Sigma^0} = [dS_{[u,s]} + uS_{[d,s]}]/\sqrt{2}, \quad \chi_V^{\Sigma^0} = [2sV_{\{u,d\}} - dV_{\{u,s\}} - uV_{\{d,s\}}]/\sqrt{6}, \tag{9}$$

$$\chi_S^{\Lambda} = [uS_{[d,s]} - dS_{[u,s]} - 2sS_{[u,d]}]/\sqrt{6}, \quad \chi_V^{\Lambda} = [uV_{\{d,s\}} - dV_{\{u,s\}}]/\sqrt{2}. \tag{10}$$

for the baryons one infers already that the three production channels we are interested in differ qualitatively in the sense that $K^+ \Lambda$ is solely produced via the $S_{[u,d]}$ diquark, $K^+ \Sigma^0$ via the $V_{\{u,d\}}$ diquark, and $\pi^+ n$ via S and V diquarks. An important consequence of this observation is that helicity amplitudes and hence spin observables which require the flip of the baryonic helicity are predicted to vanish for the K^+-Λ final state (e.g., the polarization P of the outgoing Λ). For the other two channels helicity flips may, of course, take place by means of the V diquark.

One of the qualitative features of the HSA is the fixed-angle scaling behaviour of cross sections. The pure quark HSA implies an s^{-7} decay of the photoproduction cross section. This scaling behaviour is, of course, recovered within the diquark model in the limit $s \to \infty$. However, at finite s, where the diquark form factors become operational and diquarks appear as nearly elementary particles, the s^{-7} power-law is modified. Additional deviations from the s^{-7} decay of the cross section are due to logarithmic corrections which have their origin in the running coupling constant α_s and eventually in the evolution of the DAs ϕ^H (neglected in our calculation).

3.1 The K$^+$-Λ Channel

Figure 3 (left) shows the diquark-model predictions for the (scaled) photoproduction cross section $s^7 d\sigma/dt$ along with the few existing photoproduction data at large-momentum transfer[12] and the outcome of the pure quark HSA[7] (dash-dotted curve). Whereas the DAs of proton and Λ have been kept fixed according to Eq. (2) we have varied the K$^+$ DA. The solid and the dashed line represent results for the asymptotic (Eq. (3)) and the two-humped (Eq. (4)) K$^+$ DA, respectively, evaluated at $p^{\gamma}_{\text{lab}} = 6$ GeV. The better performance of the asymptotic DA and the overshooting of the asymmetric DA is in line with the conclusion drawn from the investigation of the pion-photon transition form factor[14] where, for the case of the pion, ϕ^{π}_{CZ} leads also to an overshooting the data. Our findings have to be contrasted with those obtained within the pure quark-model calculation of photoproduction[7], where the asymptotic forms for both, baryon and meson DAs, give systematically larger results than the combination of very asymmetric DAs. However, the numerics of Ref. 7 must be taken with some provisio. For Compton scattering off nucleons it has been demonstrated[15] that the very crude treatment of propagator singularities adopted in Ref. 7, namely keeping $i\epsilon$ small but finite, may lead to deviations from the correct result which are as large as one order of magnitude. At this point we want to emphasize that we have paid special attention to a correct and numerically robust treatment of the propagator singularities which occur in the convolution integral, Eq. (1). By carefully separating the singular contributions, exploiting delta functions, rewriting principal-value integrals as ordinary integrals plus analytically solvable principle-value integrals, we are able to do all the numerical integrations by means of very fast and stable fixed-point Gaussian quadratures.

We have also examined the relative importance of various groups of Feynman graphs and found the 3-point contributions to be by far the most important (cf. Fig. 3 (left)). 4- and 5-point contributions amount to $\approx 5\%$

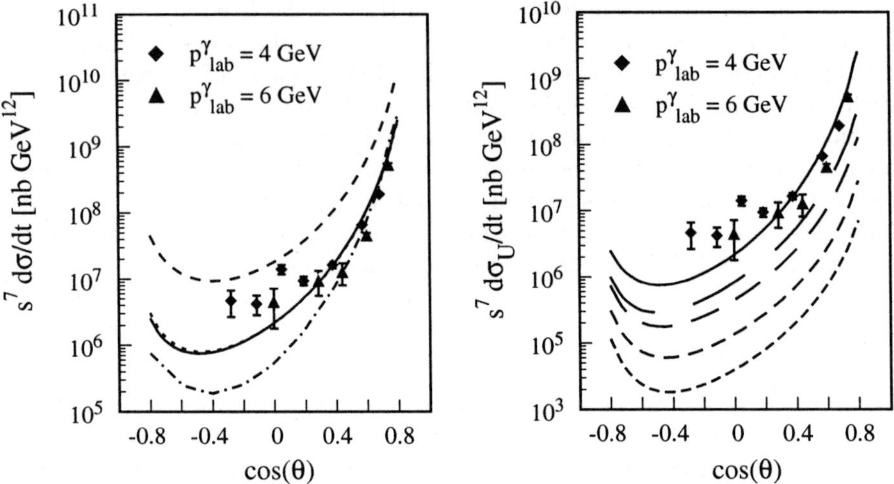

Figure 3: Diquark-model predictions for photo- (left figure) and electroproduction (right figure) of the K$^+$-Λ final state. Results are shown for fixed $p^\gamma_{\mathrm{lab}} = 6$ GeV. The experimental points are photoproduction data taken from Anderson et al.[12]. *Left figure*: differential cross section for $\gamma\,\mathrm{p} \to \mathrm{K}^+\Lambda$ scaled by s^7 vs $\cos(\theta_{\mathrm{cm}})$; solid (dashed) line: diquark-model result for p and Λ DAs chosen according to Eq. (2), K$^+$ DA according to Eq. (3) (Eq. (4)); dotted line: same as full line, but only three-point contributions taken into account; dash-dotted line: quark-model result[7] for the asymmetric p and Λ DAs of Ref. 13 and the two-humped K$^+$ DA of Eq. (4). *Right figure*: the spin-averaged cross section $d\sigma_{\mathrm{U}}/dt$ (scaled by s^7) vs $\cos(\theta_{\mathrm{cm}})$ for transversely polarized photons with virtualities $Q^2 = 0$ (photoproduction limit), 0.5, 1, 2, and 3 GeV^2; dashes become shorter with increasing Q^2. Proton and Λ DAs are chosen according to Eq. (2), the K$^+$ DA according to Eq. (3).

at $\theta_{\mathrm{cm}} = 90°$ and $E^\gamma_{\mathrm{lab}} = 6$ GeV as long as only $d\sigma/dt$ is considered. Spin observables, on the other hand, are much more affected by 4- and 5-point contributions. A more detailed discussion of $\gamma\,\mathrm{p} \to \mathrm{K}^+\Lambda$ (and also $\gamma\,\mathrm{p} \to \mathrm{K}^{*+}\Lambda$) with a full account of calculational techniques and analytical expressions for the photoproduction amplitudes can be found in Ref. 16.

The diquark-model predictions for a few electroproduction observables are depicted in Fig. 3 (right) and Fig. 4. In these plots we have concentrated on the asymptotic form, Eq. (3), of the K$^+$ DA. Like in the case of photoproduction the large-s behaviour of the four cross section contributions is s^{-7} (modified by logarithmic factors) provided that Q^2/s is kept fixed. For fixed photon virtuality Q^2 and $s \to \infty$, however, $d\sigma_{\mathrm{L}}/dt$ and $d\sigma_{\mathrm{I}}/dt$ decay like s^{-9} and s^{-8}, respectively. For fixed $p^\gamma_{\mathrm{lab}} = 6$ GeV the transverse cross-section contribution $d\sigma_{\mathrm{U}}/dt$ decreases with increasing Q^2/s (cf. Fig. 3 (right)). The same holds for

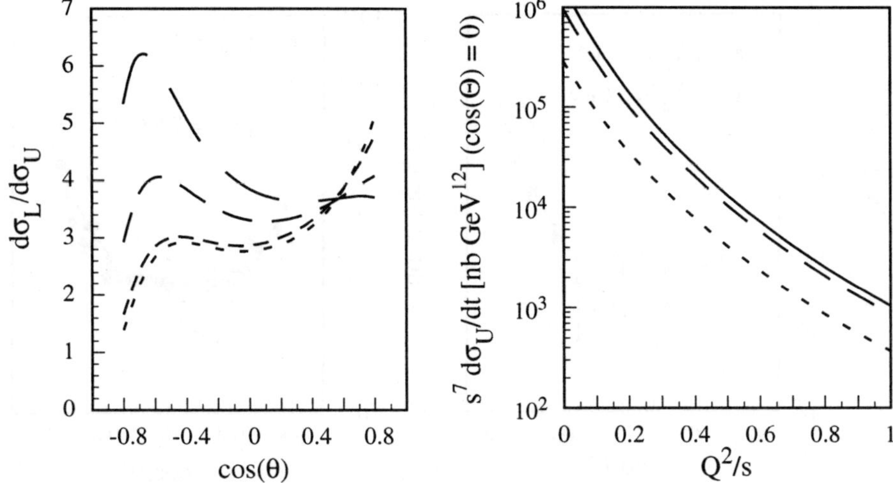

Figure 4: Diquark-model predictions for electroproduction of the K^+-Λ final state. *Left figure*: ratio of cross sections $d\sigma_L/d\sigma_U$ for longitudinally and transversely polarized photons, respectively, vs $\cos(\theta_{cm})$ for fixed $p^\gamma_{lab} = 6$ GeV; dashes become shorter with increasing photon virtuality Q^2 ($= 0.5, 1, 2, 3$ GeV^2). *Right figure*: $d\sigma_U/dt$ at fixed $\theta_{cm} = 90°$ vs Q^2/s for $s = 10$ (solid), 20 (long dashed), and 100 GeV^2 (short dashed).

$d\sigma_L/dt$ (if $Q^2/s \gtrsim 0.04$). At $Q^2/s = 0.04$ $d\sigma_L/dt$ is already larger than $d\sigma_U/dt$ (cf. Fig. 4 (left)). Since $d\sigma_L/dt$ vanishes for $Q^2 \to 0$ it thus has to rise very sharply at small values of Q^2/s. To get a better feeling for the Q^2-dependence of $d\sigma_U/dt$ we have plotted this quantity for fixed scattering angle ($\theta_{cm} = 90°$) and different values of s as function of Q^2/s (cf. Fig. 4 (right)). For exact s^{-7} scaling the curves for the three different values of s should coincide. The deviation from the s^{-7} scaling can mainly be ascribed to the running coupling constant α_s. The dependence of $d\sigma_U/dt$ (and likewise $d\sigma_L/dt$) on Q^2/s can be roughly parameterized by means of a function $\propto (1 + c\,Q^2/s)^{-6}$ with the parameter c between 2 and 2.5. The qualitative features, like scaling behaviour, Q^2/s dependence, and dominance of $d\sigma_L/dt$ as compared to $d\sigma_U/dt$ remain, of course, unaltered if the asymptotic K^+ DA is replaced by the asymmetric DA of Eq. (4). The angular dependence of cross-section ratios, like $d\sigma_L/d\sigma_U$, $d\sigma_T/d\sigma_U$, or $d\sigma_I/d\sigma_U$, however, exhibit a marked sensitivity on the choice of the K^+ DA. First experimental constraints on hard exclusive electroproduction are to be expected from CEBAF and later on at even higher momentum transfers from ELFE (at DESY).

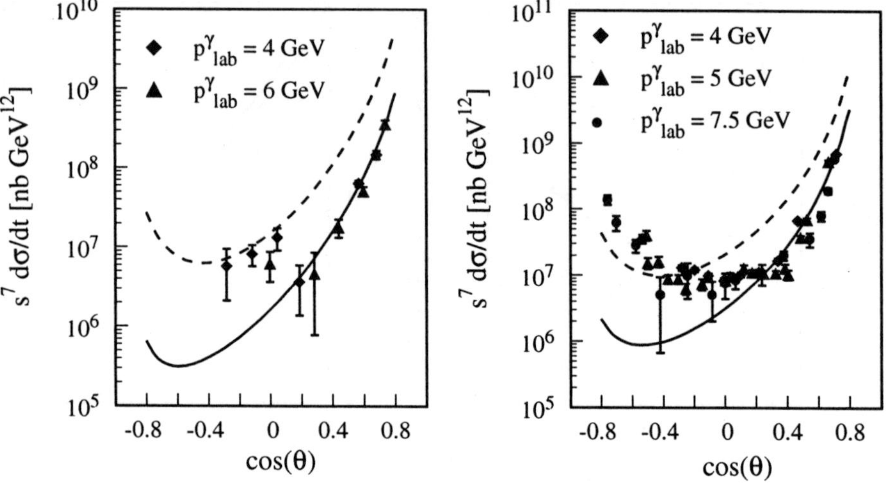

Figure 5: Diquark-model predictions for photoproduction of the K^+-Σ^0 (*left figure*) and the π^+-n (*right figure*) final states; solid (dashed) line: diquark-model results for baryon DAs chosen according to Eq. (2), K and π DAs according to Eq. (3) (Eq. (4)). Results are shown for fixed $p^\gamma_{lab} = 6$ GeV. Scalar-diquark contributions are fully taken into account, four- and five-point contributions of vector diquarks are neglected. Data are taken from Anderson et al. [12].

3.2 The K^+-Σ^0 and the π^+-n Channel

We have already mentioned that these channels differ from K^+-Λ production as far as also V-diquarks are involved. Since the treatment of V diquarks is much more intricate than that of S diquarks we have, until now, only calculated the corresponding three-point contributions. From K^+-Λ production and also from other applications of the diquark model we know, however, that spin-averaged cross sections are mainly determined by three-point contributions. Thus we do not expect a significant modification of our numerical results for the K^+-Σ^0 and the π^+-n photoproduction cross sections if four- and five-point contributions of V diquarks are included. Like in the case of K^+-Λ production we observe reasonable agreement with the K^+-Σ^0 and the π^+-n cross-section data if the asymptotic DA is taken for the K^+ and the π^+, respectively (cf. Fig. 5). The two-humped DAs, Eq. (4), seem again to be in conflict with the data. Also for these two channels the pure quark HSA fails in reproducing the data. It is remarkable that the unpolarized differential cross sections for K^+-Λ and K^+-Σ^0 production are very similar in size, although the corresponding production mechanisms (via S and V diquarks, respectively) are quite different. We expect this difference to show up more clearly in spin observables.

4 Conclusions

The predictions of the diquark model for $\gamma\,p \to K^+\,\Lambda$, $K^+\,\Sigma^0$, and $\pi^+\,n$ look rather promising if the asymptotic form ($\propto x(1-x)$) is taken for the meson DAs. To the best of our knowledge the diquark model is, as yet, the only constituent scattering model which is able to account for the large-$p_\perp$ photoproduction data. For electroproduction of the K^+-Λ final state our results are the first perturbative QCD predictions at all. With respect to future experiments it would, of course, be desirable to have more and better large momentum-transfer data on exclusive photo- and electroproduction. Polarization measurements of the recoiling particle could help to decide, whether the perturbative regime has been reached already, or whether non-perturbative effects (different from diquarks) are still at work. Spin observables, in general, could be very helpful to constrain the form of the hadron DAs.

References

1. R. Jakob, P. Kroll, M. Schürmann, and W. Schweiger, *Z. Phys.* A **347**, 109 (1993).
2. P. Kroll, Th. Pilsner, M. Schürmann, and W. Schweiger, *Phys. Lett.* B **316**, 546 (1993).
3. P. Kroll, M. Schürmann, and P. A. M. Guichon, *Nucl. Phys.* A **598**, 435 (1996).
4. E.-H. Kada and J. Parisi, *Z. Phys.* C **70**, 303 (1996).
5. R. Jakob, *Phys. Rev.* D **50**, 5647 (1994).
6. See, e.g., S. J. Brodsky and G. P. Lepage in *Perturbative Quantum Chromodynamics*, ed. A. H. Mueller (World Scientific, Singapore, 1989).
7. G. R. Farrar, K. Huleihel, and H. Zhang, *Nucl. Phys.* B **349**, 655 (1991).
8. T. Huang, *Nucl. Phys. [Proc. Suppl.]* B **7**, 320 (1989).
9. V. L. Chernyak and A. R. Zhitnitsky, *Phys. Rep.* **112**, 173 (1984).
10. I. S. Barker, A. Donnachie, and J. K. Storrow, *Nucl. Phys.* B **95**, 347 (1975).
11. R. C. E. Devenish and D. H. Lyth, *Phys. Rev.* D **5**, 47 (1972).
12. R. L. Anderson et al., *Phys. Rev.* D **14**, 679 (1976).
13. G. R. Farrar, H. Zhang, A. A. Ogloblin, and I. R. Zhitnitsky, *Nucl. Phys.* B **311**, 585 (1988).
14. R. Jakob, P. Kroll, and M. Raulfs, *J. Phys.* G **22**, 45 (1996).
15. A. S. Kronfeld and B. Nižič, *Phys. Rev.* D **44**, 3445 (1991).
16. P. Kroll, M. Schürmann, K. Passek, and W. Schweiger, preprint UNIGRAZ-UTP 15-04-96 (hep-ph/9604353), to appear in *Phys. Rev.* D.

THE $\pi\gamma$ TRANSITION FORM FACTOR AND ITS IMPACT ON CHARMONIUM DECAYS INTO TWO PIONS

P. Kroll

Fachbereich Physik, Universität Wuppertal,
D-42097 Wuppertal, Germany

The analysis of the $\pi\gamma$ transition form factor provides severe constraints on the pion's wave function. This information is used to examine charmonium decays into two pions critically. It will be argued that the standard perturbative QCD analysis of these reactions fails, i. e. the need for additional contributions can convincingly be demonstrated. Colour-octet admixtures to the charmonium states are proposed as a possible dynamical mechanisms to solve the puzzle. Consequences of the $\pi\gamma$ analysis for the electromagnetic form factor of the pion are also discussed.

1 Introduction

At large momentum transfer the hard scattering approach (HSA) [1] provides a scheme to calculate exclusive processes. Observables are described as convolutions of hadronic wave functions which embody soft non-perturbative physics, and hard scattering amplitudes T_H to be calculated from perturbative QCD. In most cases only the contribution from the lowest-order pQCD approach in the collinear approximation using valence Fock states only (termed the standard HSA) has been worked out. Applications of the standard HSA to space-like exclusive reactions, as for instance the magnetic form factor of the nucleon, the pion form factor or Compton scattering off protons revealed that the results are only in fair agreement with experiment if hadronic wave functions are used that are strongly concentrated in the end-point regions where one of the quark momentum fractions, x, tends to zero. As has been pointed out by several authors (e.g. [2,3]), the results obtained from such wave functions are dominated by contributions from the end-point regions where perturbative QCD cannot readily be applied. Hence, despite the agreement with experiment, the predictions of the standard HSA are theoretically inconsistent for such wave functions. It should also be stressed that the large momentum transfer behaviour of the helicity-flip controlled Pauli form factor of the proton remains unexplained within the standard HSA.

Applications of the HSA to time-like exclusive processes fail in most cases (e.g. G_M, F_π, $\gamma\gamma \to p\bar{p}$). The predictions for the integrated $\gamma\gamma \to \pi\pi$ cross-section ($|\cos\theta| \leq 0.6$) are in fair agreement with the data whereas the predictions for the angular distribution fails. Exclusive charmonium decays constitute another class of time-like reactions. If the end-point region concentrated wave functions are employed again, the standard HSA provides results in fair agreement with the data in many cases. It should be noted that in most calculations of exclusive charmonium decays[4] α_s values of the order of $0.2 - 0.3$ are employed. Such values do not match with α_s evaluated at the charm quark mass, the characteristic scale for these decays ($\alpha_s(m_c = 1.5\,\text{GeV}) = 0.37$ in one-loop approximation with $\Lambda_{QCD} = 200\,\text{MeV}$). Since high powers of α_s are involved in charmonium decays a large factor of uncertainty is hidden in the predictions.

Constraining the pion wave function[5,6] from the recent precise data on the $\pi\gamma$ transition form factor [7], one observes an order-of-magnitude discrepancy between data and HSA predictions for charmonium decays into two pions. In [8] contributions from the $c\bar{c}g$ Fock state are suggested as the solution of this puzzle. I am going to discuss these topics in my talk. Also I shall discuss the large momentum transfer behaviour of the pion form factor in the light of the new information on the pion's wave function.

2 The π-γ transition form factor

The apparent success of the end-point concentrated wave functions, in spite of the theoretical inconsistencies, prevented progress in understanding hard exclusive reactions for some time. Recently, with the advent of the CLEO data on the $\pi\gamma$ transition form factor $F_{\pi\gamma}$ [7], the situation has changed. The leading twist result for that form factor[a], including α_s-corrections, reads [1]

$$F_{\pi\gamma}(Q^2) = \frac{\sqrt{2}}{3} \langle x^{-1} \rangle \frac{f_\pi}{Q^2} \left[1 + \frac{\alpha_s(\mu_R)}{2\pi} K_{\pi\gamma}(Q^2, \mu_R) + \mathcal{O}(\alpha_s^2) \right]. \tag{1}$$

f_π is the usual pion decay constant (130.7 MeV) and μ_R represents the renormalization scale. The function $K_{\pi\gamma}$ has been calculated by Braaten [9] in the $\overline{MS}$ scheme. $\langle x^{-1} \rangle$ is the $1/x$ moment of the pion distribution amplitude, ϕ, which represents the light-cone wave function of the pion integrated over transverse quark momenta, $\mathbf{k}_\perp$, up to a factorization scale, μ_F, of order Q. The distribution amplitude can be expanded upon Gegenbauer polynomials, $C_n^{3/2}$, the eigenfunctions of the evolution kernel for mesons [1]

$$\phi_\pi(x, \mu_F) = \phi_{AS}(x) \left[1 + \sum_{n=2,4,\ldots}^{\infty} B_n(\mu_0) \left(\frac{\alpha_s(\mu_F)}{\alpha_s(\mu_0)} \right)^{\gamma_n} C_n^{3/2}(2x - 1) \right] \tag{2}$$

---

[a]The pion mass as well as the light current quark masses are neglected throughout.

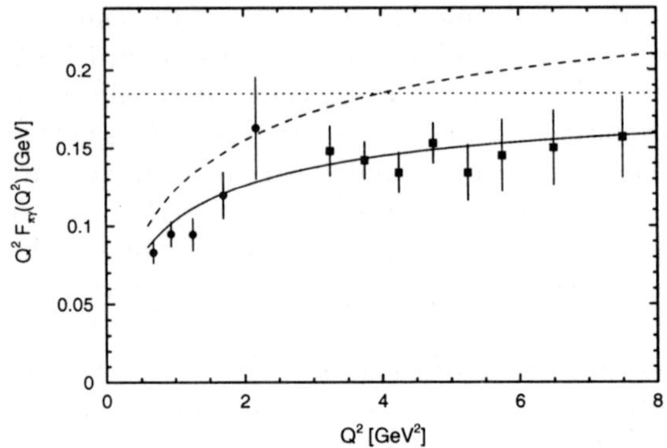

Figure 1: The scaled $\pi\gamma$ transition form factor vs. Q^2. The solid (dashed) line represents the results obtained with the modified HSA using the asymptotic (Chernyak-Zhitnitsky) wave function. The evolution of the Chernyak-Zhitnitsky wave function is taken into account. The dotted line represents the limiting behaviour $\sqrt{2}f_\pi$. Data are taken from [7,12].

where the asymptotic distribution amplitude is $\phi_{AS}(x) = 6x(1-x)$. The $1/x$ moment of the distribution amplitude reads

$$\langle x^{-1}\rangle = 3\left[1 + \sum_{n=2,4,\ldots}^{\infty} B_n(\mu_0)\left(\frac{\alpha_s(\mu_F)}{\alpha_s(\mu_0)}\right)^{\gamma_n}\right] = 3\left[1 + \sum_{n=2,4,\ldots}^{\infty} B_n(\mu_F)\right]. \quad (3)$$

The process-independent expansion coefficients B_n embody the soft physics; they are not calculable at present. μ_0 is a typical hadronic scale, actually $\mu_0 = 0.5\,\text{GeV}$. Since the anomalous dimensions, γ_n, are positive fractional numbers increasing with n (e.g. $\gamma_2 = 50/81$) any distribution amplitude evolves into the asymptotic distribution amplitude for $\ln Q^2 \to \infty$; higher order terms are gradually suppressed. Hence, the limiting behaviour of the transition form factor is

$$F_{\pi\gamma} \longrightarrow \sqrt{2}f_\pi/Q^2 \quad (4)$$

which is a parameter-free QCD prediction [11]. As comparison with the CLEO data[7] reveals, the limiting behaviour is approached from below. At 8 GeV^2 the data only deviate by about 15% from (4) (see Fig.1). In order to give a quantitative estimate of the allowed deviations from the asymptotic distribution amplitude one may assume that B_2 is the only non-zero expansion coefficient in (2). The truncated series suffices to parametrize small deviations. Moreover, it has the advantage of interpolating smoothly between the asymptotic distribution amplitude and the frequently used Chernyak-Zhitnitsky distribution amplitude [13] $(B_2 = 2/3; C_2^{3/2}(\xi) = 3/2(5\xi^2 - 1))$. For large momentum transfer the assumption on the expansion coefficients is justified by the properties of the anomalous dimensions γ_n.

In [6] it is shown that the leading twist, lowest order pQCD result (1) nicely fits the CLEO data for $B_2^{LO}(\mu_0) = -0.39 \pm 0.05$. Using Braaten's result for $K_{\pi\gamma}$ [9] that, choosing $\mu_F = \mu_R = Q$, reads

$$K_{\pi\gamma} = -\frac{10}{3} \frac{1 - 59/72\,B_2(Q^2)}{1 + B_2(Q^2)}\,, \tag{5}$$

one finds $B_2^{NLO}(\mu_0) = -0.17 \pm 0.05$ from a fit to the CLEO data. Braaten's analysis is however incomplete in so far as only the α_s corrections to the hard scattering amplitude have been considered but the corresponding corrections to the kernel of the evolution equation for the pion's distribution amplitude were ignored. As has been shown by Müller [10] recently next-to-leading order evolution provides logarithmic modifications in the end-point regions for any distribution amplitude, i.e. for the asymptotic one too. An estimate however reveals that the modifications of the evolution behaviour in next-lo-leading order are very small and can safely be neglected.

To summarize the $F_{\pi\gamma}$ form factor requires a distribution amplitude in a leading twist analysis that is narrower than the asymptotic one in the momentum transfer region of a few GeV^2. The Chernyak-Zhitnitsky distribution amplitude is in clear conflict with the data and should, therefore, be discarded[b].

Recently a modified HSA has been proposed by Botts, Li and Sterman [16] in which transverse degrees of freedom as well as Sudakov suppressions are taken into account. This approach has the advantage of strongly suppressed end-point regions. Hence, the perturbative contributions can be calculated self-consistently in the sense that the bulk of the perturbative contribution is accumulated in regions of reasonably small values of the strong coupling constant. It is to be stressed that the effects of the transverse degrees of freedom taken into account in the modified HSA represent soft contributions of higher-twist type. Still, modified HSA calculations are restricted to the

---

[b] In [14,15] a modification of the pion wave function is proposed where the distribution amplitude is multiplied by the exponential $\exp\left[-m_q^2 a_\pi^2/x(1-x)\right]$. The parameter m_q represents a constituent quark mass of, say, $330\,\text{MeV}$. Since the exponential substantially deviates from unity only in the end-point regions it leads to a strong additional suppression in the case of the Chernyak-Zhitnitsky distribution amplitude ($\langle x^{-1}\rangle$ changes from a value of 5 to 3.71 at the scale μ_0). For narrow distribution amplitudes ($B_2 \leq 0$), on the other hand, the exponential has only a minor bearing on the results for $F_{\pi\gamma}$.

dominant (valence) Fock state. Another advantage of the modified HSA is that the renormalization scale can be chosen in such a way that large logs from higher order perturbation theory are eliminated. Such a choice of the renormalization scale are accompanied by α_s singularities in the end-point regions which are, however, compensated by the Sudakov factor in the modified HSA. Singularities produced by the evolution of the wave function are also cancelled by the Sudakov factor.

Adapting the modified HSA to the case of $\pi\gamma$ transitions, one can write the corresponding form factor as [5,6]

$$F_{\pi\gamma}(Q^2) = \int dx \, \frac{d^2\mathbf{b}}{4\pi} \hat{\Psi}_\pi(x, -\mathbf{b}, \mu_F) \, \hat{T}_H(x, \mathbf{b}, Q) \, \exp\left[-S(x, b, Q)\right] \quad (6)$$

up to $\mathcal{O}(\alpha_s, k_\perp^2/Q^2)$ corrections. $\mathbf{b}$ is the quark-antiquark separation and is canonically conjugated to the usual transverse momentum $\mathbf{k}_\perp$. The use of the transverse configuration space is mandatory because the Sudakov exponent S is only known in that space [16]. The Sudakov exponent comprises those gluonic radiative corrections not taken into account in the evolution of the wave function. $\hat{T}_H$ is the Fourier transform of the lowest order momentum space hard scattering amplitude. It reads

$$\hat{T}_H(x, \mathbf{b}, Q) = \frac{2}{\sqrt{3}\pi} K_0\left(\sqrt{1-x}\, Q\, b\right) \quad (7)$$

where K_0 is the modified Bessel function of order zero. Due to the properties of the Sudakov exponent any contribution is damped asymptotically, i.e. for $\ln(Q^2/\mu_0^2) \to \infty$, except those from configurations with small quark-antiquark separations and, as can be shown, the limiting behaviour (4) emerges. b plays the role of an infrared cut-off; it sets up the interface between non-perturbative soft gluon contributions - still contained in the hadronic wave function - and perturbative soft gluon contributions accounted for by the Sudakov factor. Hence, the factorization scale μ_F is to be taken as $1/b$.

Finally, $\hat{\Psi}_\pi$ is the Fourier transform of the momentum space (light-cone) wave function of the pion for which a Gaussian $\mathbf{k}_\perp$-dependence is employed

$$\Psi_\pi(x, \mathbf{k}_\perp; \mu_F) = \frac{f_\pi}{2\sqrt{6}} \phi_\pi(x, \mu_F) N \exp\left(-a_\pi^2(\mu_F)\frac{k_\perp^2}{x(1-x)}\right). \quad (8)$$

Here $N = 16\pi^2 a_\pi^2/(x(1-x))$ and, for a distribution amplitude with $B_n = 0$ for $n \geq 4$, $a_\pi = 1/(\pi f_\pi \sqrt{8(1+B_2)})$. The $\pi^0 \to \gamma\gamma$ constraint [14] is automatically satisfied for that choice of the transverse size parameter a_π. Ψ_π represents a soft wave function, i.e. a full wave function with its pertubative tail removed from it. For $B_2 = 0$ the wave function (8) leads to a valence Fock state probability of 0.25 and a r.m.s. radius of 0.42 fm. Using the wave function (8) in a modified HSA calculation, one finds excellent agreement with the CLEO[7] and CELLO[12] data above $Q^2 \simeq 1$ GeV^2 for $B_2(\mu_0) = -0.006 \pm 0.014$[5,6] (see Fig.1). Hence, the asymptotic wave function, i.e. the asymptotic distribution amplitude combined with the Gaussian $\mathbf{k}_\perp$-dependence, works very well if the modified HSA is used.

A similar analysis of the $\eta\gamma$ and the $\eta'\gamma$ transition form factors has been carried through in[5]. It is important thereby to take into account mass corrections and the $\eta - \eta'$ mixing. The results of that analysis are in excellent agreement with the available data including the recent CLEO data[7]. The values of the $\eta - \eta'$ mixing angle and the decay constants are calculated in[5] to be $\theta_P = -18° \pm 2°$, $f_\eta = 175 \pm 10$ MeV and $f_{\eta'} = 95 \pm 6$ MeV, respectively. The $\eta_c\gamma$ transition form factor can be analyzed in the same manner. Estimates of that form factor can be found in[17].

With the advent of the CLEO data the $\pi\gamma$ transition form factor attracted much interest and, besides[5,6], many papers have been devoted to its analysis elucidating various aspects of it[15,18,19,20,21]. Particularly interesting is the generalization of (1) to the case of two virtual photons. In the standard HSA and again with $B_n = 0$ for $n \geq 4$, the $\pi\gamma^*$ transition form factor reads[9]

$$
F_{\pi\gamma^*}(Q^2, \omega) = \sqrt{2}\frac{f_\pi}{Q^2}\frac{1}{(1-\omega)^3}\Big\{[1 - \omega^2 + 2\omega\ln\omega]
$$

$$
\times[1 + \frac{1 + 28\omega + \omega^2}{(1-\omega)^2}B_2(\mu_F)] + 10\omega\ln\omega B_2(\mu_F)\Big\} \quad (9)
$$

where $\omega = Q'^2/Q^2$. The larger one of the two photon virtualities is denoted by Q^2, the smaller one by Q'^2. The factorization scale may be chosen as $\mu_F = Q\sqrt{1+\omega}$. α_s-corrections to $F_{\pi\gamma^*}$ can be found in[9] and an estimate of power corrections in[22]. The treatment of $\pi\gamma^*$ transitions within the the modified HSA is straightforward generalization of (6)[18].

Interestingly, the $F_{\pi\gamma^*}$ form factor still behaves as Q^{-2} at large Q^2. This is to be contrasted with the $Q^{-2}Q'^{-2}$ behaviour of the vector meson dominance model[23]. In the limes $\omega \to 1$ (9) simplifies to

$$
F_{\pi\gamma^*} = \frac{\sqrt{2}}{3}\frac{f_\pi}{Q^2}\left[1 + \frac{1}{2}(1-\omega)(1 - 12B_2(\mu_F))\right]. \quad (10)
$$

The limiting behaviour of the form factor for $\omega = 1$ which is strictly independent on the form of the distribution amplitude, has also been derived from QCD sum rules[24]. In[21] the triangle diagram is analyzed with the most general form of the $\pi q\bar{q}$ vertex. The result obtained for $F_{\pi\gamma^*}$ in that paper is similar to (9) provided B_2 is put to zero in (9). The differences between the two results are strongest at $\omega = 0$ (about 9%) while both the results coincide at $\omega = 1$.

3 Pionic decays of charmonium

In view of the results for $F_{\pi\gamma}$ a fresh analysis of the decays $\chi_{cJ} \to \pi\pi$ is in order. Using the information on the π wave function obtained from the analysis of $F_{\pi\gamma}$, one finds the following values for the partial widths

$$\Gamma(\chi_{c0(2)} \to \pi^+\pi^-) = 0.872\,(0.011)\ \text{keV} \tag{11}$$

within the standard HSA[8]. As usual the renormalization and the factorization scales are identified in that calculation and put equal to the c-quark mass. The parameter describing the χ_{cJ} state is the derivative $R'_P(0)$ of the non-relativistic $c\bar{c}$ wave function at the origin (in coordinate space) appropriate for the dominant Fock state of the χ_{cJ}, a $c\bar{c}$ pair in a colour-singlet state with quantum numbers $^{2S+1}L_J = {}^3P_J$. $m_c = 1.5$ GeV and, of course, the leading order standard HSA value -0.39 for $B_2(\mu_0)$ are chosen as well as $R'_P(0) = 0.22\,\text{GeV}^{5/2}$ which is consistent with a global fit of charmonium parameters[25] as well as with results for charmonium radii from potential models[26].

In[8] the modified HSA is also used to calculate the $\chi_{cJ} \to \pi\pi$ decay widths. Taking $B_2 = 0$ and the other parameters as quoted above, one finds

$$\Gamma(\chi_{c0(2)} \to \pi^+\pi^-) = 8.22\,(0.41)\ \text{keV}. \tag{12}$$

For comparison the experimental data as quoted in[27] and reported in a recent paper of the BES collaboration[28] are

$$\begin{aligned}
\Gamma(\chi_{c0} \to \pi^+\pi^-) &= 105 \pm 30 \ \ \text{keV (PDG)}, \\
&\quad\ 62.3 \pm 17.3\,\text{keV (BES)}, \\
\Gamma(\chi_{c2} \to \pi^+\pi^-) &= 3.8 \ \pm 2.0 \ \ \text{keV (PDG)}, \\
&\quad\ 3.04 \pm 0.73\,\text{keV (BES)}.
\end{aligned} \tag{13}$$

One notes that both the theoretical results, (11) and (12), fail by at least an order of magnitude. To assess the uncertainties of the theoretical results one may vary the parameters, m_c, B_2 and Λ_{QCD}. However, even if the parameters are pushed to their extreme values the predicted rates are well below data. Thus, one has to conclude that calculations based on the assumption that the χ_{cJ} is a pure $c\bar{c}$ state, are not sufficient to explain the observed rates. The necessary corrections would have to be larger than the leading terms. A new mechanism is therefore called for.

Recently, the importance of higher Fock states in understanding the production and the *inclusive* decays of charmonium has been pointed out [29]. It is therefore tempting to assume the inclusion of contributions from the $|c\bar{c}_8(^3S_1)g\rangle$ Fock state to *exclusive* χ_{cJ} decays as the solution to the failure of the HSA. The usual higher Fock state suppression by powers of $1/Q^2$ [30] where $Q = m_c$ in the present case, does not appear as a simple dimensional argument reveals: the colour-singlet and octet contributions to the decay amplitude behave as

$$M_J^{(c)} \sim f_\pi^2 f_J^{(c)} m_c^{-n_c}. \tag{14}$$

The singlet decay constant, $f_J^{(1)}$, represents the derivative of a two-particle coordinate space wave function at the origin. Hence it is of dimension GeV^2. The octet decay constant, $f_J^{(8)}$, as a three-particle coordinate space wave function at the origin, is also of dimension GeV^2. Since $M_J^{(c)}$ is of dimension GeV, $n_c = 3$ in both cases. Note that the χ_{cJ} decay constants may also depend on m_c. Obviously, the colour-octet contribution will also play an important role in the case of the χ_{bJ} decays.

In [8] the colour-octet contributions to the exclusive χ_{cJ} decays are estimated by calculating the hard scattering amplitude from the set of Feynman graphs shown in Fig. 2 and convoluting it with the asymptotic pion wave function. The colour-octet and singlet contributions are to be added coherently. The $\chi_{cJ} \to \pi\pi$ decay widths are given in terms of a single non-perturbative parameter κ which approximately accounts for the soft physics in the colour-octet contribution. A fit to the data [27,28] yields $\kappa = 0.16\,\text{GeV}^2$ (with $m_c = 1.5$ GeV; $\Lambda_{QCD} = 0.2$ GeV) and the widths

$$\Gamma(\chi_{c0(2)} \to \pi^+\pi^-) = 49.85\,(3.54)\ \text{keV}. \tag{15}$$

Comparison with (13) reveals that the inclusion of the colour-octet mechanism brings predictions and data in generally good agreement. The value found for the parameter κ has a reasonable interpretation in terms of charmonium properties and the mean transverse momentum of the quarks inside the pions. Results for the decays into pairs of uncharged pions are presented in Table 2. The quoted results for the pionic charmonium decays refer to a calculation within the standard HSA but similar good results are found when the modified HSA is used [31]. The only soft parameter appearing in the latter calculation is the octet-decay constant $f_J^{(8)}$ of the charmonium state.

Thus it seems that the colour-octet mechanism leads to a satisfactory explanation of the decay rates of the χ_{cJ} into two pions. Of course, that

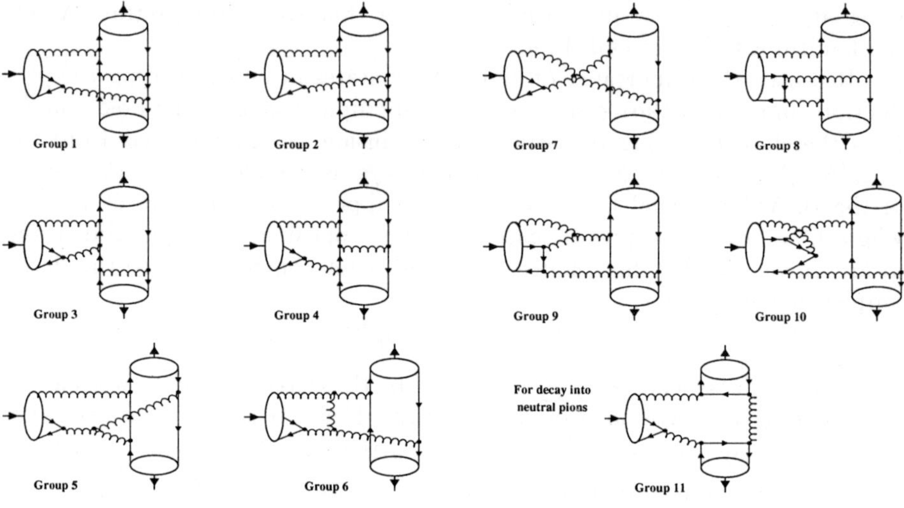

Figure 2: Representatives of the various groups of colour-octet decay graphs.

mechanism has to pass more tests in exclusive reactions before this issue can be considered as being settled.

4 The π form factor

Let us now turn to the case of the π form factor and discuss the implications of the constraints on the pion's wave function obtained from the $\pi\gamma$ analysis. The leading twist result for the pion form factor can be brought into a form similar to (1)

$$F_\pi(Q^2) = \frac{8\pi}{9}\,\langle x^{-1}\rangle^2\frac{f_\pi^2}{Q^2}\,\alpha_s(\mu_R)\,[\,1 + \frac{\alpha_s(\mu_R)}{2\pi}K_\pi(Q^2,\mu_R) + \mathcal{O}(\alpha_s^2)\,]. \quad (16)$$

	$\Gamma(\chi_{cJ} \to \pi^0\pi^0)$ [keV]		BR$(\chi_{cJ} \to \pi^0\pi^0)$ [%]	
$B_2 = 0$	25.7	1.81	0.18	0.091
Exp. PDG [27]	42 ± 18	2.2 ± 0.6	0.31 ± 0.06	0.110 ± 0.028

Table 1: Decay widths and branching ratios of $\chi_{cJ} \to \pi^0\pi^0$ (colour-octet contributions included; $m_c = 1.5\,\text{GeV}$, $\Lambda_{QCD} = 0.2\,\text{GeV}$).

Choosing $\mu_F = \mu_R$ and using the value of $B_2^{LO}(\mu_0)$ determined in the leading twist analysis of the $F_{\pi\gamma}$ as well as a value of, say, 0.4 for α_s, one obtains the result $0.097\,\text{GeV}^2/Q^2$ for F_π to lowest order pQCD. That result is much smaller than the admittedly poor experimental result[33]: $F_\pi = 0.35 \pm 0.10\,\text{GeV}^2/Q^2$. The α_s-corrections are too small to account for that discrepancy[32]. The modified HSA likewise provides too small a perturbative contribution[34]. It is important to remember at this point that, formally, the perturbative contribution to the pion form factor represents the overlap of the large momentum tails of the initial and final state wave functions. But the form factor also gets a contribution from the overlap of the soft wave functions $\hat{\Psi}_\pi$. That contribution, frequently termed the Feynman contribution, is customarily assumed to be negligible already at momentum transfers as low as a few GeV^{2c}. Examining the validity of that presumption by estimating the Feynman contribution from the asymptotic wave function (8), one finds results of appropriate magnitude to fill in the gap between the perturbative contribution and the data of [33]. The results exhibit a broad flat maximum which, for momentum transfers between 3 and about 15 GeV^2, simulates the dimensional counting behaviour. For a wave function based on the Chernyak-Zhitnitsky distribution amplitude, on the other hand, the Feynman contribution exceeds the data significantly [5,34]. Large Feynman contributions have also been found by other authors[2,35]. Thus, the small size of the perturbative contribution to the elastic form factor finds a comforting although model-dependent explanation, a fact which has been pointed out by Isgur and Llewellyn Smith[2] long time ago. Of course this line of reasoning is based on the assumption that the data of [33] are essentially correct.

A comment concerning the pion form factor in the time-like region is in order. In the standard HSA the predictions for the form factor in both the time-like and the space-like region, are identical. The experimental information on the time-like form factor comes from two sources, $e^+e^- \to \pi^+\pi^-$ and

---

^{c}At large Q^2 the Feynman contribution is suppressed by $1/Q^2$ as compared to the perturbative contribution.

$J/\Psi \to \pi^+\pi^-$, which provides, to a very good approximation[d], the form factor at $s = M^2_{J/\Psi}$. Although the e^+e^- annihilation data of Bollini et al.[36] suffer from low statistics, they agree very well with the result obtained from the J/Ψ decay. Combining both the data, one finds $|F_\pi| = 0.93 \pm 0.08\,\mathrm{GeV}^2/s$ in the momentum transfer range between 2 and 10 GeV^2, a value wich is roughly a factor of 3 larger than the space-like data. The modified HSA can account for that large ratio of the time-like over space-like form factors[38] although the perturbative contributions to the form factor in both the regions are too small.

The structure function of the pion offers another possibility to test the wave function against data. As has been pointed out in [14] the parton distribution functions are determined by the Fock state wave functions. Since each Fock state contributes through the modulus squared of its wave function integrated over transverse momenta up to Q and over all fractions x except those pertaining to the type of parton considered, the contribution from the valence Fock state should not exceed the data of the valence quark structure function. As discussed in [5,39] the asymptotic wave function respects this inequality while the Chernyak-Zhitnitsky one again fails dramatically.

---

[d]The contribution from $c\bar{c}$ transitions into the light quarks via three gluons cancel to zero if quark masses are neglected[37].

5 Summary

The study of hard exclusive reactions is an interesting and challenging subject. The standard HSA, i.e. the valence Fock state contribution in collinear approximation to lowest order perturbative QCD, while asymptotically correct (at least for form factors), does not lead to a consistent description of the data. In many cases the predicted perturbative contribution to particular exclusive reactions are much smaller then the data. The observed spin effects do not find a comforting explanation. In some reactions agreement between prediction and experiment is found although at the expense of dominant contributions from the soft end-point regions rendering the perturbative analysis inconsistent.

From a detailed analysis of the $\pi\gamma$ transition form factor it turned out that the pion distribution amplitude is close to the asymptotic form. Strongly end-point concentrated distribution amplitudes are obsolete and the ostensible successes in describing the large momentum transfer behaviour of the pion form factor in the space-like region and charmonium decays into two pions with such distribution amplitudes must therefore be dismissed.

In view of these observations it seems that higher Fock state and/or higher twist contributions have to be included in the analysis of exclusive reactions. However, not much is known about them as yet. We are lacking systematic investigations of such contributions to exclusive reactions. The colour octet model for exclusive charmonium decays is discussed in this talk as an example of such contributions.

References

1. G.P. Lepage and S.J. Brodsky, *Phys. Rev.* D **22**, 2157 (1980).
2. N. Isgur and C.H. Llewellyn Smith, *Nucl. Phys.* B **317**, 526 (1989).
3. A.V. Radyushkin, *Nucl. Phys.* A **532**, 141c (1991).
4. A. Duncan and A.H. Mueller, *Phys. Lett.* B **93**, 119 (1980);
   V.L. Chernyak et al., *Z. Phys.* C **42**, 583 (1989);
   N.G. Stefanis and M. Bergmann, *Phys. Rev.* D **47**, R3685 (1993).
5. R. Jakob, P. Kroll and M. Raulfs, *J. Phys.* G **22**, 45 (1996);
   P. Kroll, Proceedings of the PHOTON95 Workshop, Sheffield (1995), eds. D.J. Miller et al., World Scientific.
6. P. Kroll and M. Raulfs, *Phys. Lett.* B **387**, 1996 (848).
7. V. Savinov et al., CLEO collaboration, Proceedings of the PHOTON95 Workshop, Sheffield (1995), eds. D.J. Miller et al., World Scientific.
8. J. Bolz, P. Kroll and G.A. Schuler, preprint CERN-TH/96-266 (1996), hep-ph/9610265, to be published in Phys. Lett. B.

9. E. Braaten, *Phys. Rev.* D **28**, 524 (1983).

10. D. Müller, *Phys. Rev.* D **51**, 3855 (1995).

11. T.F. Walsh and P. Zerwas, *Nucl. Phys.* B **41**, 551 (1972).

12. CELLO coll., H.-J. Behrend et al., *Z. Phys.* C **49**, 401 (1991).

13. V.L. Chernyak and A.R. Zhitnitsky, *Nucl. Phys.* B **201**, 492 (1982).

14. S.J. Brodsky, T. Huang and G.P. Lepage, *Banff Summer Institute, Particles
    and Fields* 2, p. 143, A.Z. Capri and A.N. Kamal (eds.), 1983.

15. F.-G. Cao, T. Huang and B.-Q. Ma, *Phys. Rev.* D **53**, 6582 (1996).

16. J. Botts and G. Sterman, *Nucl. Phys.* B **325**, 62 (1989);
    H.N. Li and G. Sterman, *Nucl. Phys.* B **381**, 129 (1992).

17. P. Aurenche et al., $\gamma\gamma$ Physics, hep-ph/9601317.

18. S. Ong, *Phys. Rev.* D **52**, 3111 (1995).

19. A.V. Radyushkin and R.T. Ruskov, preprint hep-ph/9511270 (1995).

20. V.V. Anisovich, D.I. Melikhov and V.A. Nikonov, preprint hep-ph/9607215.

21. A. Anselm, A. Johansen, E. Leader and L. Lukaszuk,
    preprint hep-ph/9603444.

22. A.S. Gorskiĭ, *Sov. J. Nucl. Phys.* **50**, 498 (1989).

23. P. Kessler and S. Ong, *Phys. Rev.* D **48**, R2974 (1993).

24. V.A. Novikov, M.A. Shifman, A.I. Vainshtein, M.B. Voloshin and V.I.
    Zakharov, *Nucl. Phys.* B **237**, 525 (1984); V.A. Nesterenko and A.V.
    Radyushkin, *Yad. Fiz.* **38**, 476 (1983) (*Soviet J. Nucl. Phys.* **38**, 284
    (1983).

25. M.L. Mangano and A. Petrelli, *Phys. Lett.* B **352**, 445 (1995).

26. W. Buchmüller and S.-H. Tye, *Phys. Rev.* D **24**, 132 (1981).

27. Particle Data Group: Review of Particle Properties,
    Phys. Rev. D **54**, 1 (1996).

28. Y. Zhu for the BES coll., talk presented at the XXVIII Int. Conf. on High
    Energy Physics, 25-31 July 1996, Warsaw, Poland.

29. G.T. Bodwin, E. Braaten and G.P. Lepage, *Phys. Rev.* D **51**, 1125
    (1995).

30. S.J. Brodsky and G.R. Farrar, *Phys. Rev. Lett.* **31**, 1153 (1973);
    V.A. Matveev, R.M. Murradyan and A.V. Tavkheldize,
    Lett. Nuovo Cim. **7**, 719 (1973).

31. J. Bolz, P. Kroll and G.A. Schuler, in preparation.

32. R.D. Field et al., *Nucl. Phys.* B **186**, 429 (1981);
    F.M. Dittes and A.V. Radyushkin, *Sov. J. Nucl. Phys.* **34**, 293 (1981);
    E. Braaten and S.-M. Tse, *Phys. Rev.* D **35**, 2255 (1987).

33. C.J. Bebek et al., *Phys. Rev.* D **13**, 25 (1976) and D **17**, 1693 (1978).

34. R. Jakob and P. Kroll, *Phys. Lett.* B **315**, 463 (1993); B **319**, 545(E) (1993).

35. P.L. Chung, F. Coester and W.N. Polyzou, *Phys. Lett.* B **205**, 545 (1988);
     L.S. Kisslinger and S.W. Wang, *Nucl. Phys.* B **399**, 63 (1993);
     V. Braun and I. Halperin, *Phys. Lett.* B **328**, 457 (1994);
     A.E. Dorokhov, *Nuovo Cimento* A **109**, 391 (1996).

36. D. Bollini et al., *Lett. Nuovo Cim.* **14**, 418 (1975).

37. S.J. Brodsky and G.P. Lepage, *Phys. Rev.* D **24**, 2848 (1981).

38. T. Gousset and B. Pire, *Phys. Rev.* D **51**, 15 (1995).

39. T. Huang, B.-Q. Ma and Q.-X. Sheng, *Phys. Rev.* D **49**, 1490 (1994).

PRODUCTION OF MESON PAIRS INVOLVING $L \neq 0$ MESONS IN PHOTON-PHOTON COLLISIONS

L. HOURA-YAOU, P. KESSLER, J. PARISI

Laboratoire de Physique Corpusculaire, Collège de France
11, Place Marcelin Berthelot, F-75231 Paris Cedex 05, France

F. MURGIA

Istituto Nazionale di Fisica Nucleare, Sezione di Cagliari
Via Ada Negri 18, I-09127 Cagliari, Italy

J. HANSSON

Department of Physics, Luleå University of Technology,
S-95187 Luleå, Sweden

We present a formalism for studying the exclusive production or decay of mesons with any value of the internal orbital angular momentum L. As an application, we discuss the production of meson pairs (involving tensor and pseudotensor mesons) in photon-photon collisions.

1 Introduction

In a recent paper [1] we have presented a theoretical approach for the calculation of exclusive processes involving the production or decay of $(q\,\bar{q})$ mesons with any orbital angular momentum L. Starting from a bound-state model of weakly bound quarks, a formalism was derived that basically appears as a natural generalization of the usual perturbative QCD models for exclusive processes involving hadrons with zero orbital angular momentum.[2] As an interesting application, that formalism was used to study the production of meson pairs (involving $L \neq 0$ mesons) in photon-photon collisions. Predictions were given for the corresponding integrated cross sections in kinematical situations like those of LEP2 or future B-factories.

Here we will give a shortened presentation of this analysis, skipping most of the technical details of the approach and focusing on the qualitative features and on the key steps of the calculations. The interested reader can find a more detailed treatment in Ref. 1.

2   A generalization of PQCD models for exclusive processes to the production of $L \neq 0$ $(q\bar{q})$ mesons.

Before presenting our generalized approach, let us briefly recall the basic ideas underlying perturbative QCD models for exclusive processes (See Ref. 2 for more details). These models rely on factorization ideas, applied at the level of the helicity amplitudes for the physical process considered. Those amplitudes are then expressed as a convolution among: *i)* a hard-scattering amplitude involving the valence partons of all participating hadrons, assumed to be, for each hadron, collinear among themselves and with the parent hadron; *ii)* soft, non-perturbative contributions, described by means of (distribution) amplitudes for: *a)* finding the (collinear) valence partons in the incoming hadrons; *b)* the final partonic state to form the observed outgoing hadrons.

Notice that: *i)* only leading Fock states are considered (i.e., $|q\bar{q}\rangle$ for mesons, $|qqq\rangle$ for baryons); *ii)* valence partons are taken as massless and in a relative collinear configuration $(L = 0)$. A number of higher-twist effects can in principle modify the details of the models and play a relevant role at presently accessible energies.

Let us now discuss our generalized approach; consider a production process $a(\lambda_a)b(\lambda_b) \to Q(LSJ\Lambda)\,c(\lambda_c)$, where Q is a $(q\bar{q})$ meson (with quantum numbers L, S, J, and Λ) and a, b, c are any particles (with helicities resp. λ_a, λ_b, λ_c). Starting from a bound-state model of weakly bound quarks for Q, a prescription [3] relates the (hadronic-level) amplitude $\mathcal{M}_{\lambda_a\lambda_b\lambda_c}(E,\Theta)$ to the (partonic-level) amplitude $\mathcal{T}^{S\,\Lambda_S}_{\lambda_a\lambda_b\lambda_c}(E,\Theta,\mathbf{k},x)$:

$$\mathcal{M}_{\lambda_a\lambda_b\lambda_c}(E,\Theta) = \left(\frac{M_Q}{2}\right)^{1/2} \int \frac{d^3\mathbf{k}}{(2\pi)^{3/2}} \, \Psi^*(\mathbf{k}) \, \frac{\mathcal{T}^{S\,\Lambda_S}_{\lambda_a\lambda_b\lambda_c}(E,\Theta,\mathbf{k},x)}{\left[(m_q^2 + \mathbf{k}^2)(m_{\bar{q}}^2 + \mathbf{k}^2)\right]^{1/4}} \,,$$

$$(1)$$

where: E is the total energy in the c.m. frame of a,b; Θ is the c.m. scattering angle; M_Q, m_q, $m_{\bar{q}}$ are resp. the masses of the Q meson, the quark q and the antiquark $\bar{q}$; $2\mathbf{k}$ is the relative three-momentum of q, $\bar{q}$ inside the meson Q, in the meson rest-frame; $\Psi(\mathbf{k})$ is the corresponding meson wavefunction in momentum space. Notice that in Eq. (1) the q, $\bar{q}$ spinors are already combined to form a total spin state $|S,\Lambda_S\rangle$. In the spirit of PQCD models, we can now naturally generalize Eq. (1), defining the q, $\bar{q}$ 4-momenta as follows:

$$q^\mu = xQ^\mu + k^\mu \qquad\qquad \bar{q}^\mu = (1-x)Q^\mu - k^\mu \,, \qquad (2)$$

where Q^μ is the Q meson 4-momentum, and k^μ is the 4-dimensional generalization of $\mathbf{k}$. By introducing the meson distribution amplitude $\Phi_N(x)$ (normalized to unity), Eq. (1) generalizes to

$$\mathcal{M}_{\lambda_a\lambda_b\lambda_c}(E,\Theta) = \frac{1}{(2M_Q)^{1/2}}\int\frac{d^3\mathbf{k}}{(2\pi)^{3/2}}\,\Psi^*(\mathbf{k})\int_0^1\frac{dx\,\Phi_N^*(x)}{\sqrt{x(1-x)}}\,\mathcal{T}_{\lambda_a\lambda_b\lambda_c}^{S\,\Lambda_S}(E,\Theta,\mathbf{k},x)\,.$$

(3)

Using the well-known decompositions: $\Psi(\mathbf{k}) = R_L(k)\,Y_{L\Lambda_L}(\theta,\phi)$; and $|J,\Lambda\rangle = \sum C_{\Lambda_L\,\Lambda_S\,\Lambda}^{L\ S\ J}\,|L,\Lambda_L\rangle|S,\Lambda_S\rangle$ (the C's are the usual Clebsch-Gordan coefficients), we can now proceed with two crucial steps in our derivation:
i) Assuming that in the a,b c.m. frame the meson Q is extreme-relativistic, i.e. $\eta = (M_Q/E) \ll 1$, one can easily show that the partonic amplitudes $\mathcal{T}$ become independent of the azimuthal angle ϕ; as a consequence, it must be $\Lambda_L = 0$ and $\Lambda = \Lambda_S$.
ii) One can notice from Eq. (3) that the integrand with respect to the absolute value of the relative $q,\bar{q}$ 3-momentum, $k = |\mathbf{k}|$, can be expanded in increasing powers of k, the leading term being proportional to k^L; assuming that $R_L(k)$ is sharply peaked towards $k \to 0$, and keeping only the leading term in the power expansion, one gets from Eq. (3)

$$\mathcal{M}_{\lambda_a\lambda_b\lambda_c}^{LSJ\Lambda}(E,\Theta) = f_L^*\,C_{0\,\Lambda\,\Lambda}^{L\ S\ J}\lim_{\beta\to 0}\frac{1}{\beta^L}\int\frac{d(\cos\theta)}{2}\,d_{0,0}^L(\theta)$$
$$\times\int\frac{\Phi_N^*(x)dx}{\sqrt{x(1-x)}}\,\mathcal{T}_{\lambda_a\lambda_b\lambda_c}^{S\,\Lambda}(E,\Theta,\beta,\theta,x)\,,\qquad(4)$$

where we have introduced the dimensionless variable $\beta = 2k/M_Q$, and the normalization constant f_L, which is connected in the usual way to the value at the origin of the L-th derivative of the radial wave function $R_L(r)$.

This equation is the basic result of our formalism. It can easily be checked that, for $L = 0$, Eq. (4) leads exactly to the same expression as provided by the usual PQCD models.

3 An application: meson pair production in photon-photon collisions

We start this section by briefly recalling the physical interest in studying hadron production in photon-photon collisions.

3.1 *Hadron production in $\gamma\gamma$ collisions*

Photon-photon collisions represent a very useful tool for the study of hadron production. Basically, the more attractive feature is the simple, clean initial state, involving only QED interactions, which allows one to concentrates on the final, hadronic state. This way, in fact, some of the more clean tests for PQCD models were proposed.[2] It is well known that exclusive $\gamma\gamma \to$ hadron processes can be studied in e^+e^- colliders. Let us briefly recall some basic properties of $\gamma\gamma$ processes in this context.

First of all, compare the one-photon annihilation (OPA) contribution to the two-photon radiation (TPR) one. Due to the photon quantum numbers, OPA allows to investigate C-odd hadronic final states. OPA processes are of order α^2 but, due to the virtual photon propagator involved, $\sigma_{OPA}(e^-e^+ \to X) \sim 1/s$ (where s is the lab. energy squared). On the contrary, TPR processes make possible to study C-even final hadronic states (being in this sense complementary to the OPA contribution). Moreover, even though they are of order α^4, one finds that $\sigma_{TPR}(e^-e^+ \to e^-e^+X) \sim \ln^2(s/m_e^2)$: as a consequence, the TPR contribution already dominates over the OPA one at beam energies of a few GeV.

Whereas for a given beam energy the e^-e^+ kinematics for the OPA process is fixed, the continuous spectra of the photon beams in the TPR process allows simultaneous measurements at different $\gamma\gamma$ invariant masses. Notice however that this comes to the expenses of collectable statistics at a given invariant mass; moreover, due to the typical bremsstrahlung spectrum ($\sim 1/E_\gamma$) of the radiated photons, most of the TPR processes have low invariant masses.

The photon propagators in the TPR process cause the bulk of photons to be radiated nearly on mass-shell, at small angles relative to the beam. This means that e^-e^+ colliders effectively provide two colliding beams of quasi-real photons with luminosities comparable to those of the collider itself.

3.2 *The $\gamma\gamma \to M\bar{M}$ process*

Eq. (4) can be easily generalized to the process where two ($q\bar{q}$) mesons are produced, $ab \to QQ'$ (herefrom symbols without and with the "prime" are pertinent to Q and Q' mesons respectively). One needs only to substitute particle c in the original derivation with the meson Q' and repeat the same steps described in the previous section for the Q meson. We present the corresponding result directly in the case where the initial particles are two real photons:

$$\mathcal{M}_{\lambda_\gamma \lambda'_\gamma}^{LSJ\Lambda,\ L'S'J'\Lambda'}(E,\Theta) = f_L^*\ f_{L'}'^* C_{0\Lambda\Lambda}^{L\ S\ J} C_{0\Lambda'\Lambda'}^{L'\ S'\ J'}$$

$$\times\ \lim_{\beta,\ \beta' \to 0} \frac{1}{\beta^L \beta'^{L'}} \int \frac{d(\cos\theta)}{2} d_{0,0}^L(\theta) \int \frac{d(\cos\theta')}{2} d_{0,0}^{L'}(\theta')$$

$$\times\ \int \frac{\Phi_N^*(x)dx}{\sqrt{x(1-x)}} \int \frac{\Phi_N'^*(x')dx'}{\sqrt{x'(1-x')}}\ \mathcal{T}_{\lambda_\gamma\lambda'_\gamma}^{S\,\Lambda,\,S'\,\Lambda'}(E,\Theta,\beta,\beta',\theta,\theta',x,x') .$$

$$(5)$$

We can now apply our model to the case of the production, in photon-photon collisions, of meson pairs involving tensor and pseudotensor mesons. In order to get numerical results, we need three basic ingredients of Eq. (5): *i)* the hard scattering amplitudes $\mathcal{T}_{\lambda_\gamma\lambda'_\gamma}^{S\,\Lambda,\,S'\,\Lambda'}$; *ii)* the normalization constants f_L; *iii)* the meson distribution amplitudes $\Phi_N(x)$. Let us briefly describe how these quantities are fixed in our approach.

3.3 Evaluation of partonic amplitudes

Since the partonic amplitudes are of course independent of L (the valence constituents of each meson are in a collinear configuration), we can actually divide our strategy in three steps: *i)* Take the results for the partonic amplitudes as from the usual PQCD models for $L = 0$ mesons; these amplitudes have been evaluated originally by Brodsky and Lepage,[4] and are given as functions of (among other variables) x and x', the fraction of the meson momentum carried by quark q (q') inside the meson Q (Q'). *ii)* Make the substitutions: $x \to \tilde{x} = x + (\beta/2)\cos\theta$, $x' \to \tilde{x}' = x' - (\beta'/2)\cos\theta'$; it is not difficult to convince oneself that this corresponds exactly to the generalized procedure exposed in the previous section, in the limit $M/E \ll 1$ (here M indicates the Q or Q' mass). *iii)* Finally, perform a series expansion of the resulting amplitudes in powers of β, β', keeping only the physically relevant terms (i.e., those in $\beta^L \beta'^{L'}$).

Even so, the expressions of those amplitudes are quite involved and resort to numerical integration is required for performing the convolution integrals.

3.4 Normalized meson distribution amplitudes, $\Phi_N(x)$

In order to check the dependence of our results on the meson distribution amplitudes, we have considered two indicative choices: *i)* the so-called nonrelativistic DA, $\Phi_N(x) = \delta(x-1/2)$. It leads to very simple convolution integrals, and in this case we can perform analytical calculations even for the resulting differential cross sections. *ii)* A generalization of the so-called asymptotic distribution amplitude,[2] that is $\Phi_N(x) = N_L x^{L+1}(1-x)^{L+1}$, where N_L is a factor ensuring the required normalization to unity. In particular, for $L=1$, $L=2$ mesons we have respectively $\Phi_N(x) = 30x^2(1-x)^2$, $\Phi_N(x) = 140x^3(1-x)^3$. Since we have considered also the production of hybrid meson pairs (that is, pairs made of a pion plus a pseudotensor mesons), we have also to choose a DA for the pion; as an example, in the following we always use the well-known Chernyak-Zhitnitsky DA,[5] $\Phi_{N,\pi}^{CZ}(x) = 30x(1-x)(2x-1)^2$.

3.5 Normalization constants f_L

In order to make complete numerical predictions for the processes of interest, we need finally to fix the values of the normalization constants f_L appearing in Eq. (5). For the pion, we take the experimental value of the leptonic decay constant, $f_\pi \cong 93$ MeV, and use the relation $|f_0| = f_\pi/(2\sqrt{3})$.

For tensor and pseudotensor mesons we evaluate, using the same theoretical approach, the corresponding two-photon decay widths, $\Gamma(Q \to \gamma\gamma)$. Making use of the available experimental data for the masses and the two-photon decay widths of the f_2, a_2, f_2' and π_2 mesons, one gets estimates of the absolute values of the corresponding $f_{1,2}$ constants. Both the case of nonrelativistic and generalized asymptotic DA's have been taken into account.

4   Results

Once we have evaluated the helicity amplitudes for the hadronic process, $\mathcal{M}_{\lambda_\gamma \lambda_\gamma'}^{LSJ\Lambda, L'S'J'\Lambda'}(E,\Theta)$, we can give predictions for physical observables, such as the differential cross section with respect to the scattering angle Θ

$$\frac{d\sigma^{\gamma\gamma \to QQ'}(E,\Theta)}{d(\cos\Theta)} = \frac{\xi}{128\pi E^2} \sum_{\lambda_\gamma \lambda_\gamma', \Lambda\Lambda'} |\mathcal{M}_{\lambda_\gamma \lambda_\gamma'}^{LSJ\Lambda,L'S'J'\Lambda'}(E,\Theta)|^2 , \quad (6)$$

where meson masses have been neglected in the phase-space factor, and $\xi=1/2$ if Q, Q' are identical particles, $\xi=1$ otherwise.

In Fig.s 1-4 we are plotting the scaling differential cross sections $E^8 d\sigma/dt$, which is easily derived from Eq. (6) by noticing that, neglecting masses, $|t| = (E^2/2)(1-\cos\Theta)$. It is also easy to derive from Eq. (6) the differential cross section with respect to the transverse momentum p_T of the outgoing mesons, $d\sigma^{\gamma\gamma \to QQ'}(E,p_T)/dp_T$. The integrated cross section for the overall process

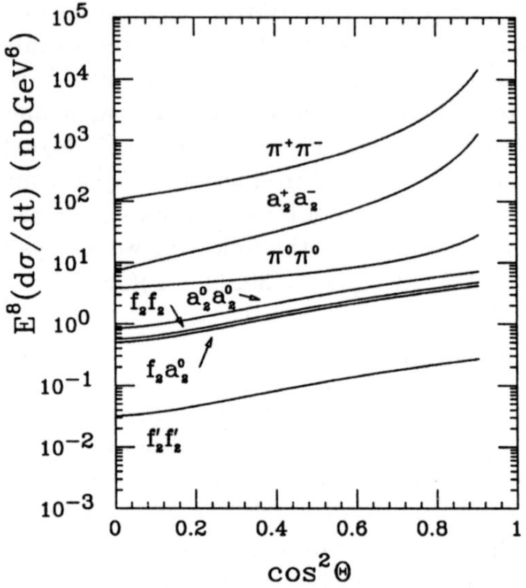

Fig. 1: Differential cross section $E^8 [d\sigma/dt]$ in nb×GeV^6, as a function of $\cos^2 \Theta$, for the process $\gamma\gamma \to QQ'$ involving the production of tensor-meson pairs; the nonrelativistic DA was used for tensor mesons; for comparison, analogous curves for $\gamma\gamma \to \pi^+\pi^-$ and $\gamma\gamma \to \pi^0\pi^0$, using the Chernyak-Zhitnitsky DA are also shown.

$(e^- e^+ \to e^- e^+ QQ')$ can then be obtained, in the equivalent-photon approximation, by convoluting $d\sigma^{\gamma\gamma \to QQ'}/dp_T$ with the equivalent-photon spectrum of the two photons, and integrating over p_T from a given minimum value of p_T, p_T^{min}. In tables 1-3 we present the results for the integrated cross section for three cases: *i)* $\sqrt{s} = 200$ GeV (LEP2 energy), $p_T > 1$ GeV ; *ii)* $\sqrt{s} = 200$ GeV, $p_T > 2$ GeV ; *iii)* $\sqrt{s} = 10$ GeV (energy of a "B factory"), $p_T > 1$ GeV.

5 Conclusions

We have presented a formalism for the study of exclusive decay or production processes involving $(q\bar{q})$ mesons having non-zero orbital angular momentum. Our approach is a generalization of the usual perturbative QCD models for exclusive processes involving $L = 0$ mesons.

As an application of our model, we have considered in detail the production of meson pairs (involving tensor, pseudotensor mesons) in photon-photon collisions. From fig.s 1-4 and tables 1-3 we can argue that the results obtained

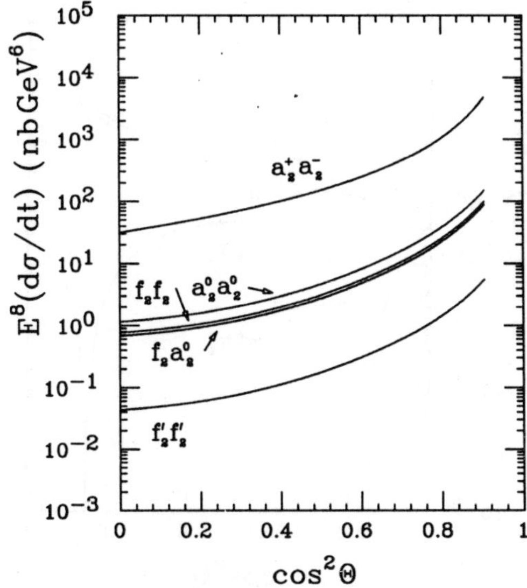

Fig. 2: The same as in Fig. 1, but the generalized asymptotic DA is used for tensor mesons.

do not depend strongly on the distribution amplitude chosen for tensor and pseudotensor mesons: apart from $\pi_2^0 \pi_2^0$ production, the generalized asymptotic DA leads to approximately equal or slightly (at most by a factor of about 3) higher values, as compared to the nonrelativistic one. One can also notice that in general the charged channels give rise to significantly higher yields than the neutral ones, being thus more favorable for experimental searches.

Finally, tables 1-3 show that, although the integrated cross sections are small, there is some hope that the production of charged-meson pairs as here considered may become measurable with high-energy $e^- e^+$ colliders of the next generation, provided integrated luminosities as high as $\approx 10^{40}$ cm^{-2} can be reached.

Acknowledgments

Three of us (J.H., F.M. and J.P.) wish to thank M. Anselmino and E. Predazzi for their kind hospitality during this very interesting workshop.

This work has been partially supported by the EU program "Human Capital and Mobility" under contract CHRX-CT94-0450.

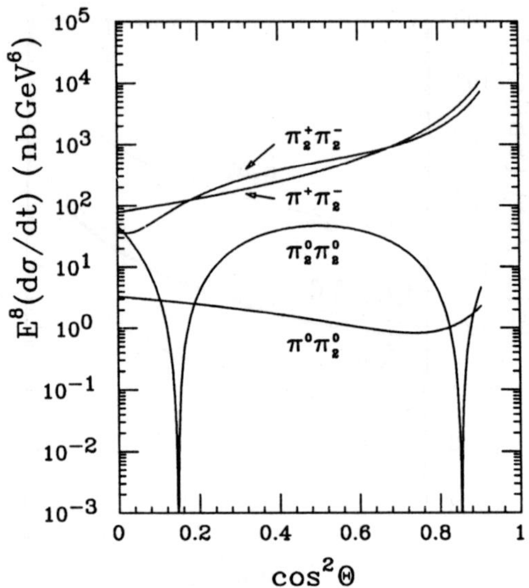

Fig. 3: Differential cross section $E^8 [d\sigma/dt]$ in nb$\times$GeV^6, as a function of $\cos^2 \Theta$, for the process $\gamma\gamma \to QQ'$ involving the production of pseudotensor-meson and hybrid (one pion plus one pseudotensor meson) pairs. The Chernyak-Zhitnitsky DA was used for pions, while the nonrelativistic DA was used for pseudotensor mesons.

References

1. L. Houra-Yaou, P. Kessler, J. Parisi, F. Murgia and J. Hansson, preprint hep-ph/9611337, LPC 96 53, INFNCA-TH9620.
2. See, e.g., S.J. Brodsky and G.P. Lepage, in *Perturbative Quantum Chromodynamics*, ed. A.H. Mueller (World Scientific, Singapore, 1989).
3. R.N. Cahn, *Phys. Rev.* D **35**, 3342 (1987).
4. S.J. Brodsky and G.P. Lepage, proceedings of *Photon-Photon Interactions 1981*, ed. G. London (World Scientific, Singapore, 1981).
5. V.L. Chernyak, A.R. Zhitnitsky, *Nucl. Phys.* B **201**, 492 (1982).

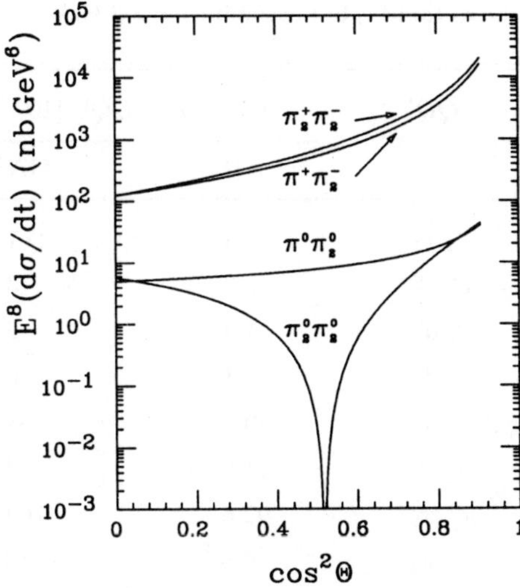

Fig. 4: The same as in Fig. 3, but the generalized asymptotic DA is used for pseudotensor mesons.

Table 1: Integrated cross sections (in 10^{-40} cm^2) of the process $ee' \to ee'QQ'$, for $\sqrt{s} = 200$ GeV, $p_T > 1$ GeV.

QQ'	$\sigma(ee' \to ee'QQ')$ $[10^{-40}$ cm$^2]$	
	NR	GASY
$f_2\, f_2$	35.2	49.1
$a_2^0\, a_2^0$	49.7	92.6
$f_2\, a_2^0$	31.1	57.5
$f_2'\, f_2'$	1.0	2.1
$a_2^+\, a_2^-$	494.6	1721.8
$\pi_2^0\, \pi_2^0$	236.1	37.7
$\pi_2^+\, \pi_2^-$	1387.6	2698.2
$\pi^0\, \pi_2^0$	165.8	389.6
$\pi^+\, \pi_2^-$	6651.7	10476.8

Table 2: Same as table 1, but assuming $p_T > 2$ GeV.

QQ'	$\sigma(ee' \to ee'QQ')$ $[10^{-40}$ cm$^2]$	
	NR	GASY
$f_2\, f_2$	1.2	1.2
$a_2^0\, a_2^0$	2.3	3.5
$f_2\, a_2^0$	1.4	2.1
$f_2'\, f_2'$	0.1	0.1
$a_2^+\, a_2^-$	20.3	68.9
$\pi_2^0\, \pi_2^0$	21.6	4.1
$\pi_2^+\, \pi_2^-$	92.6	164.2
$\pi^0\, \pi_2^0$	6.5	13.4
$\pi^+\, \pi_2^-$	189.6	298.4

Table 3: Same as table 1, but assuming: $\sqrt{s} = 10$ GeV, $p_T > 1$ GeV.

QQ'	$\sigma(ee' \to ee'QQ')$ $[10^{-40}$ cm$^2]$	
	NR	GASY
$f_2\, f_2$	2.3	2.6
$a_2^0\, a_2^0$	3.6	5.9
$f_2\, a_2^0$	2.3	3.7
$f_2'\, f_2'$	0.1	0.1
$a_2^+\, a_2^-$	33.0	113.2
$\pi_2^0\, \pi_2^0$	16.8	2.7
$\pi_2^+\, \pi_2^-$	74.0	131.8
$\pi^0\, \pi_2^0$	17.1	35.9
$\pi^+\, \pi_2^-$	524.4	825.2

COVARIANT FORMULATION OF LIGHT FRONT DYNAMICS AND RELATIVISTIC SYSTEMS

J.-F. Mathiot

Laboratoire de Physique Corpusculaire
Université Blaise Pascal, CNRS/IN2P3
F-63177 Aubière Cedex, France

We present the general properties, as well as first applications, of a covariant generalization of light-front dynamics on an arbitrary light-front surface defined by $\omega.x = 0$, where ω is an unspecified light-like four vector.

1 Light Front Dynamics

1.1 The need for a relativistic formalism

The description of few-body systems in a consistent relativistic framework has become more and more relevant in the last few years, both in nuclear and particle physics. In nuclear physics, the analysis of high momentum electromagnetic observables in order to investigate the microscopic structure of light nuclei in terms of hadronic or quark degrees of freedom has to rely on a relativistic formalism to describe both the structure of the system, and its electromagnetic interactions. A relativistic description may also be relevant at low energies (nuclei under normal conditions), but for specific observables like the spin-orbit potential or the binding energy of nuclei at normal density. In particle physics, the structure of elementary hadrons, and in particular the pion and the nucleon is, from the start, a relativistic problem since the relevant quark masses are very low.

Among the various approaches to deal with relativity in the description of bound (and scattering) states, we shall concentrate in the following on Light-Front Dynamics (LFD). In the standard formulation of LFD, the wave function of the system is defined on a plane characterized by the equation $t + z/c = 0$. This is to be compared to the usual Schrödinger formalism where the non-relativistic wave function is defined on planes of equal time (t=0 by convention), and the time evolution of the system is governed by the dynamics. For a time-independent potential, it is given for instance by the standard e^{-iEt} factor. This formulation is easily recovered in LFD by letting c going to infinity.

The formulation of relativistic systems in LFD has many advantages. May be the most important one is the absence of vacuum fluctuations. This has the important consequence that a meaningfull decomposition of the state vector describing the system under consideration in terms of Fock components of definite number of particles is possible. In the case of the deuteron for instance, one can write schematically:

$$|D\rangle = |NN\rangle + |NN\pi\rangle + |NN\pi\pi\rangle + ...\tag{1}$$

The number of Fock components to be considered in any practical calculation depends of course on the dynamics of the system, and on the kinematical regime one is interested in.

The second advantage of this formulation is the fact that any particle in the system is on its mass shell. This constraint has an immediate consequence in the fact that the formalism to calculate the wave function, and any electromagnetic observables, can be reduced to a three-dimensional one. This enables a direct and very usefull comparison with the non-relativistic phenomenology, as we shall explain later on. The price to pay is of course that the energy is not conserved at each vertex, as it is already the case in old fashioned time ordered perturbation theory. We shall come back to this point in the next subsection.

The most serious drawback of this formulation is however that the position of the light-front $t+z=0$ (with $c=1$) is not invariant in any rotation in the zx and zy plane. Since these rotations change the position of the light-front, the associated generators should depend on the dynamics and cannot be reduced to kinematical transformations [1]. This means in practice that one needs to know the complete dynamics in order to write down the general structure of a bound state of definite angular momentum [2,3]. This means also that any electromagnetic operator should have the same (dynamical) transformation properties in order to match those of the bound state wave function. This is essential in order to guarantee that any physical amplitude (or cross-section) is gauge invariant, and do not depend on the particular choice of the light-front we start with.

As a consequence, one needs an explicit procedure to exhibit in a convenient way these dynamical transformations. This is achieved in the covariant formulation of LFD which we shall concentrate on in the following.

1.2 Covariant formulation of Light-Front Dynamics

The starting point of the covariant formulation of LFD is the invariant definition of the light-front by:

$$\omega.x = 0\tag{2}$$

where ω is an (unspecified) light-like four vector ($\omega^2 = 0$). If one specifies a particular value of ω, i.e. for instance $\omega = (1,0,0,-1)$, one recovers the standard formulation of LFD we presented in the previous subsection.

This definition of the light-front is explicitely invariant by any four-dimensional rotation, or any three-dimensional rotation and Lorentz boost:

$$\omega.x = 0 \Rightarrow \omega'.x' = 0 \ . \tag{3}$$

As a consequence, these transformations become entirely kinematical, but ω-dependent, and do not necessitate the knowledge of the dynamics of the system. All the dynamics is now described by the ω dependence of *i)* the wave function, *ii)* the electromagnetic operator, *iii)* any other operator, in such a way that any physical amplitude should not depend on the particular position of the light-front, i.e. should not depend on ω.

All the details of this covariant formulation can be found in ref.[2]. We shall concentrate in the following on the main consequences for the description of physical bound state and their electromagnetic interactions.

2 General structure of bound-state wave functions

The wave functions of the system under consideration are the Fock components of the state vector defined on the light front surface $\omega.x = \sigma$, according to:

$$
\begin{aligned}
\phi(p) \ = \ & (2\pi)^{3/2} \int \Phi(k_1, k_2, p, \omega\tau) a^\dagger(\vec{k}_1) a^\dagger(\vec{k}_2)|0\rangle \\
\times \ & \delta^{(4)}(k_1 + k_2 - p - \omega\tau) \exp(i\tau\sigma) d\tau \frac{d^3 k_1}{(2\pi)^{3/2}\sqrt{2\varepsilon_{k_1}}} \frac{d^3 k_2}{(2\pi)^{3/2}\sqrt{2\varepsilon_{k_2}}} + \ . \tag{4}
\end{aligned}
$$

We left out for simplicity all spin indices, if any. As we already mentioned, the energy is not conserved at each vertex. This appears in this formulation in the delta function in eq.(4):

$$k_1 + k_2 = p + \omega\tau \ , \tag{5}$$

where p is the momentum of the system and k_1, k_2 are the momenta of the constituents. The variable τ, which has the dimension of an energy, takes care of the off-energy shell effects inherent to every bound state wave function. It is entirely determined by the on-mass shell condition for each particle, and the conservation law (5).

It is more convenient for practical calculations, and for a direct link with non-relativistic approaches, to use the variables $\vec{k}$ and $\vec{n}$ which are identical, in the system of reference where $\vec{k}_1 + \vec{k}_2 = 0$, to the momenta $\vec{k}_1$ and $\vec{\omega}$. Note that $\vec{n}$ is by definition a unit vector, and that, in this particular reference system, $\vec{p}$ is not zero, but equal to $-\vec{n}\tau$. Moreover, one can show [2] that $\vec{k}^2$ and $\vec{n}.\vec{k}$ are invariant quantities.

2.1 Spinless particles

For a system of two spinless particles in $J = 0$, the wave function $\Phi(k_1, k_2, p, \omega\tau)$ in (4) is just a scalar function of two invariants, i.e. a function of $\vec{k}^2$ and $\vec{n}.\vec{k}$, $\phi(\vec{k}^2, \vec{n}.\vec{k})$. In the non-relativistic limit, the wave function would be just a function of $\vec{k}^2$. The equation for ϕ generalizes the Schrödinger equation, and writes:

$$\left[4(\vec{k}^2 + m^2) - M^2\right] \phi(\vec{k}, \vec{n}) = -\frac{m^2}{2\pi^3} \int \phi(\vec{k}', \vec{n}) V(\vec{k}', \vec{k}, \vec{n}, M^2) \frac{d^3 k'}{\varepsilon_{k'}} , \qquad (6)$$

where V is the kernel of the equation which can be calculated for instance from the exchange of a boson. Note that this equation is three-dimsioanl, and that both ϕ and V depend on $\vec{n}$. In the Wick-Cutkosky model where the kernel corresponds to the exchange of a massless scalar boson, the wave function writes, in leading order and in the low-binding limit:

$$\phi(\vec{k}, \vec{n}) = \frac{8\sqrt{\pi m}\kappa^{5/2}}{(\vec{k}^2 + \kappa^2)^2 \left(1 + \frac{|\vec{n}.\vec{k}|}{\varepsilon_k}\right)} , \qquad (7)$$

where $\kappa = \sqrt{m|\epsilon_b|}$. The relativistic correction appears here as a factor $\frac{|\vec{n}.\vec{k}|}{\varepsilon_k}$.

For a system of two particles of spin 1/2 (the pion for instance), the wave function has two independent components. It can be decomposed for instance according to:

$$\Phi(\vec{k}, \vec{n}) = \bar{u}(k_2) \left[\frac{A_1}{m} + A_2 \frac{\omega_\mu \gamma^\mu}{\omega.p}\right] \gamma_5 v(k_2) . \qquad (8)$$

The two components A_1 and A_2 are functions of $\vec{k}^2$ and $\vec{n}.\vec{k}$. They are solution of a coupled system of equation very similar to (6).

2.2 The deuteron wave function

The general structure of the deuteron wave function can be constructed analogously in terms of the vectors at our disposal, i.e. $\vec{\sigma}, \vec{k}$ and $\vec{n}$. It has thus six components [4] instead of the usual two S- and D- non-relativistic states. The general decomposition of the wave function is written as:

$$\Psi^\lambda_{\sigma_2\sigma_1}(\vec{k}, \vec{n}) = \sqrt{m}w^\dagger_{\sigma_2}\psi^\lambda(\vec{k}, \vec{n})\sigma_y w^\dagger_{\sigma_1} , \qquad (9)$$

with

$$\vec{\psi}(\vec{k},\vec{n}) = f_1 \frac{1}{\sqrt{2}}\vec{\sigma} + f_2 \frac{1}{2}\left[\frac{3\vec{k}(\vec{k}.\vec{\sigma})}{\vec{k}^2} - \vec{\sigma}\right] + f_3 \frac{1}{2}[3\vec{n}(\vec{n}.\vec{\sigma}) - \vec{\sigma}]$$

$$+ \ f_4 \frac{1}{2k}\left[3\vec{k}(\vec{n}.\vec{\sigma}) + 3\vec{n}(\vec{k}.\vec{\sigma}) - 2(\vec{k}.\vec{n})\vec{\sigma}\right]$$

$$+ \ f_5\sqrt{\frac{3}{2}}\frac{i}{k}[\vec{k}\times\vec{n}] + f_6 \frac{\sqrt{3}}{2k}[(\vec{k}\times\vec{n})\times\vec{\sigma}]\ , \tag{10}$$

where w is the two-component nucleon spinor normalized to $w^\dagger w = 1$. The scalar functions f_{1-6} depend on the scalars $\vec{k}^2$ and $\vec{n}.\vec{k}$. The first estimate of the relativistic components of the deuteron in this formalism has been done in ref.[4]. The components f_1 and f_2 go over to the S and D states deuteron wavefunctions in the non-relativistic limit. In the momentum range from 0 to 1.5 GeV/c, the components f_3, f_4, f_6 are rather small, but the component f_5 dominates over the leading f_1, f_2 components already at momenta larger than 0.5 GeV/c. This may look at first sight rather surprising, given the success of the non-relativistic phenomenology in this kinematical domain. We shall see in the next section how one should interpret this result.

3 Electromagnetic observables

From very general considerations, any physical electromagnetic amplitude $J_\rho \equiv \langle p'|J_\rho(0)|p\rangle$ does not depend on the hypersurface where the state vector $|p\rangle$ is defined and, hence, does not contain any term dependent on ω, provided the current operator $J_\rho(0)$ is the full operator. This current operator has to contain the interaction. The consistency of the transformation properties of the current and of the state vector ensures the independence of the matrix element of the current $J_\rho(0)$ on the quantization surface orientation.

In practical calculations however, this consistency is violated twice: i) the current is usually taken in the impulse approximation, and therefore corresponds to the free current and ii) the state vector is approximated by its two-body component. Due to these reasons, the vertex J_ρ acquires spurious contributions proportional to ω_ρ. They are however completely explicited in our covariant formulation, while they are hidden in the usual formulation of the light front by $t + z = 0$.

3.1 The pion form factor

As a simple example, let us consider the pion form factor. It is easy in that case to write down the general structure of the electromagnetic amplitude:

$$J_\rho = (p + p')_\rho F(Q^2) + \omega_\rho \frac{(p+p')^2}{2\omega.p} B_1(Q^2) \ . \tag{11}$$

For the exact, on-shell physical amplitude the form factor $B_1(Q^2)$ is zero. In any approximate calculation it is not. In the relativistic regime, and in the impulse approximation for instance, it can be as large as the physical form factor $F(Q^2)$. One thus needs a definite prescription in order to extract the physical form factor from the theoretical amplitude J_ρ, at any step in the approximations which are made. For the case of $J = 0$ systems, this is rather simple since, with $\omega^2 = 0$, one has:

$$F(Q^2) = \frac{J.\omega}{2\omega.p}. \tag{12}$$

In the usual light-front formulation, with $\omega = (1,0,0,-1)$, eq.(12) corresponds to expressing the form factor through the J_+ component of the current. This is already a well known procedure. However, this procedure cannot be extended to the calculation of physical form factors of systems with total spin 1/2 and 1 [8,5], as we shall see in the next subsection.

In the asymptotical region (very high momentum transfer), the form factor is dominated by the relativistic A_2 component. This one can be calculated easily, in this domain, by one-gluon exchange interaction. The usual $1/Q^2$ behavior is recovered. Note that in the case of spinless particles (the Wick-Cutkosky model), where the wave function has only one component, the form factor behaves like $1/Q^4$ at very high momentum transfer. This formulation enables a straightforward generalization to non-leading contributions involving the transverse momentum degrees of freedom.

3.2 The nucleon form factors

The electromagnetic physical amplitude for a spin 1/2 fermion is determined by the two standard form factors:

$$J_\rho = F_1(Q^2)\bar{u}'\gamma_\rho u + \frac{iF_2(Q^2)}{2M}\bar{u}'\sigma_{\rho\nu}q^\nu u \ , \tag{13}$$

with standard notations.

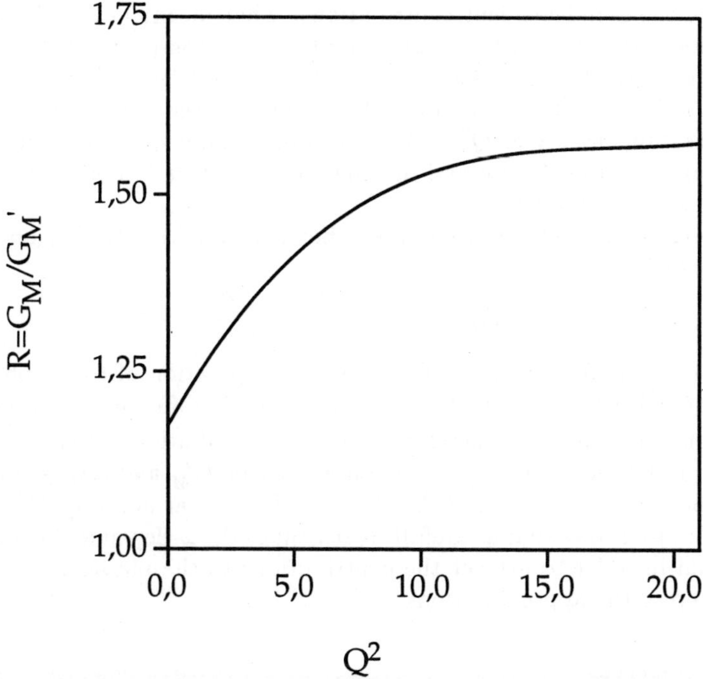

Figure 1: The ratio G_M/G'_M as a function of Q^2 in $(GeV/c)^2$.

Like in the case of spin 0 and 1, the dependence of any approximate electromagnetic amplitude on the extra four-vector ω increases the number of independent terms in the decomposition of the amplitude. Besides the two structures entering in eq.(13), one can construct the following three structures:

$$\frac{1}{\omega.p}\bar{u}'u\,\omega_\rho\;,\quad \frac{1}{(\omega.p)^2}\bar{u}'\omega_\mu\gamma^\mu u\,\omega_\rho\;,\quad \frac{1}{\omega.p}\bar{u}'\omega_\mu\gamma^\mu u\,P_\rho\;,\tag{14}$$

where $P = p + p'$. One should thus extract the ω-independent contributions from these terms. The first two one are explicitly ω-dependent, but the third one has a piece which is ω-independent after contraction with the spinors. It has therefore to be included as a contribution to the physical form factors. The exact procedure to follow in that case has been derived in ref.[8]. Note that the standard procedure to calculate the form factor in terms of the + component of the current (i.e. in our formulation from the product $J.\omega$) is not enough to get ride of the ω-independent contributions. That means in practice that the form factors extracted with this procedure acquire non-physical contributions due to the particular orientation chosen for the light-front. It turns out that the electric form factor is however the same, but the magnetic form factor differs sizeably, already at $Q^2 = 0$. To estimate these spurious contributions, we can calculate the electric and magnetic form factors in both approaches in a simplified quark model [8].

At zero momentum transfer, the difference $G'_M(0) - G_M(0)$ obtains the simple form:

$$G'_M(0) - G_M(0) = -\frac{\langle \vec{k}^2 \rangle}{3m^2} \, ,$$

where G_M is the form factor calculated after separation of the unphysical ω-dependent contributions, while G'_M is calculated from the $+$ component of the current, and includes therefore spurious contributions. Here $\langle \vec{k}^2 \rangle$ is the average quark momentum. The difference between G'_M and G_M is of course of relativistic origin. The ratio of the two form factors, at non-zero Q^2, is shown in fig.1. It also shows that a carefull treatment of the ω-dependent contribution is essential in order to extract the contributions to the physical form factors, at any step of the approximations.

4 Perspectives

We have presented the general properties of a covariant formulation of light front dynamics, and its application to relativistic few-body systems. These developments have been made particularly simple, and intuitive, by the three-dimensional nature of the formalism, and the absence of vacuum fluctuations. This enables a transparent link with the non-relativistic phenomenology developed over the last twenty years for the understanding of the microscopic structure of few-nucleon systems.

We expect many new developments in the near future along the lines developed in this contribution, both in nuclear and particle physics. In nuclear physics, the exact calculation of the two-body wave function is under investigation. This however necessitates to reparametrize the NN potential starting from the same relativistic formalism. This is essential in order to have quantitative predictions for electromagnetic observables in the few GeV range. In particle physics, the structure of the pion and nucleon can be investigated easily in this framework, and electromagnetic form factors calculated in the high momentum region. Application to virtual compton scattering is under investigation, as well as the extension to axial transitions. One other particularly interesting application is to the structure of heavy quarkonium, (the J/ψ

for instance), in order to investigate the importance of relativistic dynamical corrections, i.e. the importance of higher Fock states in the J/ψ wave function.

Acknowledgments

I acknowledge very usefull discussions with J. Carbonell, B. Desplanques and V.K. Karmanov.

References

1. P.A.M. Dirac, Rev. Mod. Phys. **21** (1949) 392
2. J. Carbonell, B. Desplanques, V.K. Karmanov and J.-F. Mathiot, prepared for Physics Reports
   V.A.Karmanov, Fiz. Elem. Chastits At. Yadra, **19** (1988) 525 (trasl.: Sov. J. Part. Nucl. **19** (1988) 228)
3. M.G.Fuda, Ann.Phys. (N.Y.) **197** (1990) 265, Phys.Rev. **D44** (1991) 1880
4. J.Carbonell and V.A.Karmanov, Nucl. Phys. **A581** (1995) 625, **A589** (1995) 713
5. V.A.Karmanov and A.V.Smirnov, Nucl. Phys. **A546** (1992) 691, **A575** (1994) 520
6. J.-F. Mathiot, Phys. Rep. **173** (1989) 63
7. B.Desplanques, V.A.Karmanov and J.-F.Mathiot, Nucl. Phys. **A589** (1995) 697
8. V.A.Karmanov and J.-F.Mathiot, Nucl. Phys. **A602** (1996) 388

Hadron structure and odderon exchange

Michael Rueter

Institut für theoretische Physik
Universität Heidelberg
Philosophenweg 16, D-69120 Heidelberg, FRG
e-mail: M.Rueter@thphys.uni-heidelberg.de

Talk given at the *Diquarks III* workshop, Torino, October 1996

supported by the Deutsche Forschungsgemeinschaft

We calculate the $C=P=-1$ contribution to high-energy scattering of hadrons in the framework of the model of the stochastic vacuum. In models, where the pomeron is generated by a two gluon exchange, this odderon contribution comes from a three gluon exchange and is much too large as compared to experimental data for pp- and $p\bar{p}$-scattering. In our model, where we have no quark-additivity, the hadron structure is very important. It is shown that a natural suppression of the odderon contribution is given by a diquark-structure of the nucleon.

In this note we summarize the results presented in a recent publication[1]. Some parameters of the used model are changed but the calculated physical quantities are rather insensitive to these changes.

1   Introduction

The possibility that the real part of the scattering amplitude increases with energy as fast as the imaginary part was first considered by Lukaszuk and Nicolescu[2]. Such a behavior would mean that a trajectory of a pole which is odd under C and P has an intercept near one. This trajectory has been called odderon. One consequence of such an odderon would be that the ratio of the real to imaginary part of the forward scattering amplitude is different for particle-particle and particle-antiparticle-scattering even at asymptotic energies. For further reference we shall use the conventional abbreviation $\Delta\rho$ for that difference:

$$\Delta\rho(s) = \rho^{\bar{p}p}(s) - \rho^{pp}(s) = \frac{\mathrm{Re}\left[T^{\bar{p}p}(s,0)\right]}{\mathrm{Im}\left[T^{\bar{p}p}(s,0)\right]} - \frac{\mathrm{Re}\left[T^{pp}(s,0)\right]}{\mathrm{Im}\left[T^{pp}(s,0)\right]} \tag{1}$$

Interest in the odderon rose again when the UA4 collaboration[3] reported a value for $\rho^{\bar{p}p}$ at $\sqrt{s} = 546$ GeV which was much larger than the one extrapolated by means of dispersion relations for proton-proton-scattering and thus seemed to indicate a large value for $|\Delta\rho|$.

The new results of the UA4/2 collaboration[4] obtained however a value $\rho^{\bar{p}p}(\sqrt{s} = 541 \text{ GeV}) = 0.135 \pm 0.015$ which is very well compatible with $\Delta\rho = 0$ at that energy[5-10] and at any rate leaves no room for a large value of that quantity. The very successful description of high-energy data by the Donnachie-Landshoff pomeron[11] also yields $\Delta\rho \approx 0$.

As far as the contribution of three *non-perturbative* gluons is concerned there is no reason for a strong suppression of the three gluon versus the two gluon exchange. In an Abelian model for non-perturbative gluon exchange[12] Donnachie and Landshoff[13] have found that the lowest order effective odderon coupling, i.e. the coupling of three non-perturbative gluons, is suppressed by a factor of two with respect to the effective pomeron coupling. Though it is very gratifying that in a non-perturbative model the three gluon coupling is smaller than the two gluon coupling (naive expectation goes in the opposite direction), this coupling still leads to a value of $|\Delta\rho| \approx .5$ which is far from consistency with the analysis of the data. In the Abelian model of Landshoff and Nachtmann, where quark additivity is a consequence of the model, the ρ-parameter for hadron-(anti)hadron-scattering is just the one for quark-(anti)quark-scattering.

In a series of papers[14] a non-Abelian model of high-energy scattering was presented which gives a good description of the experimental data and relates parameters of high-energy scattering to those of hadron spectroscopy. In this model, called the model of the stochastic vacuum[15,16] (MSV), the non-perturbative gluonic contributions to the QCD-pathintegral are approximated by a Gaussian stochastic process with a gluonic correlation length a. One of the most characteristic features of this model is that the same mechanism which leads to confinement introduces a kind of string-string interaction in high-energy scattering and leads to a marked increase of the total cross section as a function of the hadron size even if the latter is large as compared to the correlation length a. Quark-additivity does not hold in that approach. The different total cross sections for pion-nucleon-, kaon-nucleon- and nucleon-(anti)nucleon-scattering are correctly reproduced due to the different radii of the hadrons. In this note we review the evaluation of the leading $C=P=-1$ contribution of that model. We show that this contribution (and therefore also $\Delta\rho$) depends crucially on the structure of the nucleon; we especially discuss the dependence of $\Delta\rho$ on the radius of a diquark if two quarks are clustered.

2 Diffractive high-energy scattering of hadrons in the MSV

To calculate the diffractive scattering amplitude in the high-energy limit $(s \to \infty)$ one uses an eikonal approximation to separate the large energy scale s from the small momentum transfer scale t in a fixed gluon background field[17]. The resulting quark-quark-scattering amplitude (with helicity λ_i and color C_i) is very similar to the quantum-mechanical scattering in an external potential:

$$
\begin{aligned}
T(s,t) &= 2\,i\,s\,\delta_{\lambda_3\lambda_1}\delta_{\lambda_4\lambda_2}\int \mathrm{d}^2b\, e^{-i\vec{q}\vec{b}}\,\hat{j} \\
\text{where } \hat{J} &= \, < \left([\mathbf{V}[P_1]-\mathbf{1}]_{C_3C_1}\,[\mathbf{V}[P_2]-\mathbf{1}]_{C_4C_2}\right) > \\
\text{and } \mathbf{V}[P_i] &= \, \mathcal{P}\exp\left(-ig\int_{P_i}\mathbf{A}_\mu(z)\mathrm{d}z^\mu\right).
\end{aligned}
\tag{2}
$$

The quark picks up the phase $\mathbf{V}[P_i]$ by interacting with the background field on the light-like path P_i. The momentum transfer is given by $t = -\vec{q}^2$ and $\vec{b}$ is the impact-parameter. The scattering amplitude is proportional to s and we have helicity conservation. The problem is to calculate the expectation value $< \ldots >$ with respect to the background field. We do this by using the MSV. In this non-Abelian model it is crucial to respect gauge invariance. We go from quark-quark- to hadron-hadron-scattering by constructing as usual gauge invariant hadrons by connecting the constituents with Schwinger-strings. For mesons we end up with the "scattering" of Wilson-loops[14] (see fig.(1)). The meson-scattering amplitude is obtained by averaging the loop-scattering with Gaussian transversal wave function with extension parameter S:

$$
\begin{aligned}
\mathbf{W}[\partial S_i] &= \, \mathcal{P}\exp\left(-ig\oint_{\partial S_i}\mathbf{A}_\mu(z)\mathrm{d}z^\mu\right) \\
\tilde{J}(\vec{b},\vec{R}_1,\vec{R}_2) &= \, -< \frac{1}{N_C}\,\mathrm{Tr}\,[\mathbf{W}[\partial S_1]-\mathbf{1}]\cdot\frac{1}{N_C}\,\mathrm{Tr}\,[\mathbf{W}[\partial S_2]-\mathbf{1}] > \\
\hat{J}(\vec{b},S_1,S_2) &= \, \int \mathrm{d}^2\vec{R}_1\int \mathrm{d}^2\vec{R}_2\,\tilde{J}(\vec{b},\vec{R}_1,\vec{R}_2)\,|\Psi(\vec{R}_1,S_1)|^2\,|\Psi(\vec{R}_2,S_2)|^2 \\
\Psi(\vec{R}_i,S_i) &= \, \sqrt{\frac{1}{2\pi}}\frac{1}{S_i}e^{-\frac{|\vec{R}_i|^2}{4S_i^2}}.
\end{aligned}
\tag{3}
$$

In order to describe baryon-baryon-scattering, one has to start from three loops (without traces) with one common side as shown in fig.(2).
The reduced scattering amplitude for a colorless qqq-objekt is given by:

$$
\begin{aligned}
\tilde{J}(\vec{b},\vec{R}_1,\vec{R}_2) &= \, -< B_1\cdot B_2 > \\
\text{with } B_i &= \, \frac{1}{6}\epsilon_{abc}\epsilon_{a'b'c'}\left\{\mathbf{W}_{a'a}[S_{i1}]\mathbf{W}_{b'b}[S_{i2}]\mathbf{W}_{c'c}[S_{i3}]-\delta_{a'a}\delta_{b'b}\delta_{c'c}\right\} \\
\mathbf{W}_{a'a}[S_{ij}] &= \, \left[\mathcal{P}\,e^{-ig\oint_{\partial S_{ij}}\mathbf{A}_\mu(z)\,\mathrm{d}z^\mu}\right]_{a'a}.
\end{aligned}
\tag{4}
$$

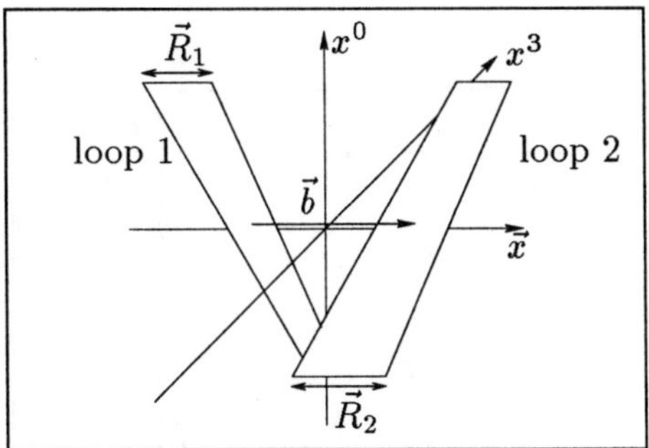

Figure 1: Two loops with transversal extension $\vec{R}_1$ and $\vec{R}_2$ and light-like sides. Loop 1 (∂S_1) describes a colorless $q\bar{q}$-pair running in negative 3-direction and loop 2 in positive 3-direction. The impact parameter $\vec{b}$ is chosen to be purely transverse and the light-like sides are considered to by infinitely long.

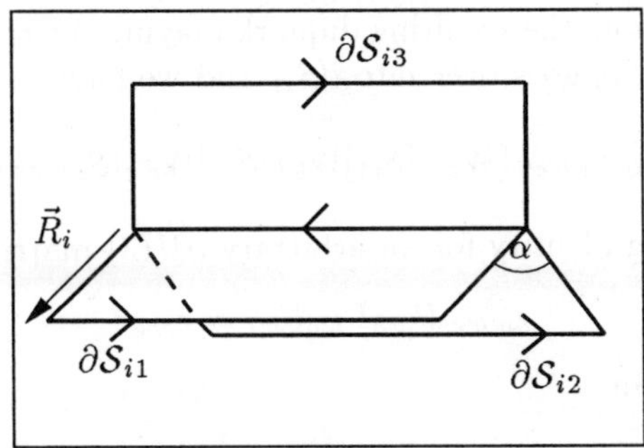

Figure 2: The colorless qqq-objekt is constructed out of 3 loops with one common line which transforms like a color singlet. Here ∂S_{ij} denotes the loop corresponding to quark j of baryon i. By varying the angle α we can consider different geometries of the baryon. With $\vec{R}_i$ we denote the radius. A baryon is obtained by averaging with a transversal wave function.

Baryon-baryon-scattering is obtained by averaging with a transversal wave function:

$$\Psi(\vec{R}_i, S_i) = \sqrt{\frac{2}{\pi}} \frac{1}{S_i} e^{-\frac{|\vec{R}_i|^2}{S_i^2}} . \tag{5}$$

We consider two classes of baryon configurations. In the first case the distances from the common line of all three loops are equal and we vary the angle α between two loops (see fig.(2)). If this angle tends to zero the two quarks together form a point-like diquark (i.e. an object transforming under the $\bar{3}$-representation of $SU(3)$). The other case we consider is a linear structure of

94

the nucleon (see fig.(3)).

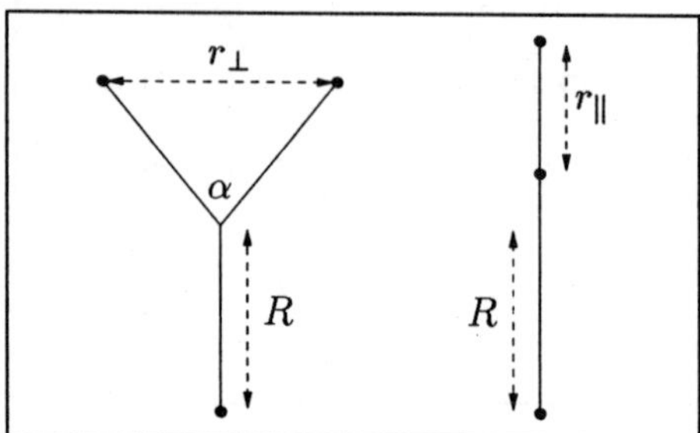

Figure 3: In this figure we show the star-like and linear geometry for the baryon. For small quark distances $r_\perp$ and $r_\parallel$ we have a diquark configuration.

In the limit of the angle α (see fig.(2)) going to zero, the baryon can effectively be treated like a meson, the resulting diquark playing the role of the antiquark. If $\alpha \to 0$ the loop ∂S_{i2} goes over into ∂S_{i1} and we have

$$B_i(\alpha = 0) = \frac{1}{6}\epsilon_{abc}\epsilon_{a'b'c'}\left\{W_{a'a}[\mathcal{S}_{i1}]W_{b'b}[\mathcal{S}_{i1}]W_{c'c}[\mathcal{S}_{i3}] - \delta_{a'a}\delta_{b'b}\delta_{c'c}\right\}. \qquad (6)$$

We use the following identity for an arbitrary $SU(3)$-matrix U

$$\epsilon_{a'b'c'}U_{a'a}U_{b'b}U_{c'c} = \epsilon_{abc}$$

from which we obtain

$$\epsilon_{a'b'c'}W_{a'a}[\mathcal{S}_{i1}]W_{b'b}[\mathcal{S}_{i1}] = \epsilon_{abh}W_{hc'}^{-1}[\mathcal{S}_{i1}] = \epsilon_{abh}W_{hc'}[\mathcal{S}_{i1}^{-1}] \qquad (7)$$

where ∂S_{i1}^{-1} is the Wilson-loop oriented in opposite direction. Inserting eq.(7) in eq.(6) we obtain

$$
\begin{aligned}
B_i(\alpha = 0) &= \frac{1}{3}\delta_{ab}\left\{W_{ac}[\mathcal{S}_{i1}^{-1}]W_{cb}[\mathcal{S}_{i3}] - \delta_{ab}\right\} = \frac{1}{3}\delta_{ab}\left\{W_{ab}[\hat{\mathcal{S}}_{i12}] - \delta_{ab}\right\} \\
&= \frac{1}{3}\,\mathrm{Tr}\left\{\mathbf{W}[\hat{\mathcal{S}}_{i12}] - \mathbf{1}\right\}
\end{aligned}
\qquad (8)
$$

where $\partial\hat{S}_{i12}$ is the union of ∂S_{i1}^{-1} and ∂S_{i3}. This is exactly the contribution of a quark traveling along line 3 and an antiquark traveling along line 1=2. The spin contribution in the high energy limit of a quark and antiquark is equal, namely $2s\delta_{\lambda\lambda'}$, where λ is the helicity in the initial and final state respectively. Thus in the limit $\alpha \to 0$ the baryon can be treated effectively as a meson, the point like diquark traveling along line 1=2 replacing the antiquark of the meson.

3 Results for the total cross-section and the slope

To calculate the vacuum expectation values in eq.(3) and eq.(4) we use the model of the stochastic vacuum. One approximates the averaging by a Gaussian stochastic process for the parallel transported fieldstrengths. This process is mainly characterized by the correlation length a and the usual gluon condensate $< g^2 FF >$. For details we have to refer to the literature[15,16,14]. The parameters are fixed by comparison with experimental data for high energy scattering and with hadron spectroscopy[14,1]. A new analysis for the different baryon geometries yields[18]:

	diquark	star-like	linear
$< g^2 FF >$	3.0 GeV^4	3.0 GeV^4	3.1 GeV^4
a	0.31 fm	0.30 fm	0.33 fm

Table 1: The parameter sets

It turns out that for the total cross section the different geometries give almost the same results. For simplicity we consider here only the diquark structure. By expanding the exponentials in eq.(3) up to second order we find the leading contribution to the scattering amplitude. This contribution is purely imaginary and even under charge parity. With the help of the optical theorem we find a total cross section which depends only on the extension parameter S and is the same for hadron-hadron- and hadron-antihadron-scattering. The numerically obtained total cross section can be parameterized by

$$\sigma^{\text{tot}} = < g^2 FF >^2 a^{10} \alpha \left(\frac{S}{a}\right)^\beta$$

where $\alpha = 5.20 * 10^{-3}$ and $\beta = 3.03$. The result is shown in fig.(4). We also calculated the slope B,

$$B = \frac{\mathrm{d}}{\mathrm{d}t} \ln \left(\frac{\mathrm{d}\sigma^{\text{el}}}{\mathrm{d}t}\right) |_{t=0}, \tag{9}$$

which in our model also only depends on the extension parameter S. It was shown[14,18] that the model reproduces the experimental data for p-p-, p-$\bar{p}$-, p-π- and p-K-scattering very well by using extension parameters which are quite similar to the electromagnetic radii of the different hadrons. By varying the extension parameter of the proton slightly with s we obtained the right energy dependency of $\sigma^{\text{tot}}_{p\bar{p}}$ and $B_{p\bar{p}}$ up to highest energies.

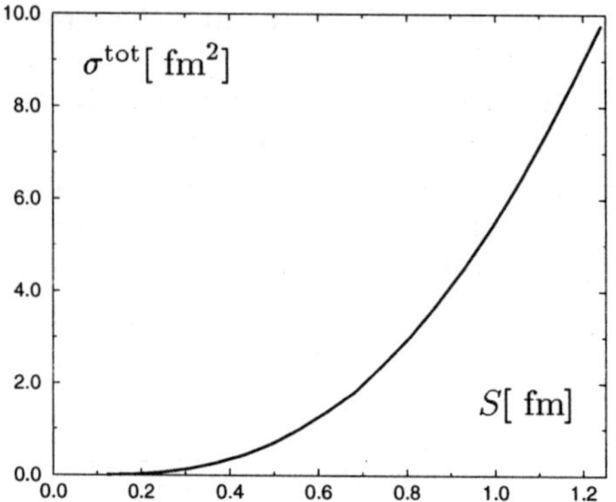

Figure 4: The total cross section of hadron-hadron-scattering with diquark geometry as a function of the extension parameter S.

4 The odderon exchange and the suppression through a diquark-structure

In this section we calculate the leading contribution to the scattering amplitude which is odd under charge parity. From eq.(3) it is easy to see that for a meson or equivalently a baryon in the diquark picture the contribution of the $C=P=-1$ exchange is appreciable for a given Wilson-loop. Constructing the hadrons by averaging the loops with wave functions ,however, cancels these contributions. So we use now the two qqq-configurations of the baryon mentioned before (see fig.(3)). By expanding the exponentials of the Wilson-loops in eq.(4) up to third order, we find[1] that the leading contribution with odd charge parity to the scattering amplitude is real. We still get the same total cross section for hadron-hadron- and hadron-antihadron-scattering but in contrast to the leading order we find a non-vanishing contribution to the rho-parameter:

$$\Delta\rho(s) = \frac{\mathrm{Re}\left[T^{\bar{p}p}(s,0)\right]}{\mathrm{Im}\left[T^{\bar{p}p}(s,0)\right]} - \frac{\mathrm{Re}\left[T^{pp}(s,0)\right]}{\mathrm{Im}\left[T^{pp}(s,0)\right]} = -2\frac{\mathrm{Re}\left[T^{pp}(s,0)\right]^{C=-}}{\mathrm{Im}\left[T^{pp}(s,0)\right]}.$$

The result for the two different baryon configurations as a function of the diquark size at UA4/2 energy ($\sqrt{s}$=541 GeV) is shown in fig.(5).
As can be seen from fig.(5) we obtain for the symmetric star-like geometry a result which is very similar to the value obtained in the Abelian non-perturbative gluon exchange model[13]. A clustering of two quarks to a diquark with an extension smaller or equal to 0.3 fm yields on the other hand already a drastic

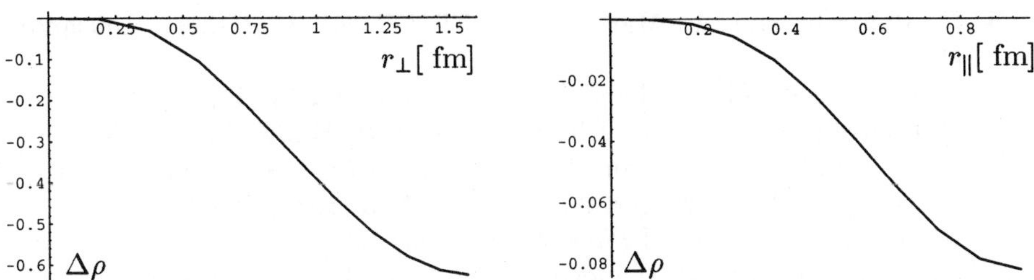

Figure 5: $\Delta\rho$ at UA4/2 energy for proton-(anti)proton scattering as a function of the quark distances $r_\perp$ and $r_\parallel$ (see fig.(3)). Left: star-like geometry (S_p=0.90 fm). Right: linear geometry (S_p=0.93 fm). Note the different scales.

suppression of $\Delta\rho$ to a value $|\Delta\rho| \leq 0.02$ which is compatible with the analysis of experiments. There is plenty of other evidence for diquark clustering in baryons: The scaling violation in nucleon structure functions[19], the strong attraction in the scalar diquark channel in the instanton vacuum[20] and the $\Delta I = \frac{1}{2}$ enhancement in semi leptonic decays of baryons[21]. Scaling violation in deep inelastic scattering is sensible to the form factors of the diquark, which is modeled by a pole fit with a pole mass of $\sqrt{3}$ to $\sqrt{10}$ GeV. This corresponds to diquark radii of 0.3 to 0.16 fm. These values are according to our model sufficiently small to give a suppression of $|\Delta\rho|$ to values below 0.02 even for the transversal extension.

Our calculation has been performed in a specific non-perturbative model. But since the limiting case of a vanishing diquark radius leads quite generally to an odderon cancelation as in the case of mesons (see eq.(8) and the discussion of it) we think that the suppression of the $C{=}P{=}{-}1$ exchange is generally due to the structure of the nucleon and cannot be seen on the quark level.

Acknowledgment

This work was done in collaboration with H.G. Dosch. The author also thanks M. Anselmino for the invitation to give this talk. Finally the author is grateful to the Deutsche Forschungsgemeinschaft and to the Bundesland Baden-Württemberg for financial support.

1. M. Rueter and H.G. Dosch, *Phys.Lett.* **B380** (1996) 177.
2. L. Lukaszuk and B. Nicolescu, *Lett.Nuov.Cim.* **8** (1973) 405.
3. D. Bernard et al., *Phys.Lett.* **B198** (1987) 583, UA4 Collaboration.
4. C. Augier et al., *Phys.Lett.* **B316** (1993) 448, UA4/2 Collaboration.

5. C. Bourrely and A. Martin, in *CERN-ECFA Workshop, Lausanne* (1984).

6. A. Donnachie and P.V. Landshoff, *Nucl.Phys.* **B244** (1984) 322.

7. C. Bourrely, J. Soffer and T.T. Wu, *Phys.Lett.* **B196** (1987) 237.

8. P. Gauron, B. Nicolescu and E. Leader, *Nucl.Phys.* **B299** (1988) 640.

9. P. Kroll and W. Schweiger, *Nucl.Phys.* **A503** (1989) 865.

10. R.J.M. Covolan et al., *Z.Phys.* **C58** (1993) 109.

11. A. Donnachie and P.V. Landshoff, *Phys.Lett.* **B296** (1992) 227.

12. P.V. Landshoff and O. Nachtmann, *Z.Phys.* **C35** (1987) 405.

13. A. Donnachie and P.V. Landshoff, *Nucl.Phys.* **B348** (1991) 297.

14. H.G. Dosch, E. Ferreira and A. Krämer, *Phys.Rev.* **D50** (1994) 1992.

15. H.G. Dosch, *Phys.Lett.* **B190** (1987) 177.

16. H.G. Dosch and Y.A. Simonov, *Phys.Lett.* **B205** (1988) 339.

17. O. Nachtmann, *Annals Phys.* **209** (1991) 436.

18. M. Rueter, *Quark-Confinement und diffraktive Hadron-Streuung im Modell des stochastischen Vakuums*, PhD-Thesis at the University of Heidelberg (1997).

19. M. Anselmino et al., *Rev.Mod.Phys.* **65** (1993) 1199.

20. T. Schäfer, E.V. Shuryak and J.J.M. Verbaarschot, *Nucl.Phys.* **B412** (1994) 143.

21. H.G. Dosch, M. Jamin and B. Stech, *Z.Phys.* **C42** (1989) 167.

COSMIC DIQUARKS

SVERKER FREDRIKSSON

Department of Physics, Luleå University of Technology, SE-971 87 LULEÅ, Sweden

This is a mini-review of the possible astrophysical roles of diquarks. References are given to the existing literature, and some old and new arguments, from ongoing work with Svante Ekelin and Johan Hansson, are presented in favour of the view that diquarks should be relevant for the history of Universe and for some of its exotic sides. This includes the ignition of supernovas, the formation of galaxies (with dark matter) and the origin of gamma-ray bursts. A collaboration with Joseph Parisi along these lines is at the planning stage within the EC-sponsored project 'The Fundamental Structure of Matter'.

1   Introduction

Eight years ago, at the very first Workshop on Diquarks here at Villa Gualino, I promised[1] to be back one day and explain, with diquarks: (i) The mystery of the Torino shroud, (ii) The extinction of dinosaurs and (iii) Big Bang. Since then the shroud has been shown to be a fake, and I have given up (for the moment) on dinosaurs.

However, Big Bang is a "must" for any diquark enthusiast with cosmic ambitions, since diquarks might appear wherever there are quarks. And the only cosmic situation for which there seems to be a consensus about the role of quarks is the period between the creation of the first giant quark-gluon plasma (QGP) and its hadronisation to normal matter.

So, my logics in discussing diquarks in cosmos is the following:

If diquarks exist as two-quark bound objects, *e.g.*, in a nucleon, they should appear also in other multi-quark systems whenever they have a chance.

If QGPs exist, they must also have a diquark component, if temperatures and densities are not too hostile.

If there are cold QGPs in space, they would be strongly influenced by diquark formation, hence providing 'cosmic diquarks'.

Would such diquark effects be non-trivial and interesting? I believe that diquark effects are not only non-trivial and interesting, but also *crucial* for the structure of galaxies, and for some everyday astrophysical phenomena. Proving this is of course a different story!

First, there is no general agreement, even among us diquark workers, about the role of diquarks inside nucleons (33 years after the first mentioning of diquarks[4]). This is obvious from reviews,[2] the literature[3] and this workshop.

There seems to be a convergence, though, toward diquarks being mostly spin-0, colour antitriplet objects, due to a spin-spin (colour-magnetic) attraction. In spin-1 pairs this force is repulsive, so that vector diquarks are either unbound, or more loosely bound than scalars.

As for the radius of the scalar (ud) there are two recent studies pointing at a value of $0.30 - 0.35$ fm; one of the non-relativistic three-quark system,[5] and the other of higher-twist effects in deep-inelastic scattering.[6] The (ud) should then have a binding of roughly 300 MeV, with two 400 MeV quarks forming a 500 MeV diquark. It should survive QGP temperatures up to a few hundred MeV and densities about ten times that of a nucleon.

Secondly, there are no clear experimental signals of QGPs in high-energy heavy-ion data, and no agreement among theorists what these signals would be, or how to estimate their magnitudes. Hence a diquark worker must borrow some phenomenological QGP methods, or invent new ones.

Thirdly, there are no observations from cosmos showing that quarks play a role in space except that inside nucleons. Even among workers who believe in QGPs inside supernovas, neutron stars etc., there are many schools, disputing both details and general principles, *e.g.*, how to extrapolate properties of simple hadrons, or small QGPs, to systems of, say, 10^{57} quarks. Therefore, a diquark worker must first adopt a model for diquarks in big QGPs, and then take a stand on some astrophysical controversies.

Nevertheless, several groups have suggested 'cosmic diquarks', and the connected issue of 'diquark matter', both in the form of rather detailed calculations, and as mere speculations. I list these in the next section.

2   Chronology

1976: Johnson and Thorn suggested that there are mesons with a diquark and an antidiquark.[7] No such states have been found, but there are still predicted heavy states to be searched for. See, *e.g.*, the contribution by Lichtenberg, Roncaglia and Predazzi at this Workshop.[8]

1981: Fredriksson and Jändel presented the idea of a 'diquark deuteron',[9] consisting of three diquarks. The name 'demon nuclei' was coined for hypothetical nuclei polarised by a three-diquark kernel, as was the concept of a 'diquark plasma', consisting of diquarks only. None of the suggestions[10,11,12] for detecting a 'demon deuteron' has given any positive results.

1985: Ekelin and Fredriksson[13] introduced the idea that a QGP could have a diquark component, and suggested some experimental signals from heavy-ion collisions, one being an excess of dibaryons after hadronisation.

1986: Ekelin [14] calculated the thermodynamic properties of a plasma consisting of quarks, gluons and diquarks. He found that the diquark component might dominate at low-enough temperatures, and there could even exist a *Bose-condensed* state of pure diquarks.

1988: At the first Workshop on Diquarks, Fredriksson took diquarks to cosmos and presented the 'supernova diquark detonator'. [1] *If* a QGP is created in the centre of a collapsing star, just before the explosive bounce, then a diquark surplus could occur, because extra diquarks can be formed out of the single quarks inside nucleons. At first this extra pairing speeds up the collapse. But then the gammas released from the diquark binding create a radiation pressure, and 'later' the diquarks repel each other at a length scale of 0.3 fm (the diquark radius). Hence, the supernova bounce should occur at this length scale, or at a density scale ten times that of a nucleon.

Donoghue and Sateesh [15] presented the first detailed account of diquarks in a QGP with a formalism developed for a nucleon gas. They included a repulsive diquark self-coupling, which, *e.g.*, prevents Bose condensation. The repulsion is sizeable, but almost pointlike. The diquark phase gives a significant lowering of the energy as compared to a gas of free quarks. The authors suggest that it is relevant for the core of a neutron star.

1991: Kastor and Traschen published the first analysis of such 'diquark stars', [16] by using the formalism of Donoghue and Sateesh. They find that such a cosmic QGP is dominated by the diquark component, which is less dense than a diquark-free plasma. They propose a detailed study of the role of diquarks for a supernova bounce, and of diquarks inside strange quark matter (diquark studies are normally limited to non-strange QGPs).

1992-93: Sateesh [17] suggested experimental signatures of a diquark phase inside a QGP as created in heavy-ion collisions, *e.g.*, the baryon ratio Λ_c/Σ_c.

Horvath *et al.* [18] refined the model of Donoghue and Sateesh, and concluded that diquark stars might exist without an external gravitational pressure (assumed by Kastor and Traschen). It is also stressed that the normal way of treating huge QGPs, namely within the MIT bag model, must be revised if there are diquarks in the plasma. All parameters must, *e.g.*, be re-fitted to baryon properties with diquarks included, in particular the 'universal' bag pressure parameter, B.

Sahu [19] studied pulsars with a quark-diquark core. He suggests an equation of state for such a core and computes the influence of diquarks on the radial pulsation frequencies and the neutrino emissivity of pulsars. None of these effects are expected to be sizeable.

1995-96: Fredriksson and Hansson claimed that diquarks are to be blamed for almost everything in Universe,[20] including galaxy formation, dark matter and cosmic gamma-ray bursts (more on this below).

Quite recently, Karn *et al.* found [21] that the size of diquarks (previously taken for pointlike) can be important for QGPs. Examples are the energy and density (both being lower with extended diquarks) and the spatial distribution of diquarks (being more uniform with extended diquarks).

There are also many inspiring papers on 'cosmic quarks', although with no mentioning of diquarks, *e.g.*, the paper where Witten [22] suggested the existence of stable quark nuggets in space, and the extensive work by Madsen *et al.* [23] and Glendenning *et al.* [24] on the quark content of pulsar cores.

3   The Crucial Idea - Diquark Repulsion

In this section I argue that there might exist pure, low-density and stable diquark plasmas, in particular in space.

Donoghue and Sateesh [15] suggested that diquarks repel each other, and describe this with a $\lambda\phi^4$ diquark self-coupling. It is due to residual spin-spin quark forces, leaking out of scalar diquarks. This interaction is short-ranged, due to the almost pointlike nature of colour-magnetism, leading to moderate corrections to a free-diquark gas. The contribution to the total energy goes like N, the number of diquarks, rather than like N^2, the number of diquark pairs.

In the same spirit, one should also analyse the leakage of *colour-electric* quark forces between diquarks. Also such forces could contribute a net repulsion to a diquark plasma, and be *long-ranged*. The reason for this belief is that there is a balance between attraction and repulsion in an infinite free-quark gas (treated essentially non-relativistically). In a colour-antitriplet pair there is an attraction, with a magnitude twice that of the repulsion in a colour-sextet pair, and there are twice as many sextet pairs as antitriplets. (In a small system, such as a nucleon or a dibaryon, the wave function is too biased by being a total colour-singlet, and the pair-forces do not cancel.)

However, when a QGP collapses to a diquark plasma, due to the attractive colour-magnetic force, the situation changes. Now, part of the attraction between quarks is absorbed in the colour-antitriplet diquarks, and the net colour forces between quarks in different diquarks should be repulsive. The potential decreases like $1/r$, and summed over N diquarks, filling a volume with radius R, the contribution to the total energy would grow like N^2/R, *i.e.* like the number of diquark pairs. This is quite different from the colour-magnetic forces and the kinetic energies, which grow like N/R, and the pressure term, which grows like R^3. Diquarks do not obey the Pauli principle, so that any number of diquarks can gather in an energy interval with a width related to the temperature and the repulsion.

There is no rigorous way to analyse a plasma with a strong internal interaction. The standard theory for QGPs is the MIT bag model.[25] Here, a spherical bag, with an external vacuum pressure ('B'), contains non-interacting quarks. The colour-electric interaction is assumed to be fully absorbed in the B value. If so, the density becomes roughly independent of the number of quarks, leading to, *e.g.*, pulsar cores about as dense as single nucleons. With a sizeable repulsive interaction in the plasma the density would decrease with the mass of the plasma, and would ultimately approach zero (would all other assumptions in the MIT bag model be correct). If so, there would be diquark plasmas with densities *lower than that of nuclear matter*. This should occur only well beyond the mass range of normal nuclei, but before (hypothetical) cosmic QGP sizes of 10^{50} quarks.

Ultimately, the diquark plasma density should decrease like $N^{-1/2}$, but it is impossible to foresee the exact dependence, as long as we cannot solve the full many-body problem. In a correct solution, it is possible that the spread in diquark momentum couples to the strength of the repulsion in such a way that also the total kinetic energy grows like N^2.

In addition, the parameters in the MIT bag model must be reconsidered with diquarks properly taken into account also in normal baryons. Treating the proton as a quark-diquark system should require a smaller than normal B value, which would lower the vacuum pressure and the density.

Observe that also Kastor and Traschen[16] found that 'diquark stars' could be less dense than normal nuclear matter in special cases, but that they seemingly dismiss this as a physical reality.

In the future, we will look more deeply into this, *e.g.*, by trying to derive a value for the diquark repulsion, by estimating the N dependence of the kinetic and repulsive energies, and by re-fitting the MIT bag model to a quark-diquark model of the nucleon. One should also study strange quarks in the diquark plasma, with equally many (ud), (us) and (ds) diquarks, in the spirit of the 'strangelet' idea by Witten and others.

There remains to ask why a diquark plasma would not hadronise once its density falls below that of free nucleons. QCD is commonly believed to have an inherent length scale of '1 fm', which prevents interquark distances from growing beyond this scale, whatever cause they may have. The punishment for such a violation is hadronisation.

My own attitude is that it is impossible to prove anything rigorously in QCD, in particular the alleged 1 fm length scale. There is, for instance, no such number in the QCD lagrangian. Rather, it happens to be the size of the proton, which could equally well mirror a delicate combination of quark masses, the number of quarks and their interactions, $e.g.$, the J/Ψ having a radius of only 0.3 fm, which reflects the high c quark mass, and not a 'QCD length scale'.

Confinement could be a macroscopic property of a full colour-neutral quark system rather than a microscopic one related to a length-scale of 1 fm. A QGP or diquark plasma would then hadronise (or decay) *if and only if* the smaller systems would get a lower total energy than the original plasma. This is the practical way 'hadronisation' is treated in the MIT bag model, which does not keep track of phase transitions as such. Here one compares the total energies of competing systems, $e.g.$, that of a dibaryon with those of two baryons, in order to explore the stability.

Localised system that happen to be colour-neutral would be 'screened' from hadronising by the global confinement field. Such a global screening should work as well for the plasma's own constituents, as in the school-book case of an 'extra' quark-antiquark pair in a QGP. [26]

4   The Diquark Big Bang Scenario

Once upon a time, Universe was a QGP. Shortly after Big Bang it was split up into smaller pieces by (presumably) quantum fluctuations. These pieces might have been the embryos of galaxies, or clusters of galaxies.

Would the pairing of quarks into diquarks be as important as we believe, these proto-galaxies should have turned into diquark plasmas as soon as the interquark distance exceeded, say, 0.3 fm, $i.e.$, at a fairly early time, and well before the time-scale of the hadronisation in the standard Big Bang scenario.

Suppose that an average proto-galaxy was so heavy that a diluted diquark plasma was its absolute ground state. Then the plasma must have continued to expand beyond the density of normal nuclear matter. And at a time not too long after Big Bang, the internal expansion of the plasma should have stopped to mimic the general expansion of Universe, once its equilibrium density was reached.

But such a galactic plasma could not possibly be stable! It would by far fulfil the requirement for becoming a *black hole*, even if its density would be many orders of magnitude lower than that of nuclear matter. The time-scale for the galaxy to 'discover' this fact is not easy to estimate, which is obvious from the unclear situation regarding primordial black holes in the normal Big Bang scenario. It is, *e.g.*, not clear whether the relevant time scale is that of the slow-down of the internal expansion, or that of light crossing the proto-galaxy.

The plasma density was probably lowest in the centre, due to the overall diquark repulsion, which means that the black hole could not swallow more than a minor piece of the total proto-galaxy.

It should however, have caused a detonation in the galactic centre, which sent a pressure/density wave through the plasma. So, starting from the black hole, the plasma was pushed into a higher density, and ripped into smaller pieces. This triggered local hadronisation, starting near the galactic centre. The energy released from this hadronisation enforced the detonation, until the effect died out at larger distances. Beyond that point, the plasma was just split up into smaller chunks, for which a diquark plasma is still the absolute ground state. Ultimately, the galaxy consisted of a hadronic central part, and a halo of diquark matter of various sizes.

One might ask why the visible galaxies are mostly disc-like. This must be due to the rotation of the original system, giving a rotating black hole, and a two-dimensional pressure wave. Once the disc-like hadronisation started, its own radiation became more spherical, and triggered a spherical expansion of the galactic halo.

5 Dark Matter, Gamma-Ray Bursts and All That

What I suggest is, of course, that the dark matter in the galactic halos are diquark plasmas with a range of masses, and that these are the MACHOs observed through micro-lensing.

These speculations at least explain why dark matter is more peripheral than visible matter, and also why it exists in such huge quantities.

Also, the model does not refer to hitherto undiscovered, exotic particles. Rather, it suggests that 'normal' matter is an 'accidental perturbation' of the truly normal and original state of Universe, while dark matter still represents that virgin state.

Interestingly, the QGP has never been a popular candidate for dark matter, in spite of Witten's much-quoted idea. This seems to be due to an argument by Madsen,[27] that all neutron stars would have turned into quark stars, would there be too much 'catalysing' quark matter in the galaxy. Observations of so-called pulsar glitches contradict such high core densities (according to Madsen). However, this idea relies on many delicate assumptions (as do all ideas presented here!). Also, several groups indeed claim that most pulsars have quark cores, and that glitches can occur anyway (or even *because* of the QGP).

The detonation following the creation of the black hole, and the subsequent hadronisation into visible matter, should release an enormous amount of radiation, typically a hundred MeV per produced nucleon. Most of these gammas are consumed when ripping the diquark matter apart. After the halo has been created, there should, however, remain a spherical shell of gammas expanding from the newborn galaxy. Its optical thickness would reflect the thickness of the outermost plasma layer at the moment when the shockwave breaks it up. This radiation could be exactly what we witness almost daily from the sky in the form of the celebrated and mystical cosmic gamma-ray bursts.

I will not describe those observations in detail, but rather refer the reader to one of the many reviews. [28] The most important features of the bursts are that they seem to be extra-galactic (due to the isotropic distribution), extremely energetic, and up to now uncorrelated to visible objects in the sky. A typical burst lasts for a fraction of a second up to a few minutes. There is no hint of a distance scale, but *if* the sources are evenly distributed in Universe the bursts have energies typical for what is expected from neutron star mergers, which is currently the most popular explanation.

There is, however, no observations that contradict that they come from a very distant shell, *i.e.* from the early stage of Universe. Then, of course, they would be much more energetic than if they are from all over Universe.

So, what I suggest is that gamma-ray bursts are the fireworks that celebrate the birth of galaxies as we know them.

Strangely enough, there is, to my knowledge, no model on the market that couples gamma-ray bursts to the birth of galaxies. There are, on the other hand, ideas [29] that the bursts are related to a QGP phase transition, although in neutron stars, and going in the opposite direction, *i.e.* from normal to quark matter.

6 Conclusions

All these fantastic scenarios stand and fall with the assumption of an intrinsic repulsion in a diquark plasma, as well as with the main ideas that diquarks will form in a QGP, and that the early temperatures were low enough so that quarks had time to create diquarks after Big Bang.

These speculations need to be underbuilt in a more rigorous way than just through hand-waving, *i.e.* from either basic QCD or at least some form of the MIT bag model.

There are no easy ways to check any of the crucial assumptions in experiments. Of course, one can try to find clear diquark effects in nucleons, or even in possible QGPs from heavy-ion collisions. But it is not likely that a small diquark plasma will be even quasi-stable, or that any of the many predictions for a 'normal' QGP will be valid in real life. Most models take it for granted that an equilibrium state has time to settle in a high-energy heavy-ion collision, and that seems highly unlikely to me! It should be easier to *disprove* our ideas, one way being to find that gamma-ray bursts come from existing galaxies, or that they repeat at the same locations. Another would be to prove that they come from the full volume of Universe. On the theoretical side, one might prove that no diquark repulsion exists, or that it will be screened from becoming sizeable.

In that case I will try diquarks on dinosaurs instead, and report the outcome at the next Torino Worskhop on Diquarks in the year 2000!

Acknowledgements

I am grateful to the Organisers of these Workshops, Mauro Anselmino, Enrico Predazzi and collaborators, for enlightening three consecutive (summer) Olympic years with such thrilling activities, and for giving old diquark friends another chance to meet. I appreciate fruitful discussions with Gerhard Fischer, Don Lichtenberg and Helmut Satz about quarks, diquarks, QCD, screening and all that. I also thank Fischer and Satz for hospitality during two visits to CERN. This work is supported by the European Commission under contract No. CHRX-CT94-0450.

References

1. S. Fredriksson, in *Workshop on Diquarks*, eds. M. Anselmino and E. Predazzi (World Scientific, Singapore, 1989), p. 22.
2. M. Anselmino, E. Predazzi, S. Ekelin, S. Fredriksson and D.B. Lichtenberg, *Rev. Mod. Phys.* **65**, 1199 (1993).
3. S. Fredriksson and J. Hansson, *Diquark '97 - a Literature Survey*, Luleå University of Technology report (1997), unpublished.
4. M. Gell-Mann, *Phys. Lett.* B **8**, 214 (1964).
5. S. Fredriksson, J. Hansson and S. Ekelin, *Z. Phys.* C, in press.
6. M. Anselmino *et al.*, *Z. Phys.* C **71**, 625 (1996).
7. K. Johnson and C.B. Thorn, *Phys. Rev.* D **13**, 1934 (1976).
8. D.B. Lichtenberg, R. Roncaglia and E. Predazzi, these Proceedings.

9. S. Fredriksson and M. Jändel, *Phys. Rev. Lett.* **48**, 14 (1982); *Nature* **301**, 564 (1983); see also *5th High Energy Heavy Ion Study*, ed. G.S. Pappas (Lawrence Berkeley Laboratory, 1981), p. 432, and *VI International Seminar on High Energy Physics Problems*, ed. A.M. Baldin (JINR-Dubna, 1981), p. 250.

10. K. Seth *et al.*, *To catch a demon*, proposal to the Los Alamos National Laboratory (1983).

11. S. Fredriksson, in *Workshop on the Physics Program at CELSIUS*, eds. B.R. Karlsson and G. Tibell (Uppsala University, 1983).

12. P. Pal and P. Dasgupta, *Phys. Rev.* D **31**, 475 (1985).

13. S. Ekelin and S. Fredriksson, in *Nucleus-Nucleus Collisions II, Vol. 1*, eds. B. Jakobsson and K. Aleklett (Lund University, 1985), p. 132.

14. S. Ekelin, Royal Institute of Technology (Stockholm) PhD thesis TRITA-TFY-86-17 (1986), unpublished, and in *Strong Interactions and Gauge Theories*, ed. J. Tran Thanh Van (Editions Frontières, Gif-sur-Yvette, 1986), p. 132.

15. J.F. Donoghue and K.S. Sateesh, *Phys. Rev.* D **38**, 360 (1988).

16. D. Kastor and J. Traschen, *Phys. Rev.* D **44**, 3791 (1991).

17. K.S. Sateesh, *Phys. Rev.* D **45**, 866 (1992).

18. J.E. Horvath, *Phys. Lett.* B **294**, 412 (1992); J.E. Horvath, J.A. Freitas Pacheco and J.C.N. de Araújo, *Phys. Rev.* D **46**, 4754 (1992).

19. P. Sahu, *Mod. Phys. Lett.* **8**, 583 (1993); *Mod. Phys. Lett.* **8**, 3435 (1993); *Int. J. Mod. Phys.* E **2**, 647 (1993).

20. S. Fredriksson and J. Hansson, poster at *Quark Matter '95*, Monterey (1995).

21. S.K. Karn, R.S. Kaushal and Y.K. Mathur, *Z. Phys.* C **72**, 297 (1996).

22. E. Witten, *Phys. Rev.* D **30**, 272 (1984).

23. J. Madsen, in *Physics and Astrophysics of Quark-Gluon Plasma*, eds. B. Sinha *et al.* (World Scientific, Singapore, 1994), p. 186.

24. F. Weber and N.K. Glendenning, Lawrence Berkeley National Laboratory report LBNL-39268 (1996), to appear in Brazilian Journal of Teaching Physics.

25. See, *e.g.*, T.A. De Grand and R.L. Jaffe, *Ann. Phys.* **100**, 425 (1976).

26. T. Matsui and H. Satz, *Phys. Lett.* B **178**, 416 (1986).

27. J. Madsen, *Phys. Rev. Lett.* **61**, 2909 (1988).

28. See, *e.g.*, G.J. Fishman, J.J. Brainerd and K. Hurley (eds.), *Gamma-Ray Bursts* (AIP, New York, 1994).

29. F. Ma and B. Xie, Astrophys. J. **462**, L63 (1996), and references therein.

The Search for Chargino-Neutralino Pairs with the ATLAS Detector at the LHC

Steve Muanza

IPN Lyon, CNRS-IN2P3. Université Claude Bernard

43, Bd du 11 Novembre 1918   F69622 Villeurbanne Cédex.

E-mail address: muanza@lyohp5.in2p3.fr

We report hereafter on particle level simualtions done in the framework of the MSSM to search for the $\tilde{\chi}_1^{\pm}\tilde{\chi}_2^0 \to 3\ell + X$ at LHC using a parametrized response of the ATLAS detector. We first establish the observability of an excess of trilepton events with respect to the SM predictions. Then we show the particular process cited above can be extracted from this trilepton excess. We propose an original method to evaluate the MSSM main parameter, namely $m_{\tilde{g}}$, using the above signal. And finally we apply the lattest to also approach the charginos and neutralinos masses.

1   Introducing the MSSM

The MSSM [1] is the simplest supersymmetric extension of the SM. It has a single SUSY generator (N=1) and has the same gauge group as the SM. It has a minimal particle content displayed in the following tables.

S=1	S=1/2	S=0
g	$\tilde{g}$	
	q	$\tilde{q}_{L,R}$
	ℓ	$\tilde{\ell}_{L,R}$
B_μ	$\tilde{B}_\mu$	
	$\tilde{\Phi}_1,\ \tilde{\Phi}_2$	$\Phi_1,\ \Phi_2$
W^i_μ	$\tilde{W}^i_\mu$	

(Massless particles)

S=1	S=1/2	S=0
g	$\tilde{g}$	
	q	$\tilde{q}_{L,R}$
	ℓ	$\tilde{\ell}_{L,R}$
$\gamma,\ Z$	$\tilde{\chi}_3^0,\ \tilde{\chi}_4^0$	$h^0,\ A^0$
	$\tilde{\chi}_2^0,\ \tilde{\chi}_1^0$	H^0
$W^{\pm}$	$\tilde{\chi}_2^{\pm}$	
	$\tilde{\chi}_1^{\pm}$	$H^{\pm}$

(Massive particles)

Tables 1,2: MSSM particle content

The left table doesn't account for $SU(2)_L \otimes U(1)_Y$ and SUSY breaking, contrarily to the right one. The electroweak (EW) symmetry breaking provides masses for the SM particles (except ν_s) whilst the sparticles masses are mainly due to the SUSY breaking. These symmetries breaking cause a mixing in the EW gaugino sector, on one hand between the zino, photino and neutral hig-

110

gsinos to form the 4 mass eigenstates called neutralinos $(\tilde{\chi}_j^0)$ and on the other hand between the wino and charged higgsinos to form the 2 mass eigenstates called charginos $(\tilde{\chi}_i^\pm)$. The MSSM depends essentially on the five following parameters:

$$
\begin{cases}
m_{\tilde{g}} : & \text{the gluino mass,} \\
m_{\tilde{q}} : & \text{the mean squark mass,} \\
\mu : & \text{the higgsino mass term,} \\
m_A : & \text{the pseudo-scalar higgs mass and} \\
\tan\beta : & \text{the ratio of the two Higgs doublets vev;}
\end{cases}
$$

the MSSM gaugino sector depending only on 3 of them: $m_{\tilde{g}}$, μ and $\tan\beta$. The mass matrix of the charginos is:

$$
\begin{pmatrix}
M_2 & \sqrt{2} M_w \sin\beta \\
\sqrt{2} M_w \cos\beta & \mu
\end{pmatrix},
$$

its eigenvectors are the two charginos and its eigenvalues are their masses simply related to the 3 sector parameters by:

$$
m^2_{\tilde{\chi}^\pm_{1/2}} = \frac{1}{2}[(M_2^2+\mu^2+2m_w^2) \mp \sqrt{(M_2^2 + \mu^2 + 2m_w^2)^2 - 4(\mu M_2 - m_w^2 \sin(2\beta))^2}].
$$

The mass matrix of the neutralinos is:

$$
\begin{pmatrix}
M_1 & 0 & -M_z \sin\theta_w \cos\beta & M_z \sin\theta_w \sin\beta \\
0 & M_2 & M_z \cos\theta_w \cos\beta & -M_z \cos\theta_w \sin\beta \\
-M_z \sin\theta_w \cos\beta & M_z \cos\theta_w \cos\beta & 0 & -\mu \\
M_z \sin\theta_w \sin\beta & -M_z \cos\theta_w \sin\beta & -\mu & 0
\end{pmatrix}
$$

its eigenvectors are the four neutralinos and its eigenvalues are their masses that are very complicated functions of the 3 sector parameters [2].
The gaugino masses M_i are all proportionnal to $m_{\tilde{g}}$ [a].

2 The $\tilde{\chi}_1^\pm \tilde{\chi}_2^0 \to 3\ell + \mathbf{X}$ channel

2.1 The signal

As it has been shown before [3], the optimal way to look for a signature from charginos and neutralinos decay is to focus on their leptonic modes in order not to be overwhelmed with the standard and SUSY QCD productions. The

---

[a] $M_1 = (5\alpha_1/3\alpha_3)m_{\tilde{g}}$, $M_2 = (\alpha_2/\alpha_3)m_{\tilde{g}}$ and $M_3 = m_{\tilde{g}}$. The α_i are the coupling constants associated to the $SU(3)_C$, $SU(2)_L$ and $U(1)_Y$ gauge sub-groups

$\tilde{\chi}_1^\pm \tilde{\chi}_2^0$ pairs are produced by Drell-Yan in the s-channel and by squark exchange in the t-channel. They have sufficiently high production cross-section (O(10 pb)) and their trilepton mode turns out to be the best channel for the $\tilde{\chi}_i^\pm \tilde{\chi}_j^0$ pairs searches at hadron colliders.

2.2   The background

Of course there are many other processes leading to the same final state in the SM as well as in the MSSM. As SM backgrounds we've considered the $t\bar{t}$, WZ, ZZ, $b\bar{b}$ and $c\bar{c}$. These processes were generated with PYHTIA 5.7-JETSET 7.4 for a top mass of 175 GeV. The MSSM backgrounds counted SUSY QCD production ($\tilde{g}\tilde{g}$, $\tilde{q}\tilde{g}$, $\tilde{q}\tilde{q}$), associated production of a chargino or a neutralino with a gluino or a squark (noted $\tilde{G}\tilde{g}$, $\tilde{G}\tilde{q}$), production of slepton pairs as well as other combinations of chargino-neutralino pairs. We generated all the MSSM processes having a $\sigma_{prod} > 10$ fb with ISAJET 7.08 and we scanned the parameter space as follows:

$$\begin{cases} m_{\tilde{g}} = 200, 300, 400, 500, 600 \text{ GeV}, \\ m_{\tilde{q}} = m_{\tilde{g}} + 20 \text{ GeV or } m_{\tilde{q}} = 2 \times m_{\tilde{g}}, \\ \mu = -m_{\tilde{g}}, \tan\beta = 2, m_A = 500 \text{ GeV}. \end{cases}$$

We produced 10^4 pb^{-1} for all processes except the MSSM ones with $m_{\tilde{g}} =200$ GeV for which only 10^3 pb^{-1} was produced.

2.3   Events reconstruction

To simulate ATLAS response we used a toy detector with an pseudo-rapidity coverage of $|\eta| <5$ and a granularity of $\Delta\eta \times \Delta\phi = 0.05 \times 0.05$. We used a fixed cone algorithm to reconstruct the jets and the p_T^{miss} was calculated as the vector sum, within the acceptance, of all stable particle momenta except for the ν_s and LSP$_s$.

2.4   Analysis

We performed the analysis [5] in two steps: first we wanted to see wether the MSSM would predict an observable excess of trilepton with respect to the SM predictions, then we made sure the $\tilde{\chi}_1^\pm \tilde{\chi}_2^0 \to 3\ell+$X could be extracted from this excess. Hence we started the analysis at low luminosity ($\mathcal{L} = 10^{33} cm^{-2}s^{-1}$) neglecting the pile-up phenomenom, then we redid it at high luminosity ($\mathcal{L} = 10^{34} cm^{-2}s^{-1}$) adding a mean of 36 minimum bias events per bunch crossing.

Here are the details of the cuts we applied to separate the signal from the backgrounds, we required:

$$\left\{ \begin{array}{l} (1)\ 3\ell \text{ with } |\eta| < 2.5, \\ (2)\ p_T(\ell_1, \ell_2) > 20 \text{ GeV and } p_T(\ell_3) > 10 \text{ GeV}, \\ (3)\ 1 \text{ opposite} - \text{sign and same} - \text{flavor lepton pair per event}, \\ (4)\ |m_{\ell^+\ell^-} - m_z| > 10\ (7.5) \text{ GeV}, \\ (5)\ \text{lepton isolation}: E_T^{\text{HAD}} < 5 \text{ GeV in a cone of} \\ \quad \Delta R = 0.3\ (0.2) \text{ around the lepton direction}, \\ (6)\ \text{central jets rejection}: \forall |\eta_{\text{jet}}| < 3,\ p_T < 25 \text{ or } 10\ (50) \text{ GeV, the} \\ \quad \text{size of the jet cone was } \Delta R = 0.6\ (0.5) \text{ and } p_T^{\min} = 10 \text{ GeV}; \end{array} \right.$$

the values between brackets are the high luminosity thresholds. We didn't apply any cut on the p_T^{miss}, we took an overall lepton identification efficiency of 90% and considered a charge misassignment of 1%.

The second cut completely removes backgrounds like $b\bar{b}$ and $c\bar{c}$ because they yield too soft trileptons, the fourth cut removes the Z peak and the fifth and sixth reject the hadronic activity in the trilepton events. Indeed, the signal's only source of hadronic activity is a possible gluon radiation in the initial state whereas for backgrounds like $\tilde{g}\tilde{g}$ or $t\bar{t}$ for instance there are many sources of hadronic activity leading to high jet multiplicities and rather hard jet p_T spectra. So the signal is essentially distinguished from the backgrounds using cuts (5) and (6).

The results of this analysis are summarized in the tables 3-6 giving the statistical significances ($\sigma = S/\sqrt{S+B}$, where S is the signal and B the background) as functions of $m_{\tilde{g}}$ for $\mu = -m_{\tilde{g}}$, $\tan\beta = 2$ and $m_t = 175$ GeV. We remind here that the discovery criterion is $\sigma \geq 5$.

Tables 3 and 4 give the statistical significances of an excess of trilepton events with respect to the SM predictions for an integrated luminosity of $\int \mathcal{L}dt = 10^4\ pb^{-1}$ (one year of LHC run at low luminosity). The first line corresponds to $m_{\tilde{q}} = m_{\tilde{g}} + 20$ GeV and the second to $m_{\tilde{q}} = 2 \times m_{\tilde{g}}$. Table 3 corresponds to a low energy run ($\sqrt{s} = 9.3$ TeV) whilst table 4 corresponds to the LHC nominal energy ($\sqrt{s} = 14$ TeV).

$m_{\tilde{g}}$ (GeV)	200	300	400	500	600
$m_{\tilde{q}} = m_{\tilde{g}} + 20$ GeV	$\sigma = 51.9$	$\sigma = 21.7$	$\sigma = 13.5$	$\sigma = 8.2$	$\sigma = 2.6$
$m_{\tilde{q}} = 2 \times m_{\tilde{g}}$	$\sigma = 14.6$	$\sigma = 6.4$	$\sigma = 3.3$	$\sigma = 2.3$	$\sigma = 1.7$

Table 3: The statistical significances of the trilepton excess due to the MSSM production at low energy. $\mu = -m_{\tilde{g}}$, $\tan\beta = 2$, $\int \mathcal{L}dt = 10^4\ pb^{-1}$.

$m_{\tilde{g}}$ (GeV)	200	300	400	500	600
$m_{\tilde{q}} = m_{\tilde{g}} + 20$ GeV	$\sigma = 76.9$	$\sigma = 32.0$	$\sigma = 20.9$	$\sigma = 13.5$	$\sigma = 4.6$
$m_{\tilde{q}} = 2 \times m_{\tilde{g}}$	$\sigma = 21.8$	$\sigma = 9.6$	$\sigma = 5.4$	$\sigma = 3.9$	$\sigma = 2.9$

Table 4: The statistical significances of the trilepton excess due to the MSSM production at high energy. $\mu = -m_{\tilde{g}}$, $\tan\beta = 2$, $\int \mathcal{L}dt = 10^4 \ pb^{-1}$.

$m_{\tilde{g}}$ (GeV)	200	300	400	500	600
$m_{\tilde{q}} = m_{\tilde{g}} + 20$ GeV	$\sigma = 19.5$	$\sigma = 17.5$	$\sigma = 14.2$	$\sigma = 9.4$	$\sigma = 3.4$
$m_{\tilde{q}} = 2 \times m_{\tilde{g}}$	$\sigma = 11.8$	$\sigma = 6.8$	$\sigma = 4.0$	$\sigma = 2.9$	$\sigma = 2.2$

Table 5: The statistical significances of the $\tilde{\chi}_1^\pm \tilde{\chi}_2^0 \to 3\ell$ process with respect to the SM and MSSM background (no pile-up). $\sqrt{s} = 14$ TeV, $\mu = -m_{\tilde{g}}$, $\tan\beta = 2$, $\int \mathcal{L}dt = 10^4 \ pb^{-1}$.

$m_{\tilde{g}}$ (GeV)	200	300	400	500	600
$m_{\tilde{q}} = m_{\tilde{g}} + 20$ GeV	$\sigma = 33.4$	$\sigma = 57.3$	$\sigma = 48.8$	$\sigma = 31.3$	$\sigma = 8.7$
$m_{\tilde{q}} = 2 \times m_{\tilde{g}}$	$\sigma = 106.3$	$\sigma = 149.3$	$\sigma = 71.1$	$\sigma = 28.0$	$\sigma = 16.8$

Table 6: The statistical significances of the $\tilde{\chi}_1^\pm \tilde{\chi}_2^0 \to 3\ell$ process with respect to the SM and MSSM background (pile-up included). $\sqrt{s} = 14$ TeV, $\mu = -m_{\tilde{g}}$, $\tan\beta = 2$, $\int \mathcal{L}dt = 10^5 \ pb^{-1}$.

3   A new method to evaluate $m_{\tilde{g}}$

3.1   The $m_{\tilde{\chi}_2^0} - m_{\tilde{\chi}_1^0}$ measurement

The $\tilde{\chi}_2^0$ and $\tilde{\chi}_1^0$ mass splitting can be measured by looking at the $\tilde{\chi}_2^0$ decay: $\tilde{\chi}_2^0 \to \ell^+\ell^- \tilde{\chi}_1^0$. These dilepton events can be selected by choosing $\ell^+\ell^-\ell'$ events (like $\mu^+\mu^- e^\pm$ for instance) so as to be sure the dilepton come from the $\tilde{\chi}_2^0$ decay and the third lepton from the $\tilde{\chi}_1^\pm$. The dilepton invariant mass is limited by: $m_{\ell^+\ell^-} \leq m_{\tilde{\chi}_2^0} - m_{\tilde{\chi}_1^0}$ and if there's a sufficient statistics then $m_{\ell^+\ell^-}^{max}$ constitutes an experimental measurement of $m_{\tilde{\chi}_2^0} - m_{\tilde{\chi}_1^0}$.

3.2   The Gunion-Haber parametrization

In order to relate this measurement to the sector parameters one clearly needs to have a simple relation between the neutralino masses and the triplet ($m_{\tilde{g}}$,

114

μ, $\tan\beta$). As this isn't available with the exact neutralino matrix eigenvalues, I propose [5] to use a parametrization due to J. Gunion and H. Haber (G-H) [4]. This parametrization is valid if $|M_1 \pm \mu| >> m_z$ or $|M_2 \pm \mu| >> m_z$ and gives:

$$\begin{cases} m_{\tilde{\chi}_1^\pm} \approx M_2 + m_w^2 (M_2 + \mu \sin(2\beta))/(M_2^2 - \mu^2) \\ m_{\tilde{\chi}_2^\pm} \approx |\mu| + m_w^2 Sign(\mu)(\mu + M_2\sin(2\beta))/(\mu^2 - M_2^2), \end{cases}$$

$$\begin{cases} m_{\tilde{\chi}_1^0} \approx M_1 + m_z^2 \sin^2\theta_w (M_1 + \mu \sin(2\beta))/(M_1^2 - \mu^2) \\ m_{\tilde{\chi}_2^0} \approx M_2 + m_z^2 \cos^2\theta_w (M_1 + \mu \sin(2\beta))/(M_2^2 - \mu^2) \\ m_{\tilde{\chi}_3^0} \approx |\mu| + \frac{m_z^2}{2} Sign(\mu)(1 + \sin(2\beta)) \\ \qquad (\mu + M_2\sin^2\theta_w + M_1\cos^2\theta_w)/(\mu + M_2)(\mu + M_1) \\ m_{\tilde{\chi}_4^0} \approx |\mu| + \frac{m_z^2}{2} Sign(\mu)(1 + \sin(2\beta)) \\ \qquad (\mu - M_2\sin^2\theta_w - M_1\cos^2\theta_w)/(\mu - M_2)(\mu - M_1), \end{cases}$$

with $\begin{cases} M_1 \approx \frac{5\alpha_1}{3\alpha_3} m_{\tilde{g}} \\ M_2 \approx \frac{\alpha_2}{\alpha_3} m_{\tilde{g}} \end{cases}$ and $\begin{cases} \alpha_1(m_z) = 1/128 \\ \alpha_2(m_z) = 0.02615 \quad \Leftarrow \sin^2\theta_w = 0.23. \\ \alpha_3(m_z) = 0.10 \end{cases}$

Comparing $m_{\tilde{\chi}_2^0} - m_{\tilde{\chi}_1^0}$ in ISAJET 7.08 and in the G-H parametrization one can see that this quantity is linearly dependent on $m_{\tilde{g}}$ and not much on μ for large $|\mu|$. The third parameter $\tan\beta$ has only a moderate influence on $m_{\tilde{\chi}_2^0} - m_{\tilde{\chi}_1^0}$.

3.3 The gauginos mass evaluation

To evaluate $m_{\tilde{g}}$ one writes down $m_{\tilde{\chi}_2^0} - m_{\tilde{\chi}_1^0}$ in the G-H parametrization then solves numerically the 5th degree equation in $m_{\tilde{g}}$. The problem then is the unknown values of μ and $\tan\beta$. To circumvent it, one takes the gluino masses obtained with extreme values of $\tan\beta$ ($\tan\beta = 2$ and 40, for example) and take an arbitrary high value of $|\mu|$ ($|\mu| = 400$GeV, for instance). The terms depending on these two parameters in the G-H expression of $m_{\tilde{\chi}_2^0} - m_{\tilde{\chi}_1^0}$ are two (independent) monotonous functions of them. In these conditions one can interpolate between the extreme values of $\tan\beta$ and the two signs of the large $|\mu|$. Here are typical results obtained with this method:

$$\begin{cases} \text{if } \Delta(m_{\tilde{\chi}_2^0} - m_{\tilde{\chi}_1^0}) = 1 \text{ GeV}, & \frac{\Delta m_{\tilde{g}}}{m_{\tilde{g}}} < 10\% \\ \qquad\qquad\qquad = 2 \text{ GeV}, & \frac{\Delta m_{\tilde{g}}}{m_{\tilde{g}}} \leq 10 - 15\% \\ \qquad\qquad\qquad = 4 \text{ GeV}, & \frac{\Delta m_{\tilde{g}}}{m_{\tilde{g}}} \leq 25 - 30\% \end{cases}$$

Once one has an estimation of $m_{\tilde{g}}$ one can plug it in the G-H parametrization to evaluate all the charginos and neutralinos masses:

$$\text{if } \Delta(m_{\tilde{\chi}_2^0} - m_{\tilde{\chi}_1^0}) = 1 \text{ GeV}, \qquad
\begin{cases}
\dfrac{\Delta m_{\tilde{\chi}_1^\pm}}{m_{\tilde{\chi}_1^\pm}} \leq 10\% \\[2ex]
\dfrac{\Delta m_{\tilde{\chi}_2^\pm}}{m_{\tilde{\chi}_2^\pm}} \leq 20\%
\end{cases}
\qquad
\begin{cases}
\dfrac{\Delta m_{\tilde{\chi}_1^0}}{m_{\tilde{\chi}_1^0}} \approx 50\% \\[2ex]
\dfrac{\Delta m_{\tilde{\chi}_2^0}}{m_{\tilde{\chi}_2^0}} \approx 25\% \\[2ex]
\dfrac{\Delta m_{\tilde{\chi}_3^0}}{m_{\tilde{\chi}_3^0}} \approx 20\% \\[2ex]
\dfrac{\Delta m_{\tilde{\chi}_4^0}}{m_{\tilde{\chi}_4^0}} \leq 20\%
\end{cases}$$

References

1. For a MSSM review see: H. Nilles, Phys. Rep. 110 (1984).
2. M. Guchait, Z. Phys. C57 (1993) 157-163.
3. R. Barbieri et al., Nucl. Phys. B367 (1991) 28-59,
   H. Baer and X. Tata, Phys. Rev. D47 (1993) 2739.
4. J. Gunion and H. Haber, Phys. Rev. D37 (1988) 2515-2532.
5. In this very short report a lot of details, justifications and illustrations lack, the reader can find them in:
   G. Montarou and S. Muanza, Atlas Internal Note, PHYS-No-61 (1994),
   S. Muanza, "La recherche des charginos et des neutralinos avec le détecteur ATLAS au LHC", Ph.D. thesis from L'Université Blaise Pascal in Clermont-Ferrand, France (1996) this thesis is available on 3W at http://atlasinfo.cern.ch/Atlas/GROUPS/THESIS/thesis.html

Physics at LISS

L.C. Bland

Department of Physics, Indiana University, Bloomington, IN 47405 USA

The Indiana University Cyclotron Facility (IUCF) is in the process of preparing a proposal for the construction of a new research facility – the Light-Ion Spin Synchrotron (LISS) – a 21 GeV/c synchrotron/storage ring. The scientific motivations for the proposed facility, including a broad outline of the envisioned experimental program and specific examples that illustrate the unique features of LISS, and the design of the accelerator and experimental facilities are presented.

It is my great pleasure that the conference organizers invited me to present some of the ideas we, at the Indiana University Cyclotron Facility (IUCF), have been developing for the Light-Ion Spin Synchrotron (LISS) project over the past two years. As you will see, LISS is envisioned as a new facility that would be devoted to the study of the strong interaction in the non- perturbative regime. It is appropriate to present these ideas at this workshop since models involving diquarks provide one way of dealing with the difficulties of non-perturbative QCD. Indeed, many of the ideas presented at this workshop could lead in the future to specific predictions that we could experimentally test at LISS!

A fundamental theory of the strong interaction (QCD) has been extremely successful at describing many phenomena at large Q^2 where the coupling constant of the theory is small. Since the force carriers of the strong interaction themselves interact strongly, the QCD coupling constant increases as Q^2 decreases. Hence, perturbative methods are not applicable at low Q^2 – and the correspondingly longer distance scales – as required to understand the bulk of matter in the universe and their low-energy strong interactions responsible for binding hadrons to form atomic nuclei. Further understanding of the strong interaction will necessitate dealing with QCD in the non-perturbative regime. It is likely that new approaches are required, possibly dealing with other (effective) degrees of freedom, to gain some intuition into the complexities of the strong interaction. As was the case for low-energy strong-interaction studies, experimental results will provide critical tests and help guide the development of new models.

At the Indiana University Cyclotron Facility (IUCF), we have been developing a proposal over the past two years for a new application of *storage ring* technology to the experimental study of outstanding problems in non-perturbative QCD. The Light-Ion Spin Synchrotron (LISS) project represents a natural extension of both the physics and technical capabilities of IUCF to the study of strong interactions using proton and other light-ion beams in the momentum range, $1 \leq p_{beam} \leq 21$ GeV/c. At the heart of the scientific justification for LISS, we are relying on our experience of using polarization as a tool, developed in the study of strong-interaction phenomena at low energies, to gain a better understanding of the forces between baryons at small distances. We believe that understanding of these short-distance phenomena (intermediate between the low-energy regime where effective meson-exchange theories work extremely well and the very high-energy regime where perturbative QCD can be used to describe phenomena) will be enhanced by the spin and isospin probes provided by LISS. The hadron beams provided by LISS will be an *essential complement* to the electromagnetic probes available at CEBAF and proposed higher energy electron accelerators (CEBAF upgrade or ELFE) for the study of the strong interaction at short distances.

In this talk, I will outline the general physics program we currently envision for LISS and then take a few specific examples from that outline to illustrate the unique attributes of a storage ring and a polarized proton beam for increased understanding of strong-interaction phenomena. I will also, by necessity, discuss some of the general technical features that will be unique to the LISS facility. Unlike the case with conventional accelerators where beams are extracted and then directed onto a target (effectively decoupling the accelerator from the experiment), for storage-ring experiments it is essential to understand the accelerator itself since the experiment is an integral part of the machine.

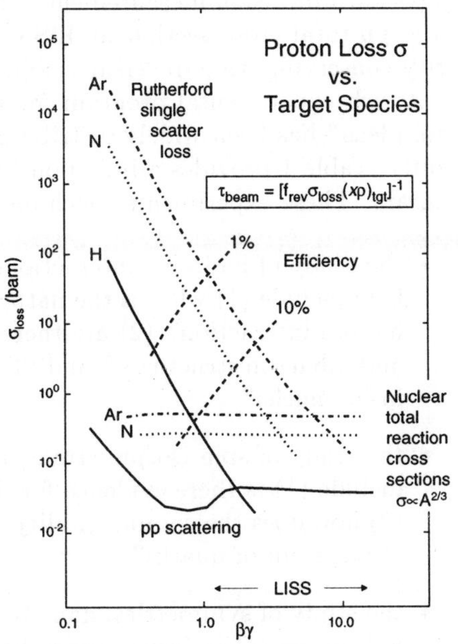

Figure 1: Illustrated is the 'loss cross section', defined in terms of the beam lifetime (τ_{beam}) in a storage ring in the presence of a target of thickness $x\rho$, vs. p/m. The revolution frequency of the beam in the ring is f_{rev}.

Many of the principal techniques to be exploited at LISS – including Siberian Snakes for maintaining polarization of a stored beam during acceleration[1] and pure polarized targets for internal target experiments[2] – have been developed at the IUCF Cooler ring. Some of these techniques are more natural to apply at higher energies. Perhaps the best example of this is illustrated in Fig. 1. At low energies, the primary loss mechanism of a proton beam in a storage ring with an internal target is through Rutherford scattering. At higher energies, the nuclear cross section associated with the target nuclei becomes the predominant source of beam loss. Since nuclear reactions are the ones of interest, the *efficiency of beam use* is much greater at higher energies than for lower energy storage rings. Not only does this suggest an economic benefit presented by LISS (or, more generally for any high-energy storage ring) for all experiments in the momentum range, $1 \leq p_{beam} \leq 21$ GeV/c, but the dominant influence of nuclear reaction cross sections on the beam lifetime also suggests a novel experimental technique. Careful measurement of the beam lifetime in a storage ring with an internal target can be used to deduce the dependence of the nuclear reaction cross section on the polarization state of the beam. This method is proposed to provide a precision measurement of the parity-violating helicity dependence of the *pp* total cross section at LISS [3] and also as a means of searching for parity-conserving, time-reversal violating interactions.

A rich experimental program for LISS has been developed. A report of these ideas [4] has been widely distributed throughout the nuclear physics community. Table 1 provides a broad outline of the envisioned LISS experimental program. These experiments touch on a few central physics themes:

- the study of hadronic interactions in the confinement regime. Questions here include (1) what is the nature of the short-range ($r \leq 1$ fm) hadron-hadron interaction? (2) are there anomalous mesons, baryon resonances and dibaryon structures? and (3) how are hadronic interactions modified inside nuclei?

- the study of spin-isospin-strangeness response functions. Questions here include (1) is there evidence for 'kaonic collectivity' in hypernuclei? and (2) how does the compressibility of a system of nucleons compare to that of a system of quarks?

- the study of symmetries; namely, what is the fate of symmetries that are well preserved in lower-energy hadronic interactions?

A detailed design of the LISS ring has been completed. The ring would be used to store in excess of 10^{11} protons (either polarized or unpolarized) and other light ions (deuterons and ^{3,4}He ions) and to accelerate them to a continuously variable final momentum in the range $1 \leq p_{beam} \leq 21$ GeV/c. The light-ion capabilities, such as deuteron beams, are important for some

Table 1: Outline of the LISS experimental program.

TOPIC	EXPERIMENTS
SYMMETRY TESTS	· Parity violation in NN scattering · Time-reversal invariance tests
NN STRONG INTERACTIONS	· NN elastic scattering at large $p_\perp$ — *emphasis on single and two-spin asymmetries* · Precision measurement of pp elastic scattering in the Coulomb-nuclear interference regime

HADRONIC INTERACTIONS IN NUCLEI	· Medium modifications of NN interaction
	    — *Color Transparency Effects in $(\vec{p}, 2p)$ Reactions*
	· Response functions of nuclei
	    — *Isospin–Spin–Strangeness Responses*

SEARCHES FOR QCD EXOTICA	· Search for exotic mesons in central pp collisions · Dibaryon resonance searches — $\Delta\sigma_L$ and $\Delta\sigma_T$ *measurements* · $\Delta\Delta$ interaction in the S=3 state

HADRON STRUCTURE	· Spin filtering of isoscalar nucleon resonances
	    — *E0 response of the nucleon*
	· Study of the reaction $\vec{p}p \to pp\vec{\phi}$
	    — *Probing the strangeness content of the proton*

POLARIZATION PHENOMENA IN FLAVOR PRODUCTION	· Associated hyperon production in pp collisions · Charm production near threshold · Spin-dependent ΛN scattering length measurement · Few-body hypernucleus studies — *hypernucleus weak decay asymmetries*
MANY-BODY DYNAMICS	· Study of light-ion induced multi- fragmentation

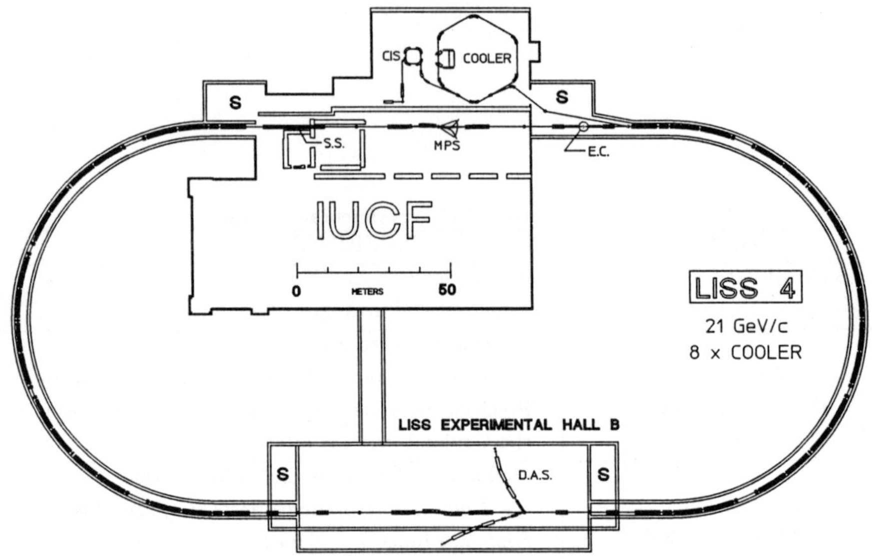

Figure 2: A layout, to scale, of the proposed LISS ring. At the top of the drawing is the existing Cooler ring. LISS would be 8× the Cooler circumference. The CIS ring is presently under construction. The CIS/Cooler combination would serve as an injector for LISS. A new experimental hall (hall B) would be required to house experimental stations.

experiments either directly or indirectly, for example by the production of secondary neutron beams. Experiments at LISS would be conducted by allowing the stored beam to interact with internal targets. No immediate plans for constructing costly external beam lines or beam extraction systems are included in the initial phase of LISS operation. With $\sim 10^{11}$ protons stored in LISS, internal target experiments can be conducted at luminosities of $10^{33} \mathrm{cm}^{-2}\mathrm{s}^{-1}$ for unpolarized hydrogen targets and $\sim 10^{32}\mathrm{cm}^{-2}\mathrm{s}^{-1}$ for polarized targets, the latter limited by fluxes from sources of polarized atoms that would feed storage cells in the ring.

Fig. 2 shows a plan view of the present design of the LISS ring. The magnet lattice of the ring is a 'racetrack', consisting of two long straight sections and two arc sections. This configuration was chosen over a more circular ring because it provides for greater space for mounting apparatus required for internal target experiments. The long straight sections have a length of ~ 150 m and are dispersion free. In parts of the straight sections the beam size is appropriate for the installation of internal polarized targets with an accompanying storage cell.

The arc sections of the LISS ring are designed to give an imaginary value for the transition energy thereby avoiding potential beam instabilities or loss during acceleration. This is achieved by constructing each arc from a number of basic 'flexible momentum compaction' (FMC) modules.[5] Each of the FMC modules in the arcs provides a straight section 4-6 m long with reasonably low dispersion that may prove useful for experiments. The ones closest to the arc ends, in particular, can serve to detect reaction products with near-beam momenta, using the final pair of arc dipoles as separation magnets.

A unique and crucial feature of the LISS ring is its design regarding the storage and acceleration of polarized beam. Building on the experience gained at IUCF in the development of Siberian Snakes used for overcoming depolarizing resonances encountered during acceleration in synchrotrons, the LISS ring has been specifically designed to include such devices. For final momenta larger than 5 GeV/c, a full, helical-dipole Siberian Snake (indicated as S.S. in Fig. 2) would be used to overcome both intrinsic and imperfection depolarizing resonances encountered during acceleration. At lower momenta, a partial Snake would be used to overcome imperfection resonances and the single remaining intrinsic resonance would be crossed via adiabatic or Snake-induced spin flip. Following acceleration, the stable polarization direction of the beam directly opposite to the Snake would be longitudinal. This means that on subsequent passes around the ring, transverse polarization components of the beam would change sign. This cancelation of unwanted polarization components is an important consideration for a precision measurement of parity violation in pp scattering.[3]

Electron cooling (E.C. in Fig. 2) will be employed to reduce the transverse emittance of the stored beams in LISS at an energy of $\sim$3.7 GeV. Cooling 10^{11} ions in LISS below this energy would lead to unacceptably large space-charge tune shifts; cooling at higher energies would be impractical because of the long cooling times and the high energy of the required electron beam. The present scheme is to inject LISS using fast-extracted beam from the existing IUCF Cooler. The Cooler would be filled by the Cooler Injector Synchrotron (CIS) ring (see Fig. 2), presently under construction. The proton beam in the Cooler would be extracted after acceleration to 400 MeV. The slow-ramping acceleration system in LISS would require $\sim$5 seconds (small compared to the long lifetime of the beam) for protons to reach full energy. The LISS acceleration cycle would then consist of partially ramping the stored beam from the injection energy to $\sim$3.7 GeV, cooling the stored beam to reduce its transverse emittance, and then completing the ramp to the final energy.

At present, two major experimental facilities are expected to be part of the LISS proposal. The first is the dual-arm spectrometer (D.A.S. in Fig. 2). This device would be used primarily for very precise measurements of NN elastic scattering at large p_T (with the primary emphasis on the measurement of single-spin and two-spin polarization observables) and for experiments aimed at understanding how the NN interaction is modified in the nuclear medium via study of $(\vec{p}, 2p)$ reactions from nuclei. Even though both of these subjects are of great interest, I won't talk about them in any more detail. Instead, I refer you to the document [4] for those interested. The other major facility is a large-acceptance spectrometer (MPS in Fig. 2) that would be equipped with tracking detectors and plastic scintillator and Cerenkov hodoscopes. This device would enable measurements of high-multiplicity charged particle final states such as those encountered in the $\vec{p}p \to pK^+\vec{\Lambda}(\Sigma^0)$ reaction (discussed in more detail below), detecting the charged decay products of the hyperon in the final state in addition to the primary proton and kaon; or, in the $\vec{p}p \to pp\phi$ reaction, with the ϕ subsequently decaying into K^+K^- pairs. Detailed simulations of the large-acceptance spectrometer are underway to finalize its design. In general, the focus of both facilities would be on experiments at the higher energy end of LISS. This recognizes the fact that the COSY ring is presently operational and provides many of the same capabilities of LISS, up to a maximum proton momentum of 3.5 GeV/c.

As an example of an experiment that would utilize unique features of a ring devoted to the acceleration and storage of polarized proton beams used for internal target experiments, it is useful to consider how one would probe the 'E0 response function' of the nucleon. By analogy to the giant monopole resonance in nuclei, E0 excitations of the nucleon involve radial exciations of the quarks – so-called 'breathing mode' excitations. A reliable determination of the energy distribution of E0 excitations would provide stringent tests of different models of nucleon structure. As is the case in nuclear physics, isolation of nucleon E0 excitations from amongst the many overlapping resonances that are excited by electromagnetic probes is quite difficult. An alternative approach is suggested from the results of an experiment [6] at Laboratoire National Saturne, shown in Fig. 3. Unlike the case for inelastic electron scattering, an experiment measuring the differential cross section for inelastic α scattering from the nucleon appears to provide a *direct observation* of excitation of the $P_{11}(1440)$, the lowest energy P_{11} resonance, also known as the Roper resonance. (The notation $L_{2I,2J}$ is a label used for non-strange baryon resonances having isospin, I, and total angular momentum, J. L refers to the orbital angular momentum of πN scattering states.) Various theoretical models have identified the $P_{11}(1440)$ as the 'breathing mode excitation' of the proton.[6,7] This interpretation is not accepted universally.[8,9] The enhancement of $P_{11}(1440)$ excitation relative to other baryon resonances of similar mass arises because inelastic α scattering by necessity must result in scaler, isoscaler excitations. The principal drawback of this method is the sizable background identified as projectile excitation. If this background could be eliminated, then inelastic α scattering could be used to map out the complete E0 response of the proton!

The measurement of *polarization observables* at a machine like LISS could be used to distinguish the projectile excitation background from direct excitation of P_{11} baryon resonances. As suggested in Fig. 3, the background in the spectrum is thought to be due to the excitation of one of the nucleons in the incident α particle to the $\Delta_{33}(1232)$ resonance. Subsequently, the excited α_Δ projectile coherently decays to an $\alpha + \pi$ final state. This process must involve

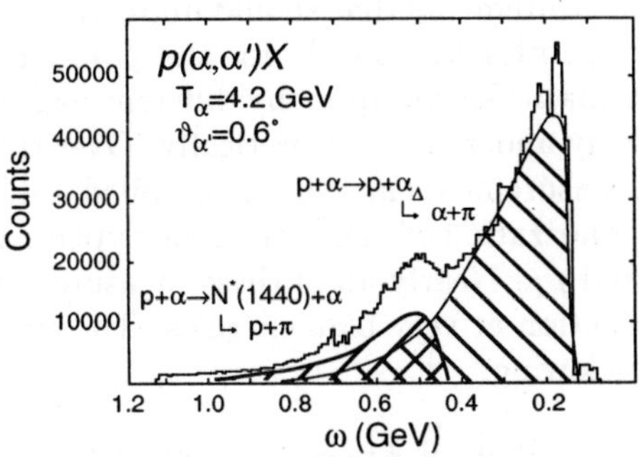

Figure 3: Results of an experiment studying inelastic α- particle scattering from the proton.

a spin and isospin excitation within the α particle, and is known [10] to have a normal-component polarization transfer coefficient, $D_{NN} \approx -1/3$. On the other hand, by using nothing more than general symmetry principles, excitation of a P_{11} resonance via α inelastic scattering must result in D_{NN} being identically $+1$ (since both the initial and final states involve identical spin/parity combinations, $0^+ + \frac{1}{2}^+$). Hence, weighting the measured differential cross section by the factor $(D_{NN} + \frac{1}{3})$ would suppress the projectile excitation background, producing a spectrum showing the energy distribution of P_{11} baryon resonances.

A potential LISS experiment to isolate the distribution of P_{11} baryon resonances must measure both the differential cross section and the polarization observable, D_{NN}. The experiment would be conducted in 'reverse kinematics', probing the same center-of-mass scattering angles as in Fig. 3, but using a polarized beam incident on an internal ^{4}He gas- jet target. The thin target allows the measurement of α particles with energies as low as $\sim$10 MeV. Observing the α particle in the final state is essential to ensure that only scaler, isoscaler excitations of the nucleon are selected. The N* polarization (P) is deduced in the final state by determining the polarization of the daughter hadrons that are emitted either parallel or antiparallel to the axis parallel to $\vec{k}_{beam} \times \vec{k}_{N*}$. If decay branches of the N* involving Λ are studied, then the parity-violating weak decay asymmetry of the $\Lambda \to p\pi^-$ decay can be used to deduce the N* polarization. The two features of a variable-energy polarized proton beam and thin internal targets makes such an experiment ideally suited to a dedicated (*polarized beam*) storage ring facility. More generally, the isolation of the E0 response function using polarization observables and the spin/isospin selectivity of hadronic probes illustrates how such probes provide an essential complement to electromagnetic probes to identify baryon resonances predicted by theory.[7]

A second example that illustrates the capabilities of LISS, is a program of measurements that should improve our understanding of the strong-interaction dynamics responsible for the polarization of hyperons observed to be a systematic feature in many different experiments, and which has, so far, not been fully understood theoretically. Over the past twenty years, an impressive set of measurements have been compiled, establishing a systematic dependence of the polarization of many different hyperons with the kinematic variables $x_{Feynman}$ and p_T. Nearly all of these measurements have been of *inclusive* hyperon production at very high energies. The systematics [11,12] can be summarized as the following:

- at fixed $x_{Feynman}$, the hyperon polarization linearly increases with p_T up to 1 GeV/c and for larger values has a constant value.

- at fixed p_T, the hyperon polarization increases with increasing $x_{Feynman}$.

- the polarization is found to be essentially independent of $\sqrt{s}$.

Some of these attributes are illustrated in Fig. 4 which shows the momentum dependence for different hyperons produced in *inclusive* reactions initiated by protons with an incident energy of 400 GeV. One thing immediately clear from this figure are the different *sign* polarizations for $\Sigma^{\pm}$ and Λ. This feature, along with most of the other systematics, is consistent with having the s quark in the outgoing hyperon (boosted onto the mass shell from the quark sea) produced with its polarization anti-parallel to the axis lying along $\vec{k}_{beam} \times \vec{k}_{hyperon}$, *i.e.* the s quark is

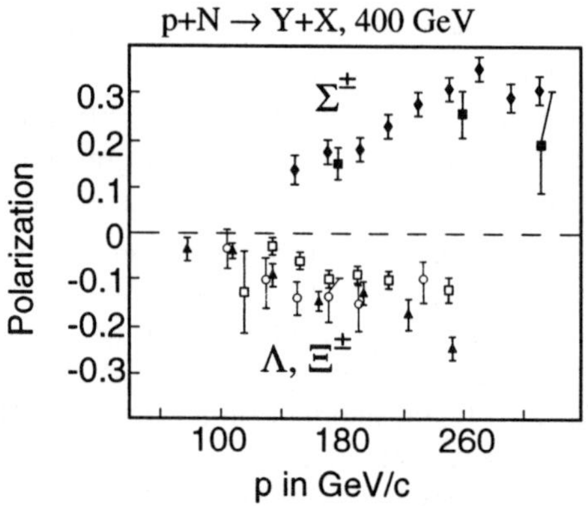

Figure 4: Some systematic trends of hyperon polarization observed in *inclusive* experiments with incident high-energy protons.

preferentially produced with its spin pointing down. A (probably overly) simple model,[13,14] that uses SU(6) wave functions to describe both the incident proton and the outgoing hyperon and has the common valence quarks between the two hadrons viewed as spectators, is capable of describing nearly all of the systematic features. Viewed as a 'leading particle' phenomenon, there is expected to be no polarization when the hyperon shares none of the valence quarks from the incident proton. Unfortunately, recent measurements show that these expectations aren't met in all cases, since significant polarization is observed for $\bar{\Xi}^+$ and $\bar{\Sigma}^-$ production,[15] in contrast to the near zero polarization observed for Ω^- and $\bar{\Lambda}$ that is consistent with the 'leading-particle' expectations.[11,12] The bottom line is that the phenomena of hyperon polarizations, although apparently simple based on a subset of the available systematics, shares many of the same complexities as other problems in *non-perturbative* QCD. It's appeal lies in the ability to use the *strange* quark as a probe of the underlying strong-interaction dynamics.

A LISS program of measurements directed at this problem would, first of all, emphasize *exclusive* hyperon-production measurements. Such measurements are essential to avoid the 'feeding' of Λ observed in the final state from the decays of other hyperons produced in the reactions; particularly from the decay of Σ^0 which has a branching ratio of $\sim$100% to $\Lambda\gamma$, and, as is clear from Fig. 4, has very different values for the polarization observables. A second feature of a LISS program would be on the measurement of additional polarization observables accessible when both the beam and target are polarized, including the normal-component polarization transfer coefficient (D_{NN}) and the beam and target analyzing powers (A_y). One advantage of having both entrance channel particles polarized is to attempt and correlate the hyperon's polarization with either the beam or target spin orientation. Such measurements would provide stringent tests of theoretical models (specifically those based on 'spectator' quarks or diquarks transferred from the projectile to the hyperon) designed to explain hyperon polarization systematics.

To accomplish *exclusive* measurements at LISS, a multi-particle magnetic spectrometer (MPS) would be employed. Since the device must have a large acceptance to cover a significant fraction of the phase space, it is unlikely that momentum resolution better than $\delta p/p \sim 1\%$ could be attained, hence making it impossible to achieve the requisite mass resolution to distinguish Λ from Σ^0 (separated by only 77 MeV/c^2) by detecting only charged particles. Instead, a large acceptance, high-energy photon detector (similar to the lead-glass wall employed in the Brookhaven experiment, E852 [16]) would be an essential complement to the MPS. Phase space distributions show that at the maximum incident energy that could be studied at LISS and with cuts imposed corresponding to where the hyperon polarization is known to be large ($x_{Feynman} > 0.2$ and $p_T > 1$ GeV/c), the average energy of the produced hyperon in pKY final states is $\sim$12 GeV, meaning there is a substantial 'Doppler shift' of the photons from the $\Sigma^0 \to \Lambda\gamma$ decay. Hence, the photons (which are emitted isotropically in the Σ^0 rest frame with an energy of 75 MeV) are constrained to a narrow cone around the Σ^0 direction and have energies of nearly 1 GeV, making their detection feasible.

The real advantage of LISS for such measurements is that, unlike at existing accelerators, *polarized* proton beams and pure *polarized* proton targets would be the primary focus of this new facility and would readily be available for such experiments. This availability makes it conceivable to consider new measurements that could produce substantially greater understanding of long-standing mysteries. Similar considerations exist for all of the experiments that have been conceived for LISS.[4]

The estimated total cost of the LISS facility including the accelerator hardware, major experimental facilities and the civil construction is approximately $55M. The estimated cost for the two major experimental facilities envisioned for the first round of experiments at LISS is ~$15M. At present, there is strong support from Indiana University, suggesting that ~$20M of the costs (primarily for the civil construction, including the ring tunnel and the second experimental hall) would be provided by the State of Indiana. A proposal to the National Science Foundation for funding the remaining ~$35M for construction of the accelerator and the major experimental facilities will be submitted in the 1997 calendar year.

Acknowledgments

I would like to acknowledge the efforts of all of the people at IUCF and other institutions who have helped to develop the ideas for the LISS project. There are too many people to list individually. There are two people who need to be singled out for their contributions to the project: Peter Schwandt has done nearly all of the design work for the accelerator and Steve Vigdor conceived most of the physics program.

References

1. A.D. Krisch *et al.* Phys. Rev. Lett. **63**, 1137 (1989).
2. T. Wise, *et al.*, Nucl. Inst. Meth. **A336** (1993); M. A. Ross, *et al.*, Nucl. Inst. Meth. **A344** (1994).
3. S.E. Vigdor, in *Future Directions in Part. and Nucl. Phys. at Multi-GeV Hadron Beam Fac.*, ed. D.F. Geesaman (BNL rep. #BNL-52389, 1993), p. 171.
4. *The LISS White Paper*, eds. L.C. Bland, *et al.*, IUCF Internal Report, 1996.
5. S.Y. Lee, *et al.*, Phys. Rev. E **48**, 3040 (1993)

6. H.P. Morsch, *et al.*, Phys. Rev. Lett. **69**, 1236 (1992).

7. R. Bijker, F. Iachello and A. Leviatan, Ann. Phys. **236**, 69 (1994).

8. C. Schütz, K. Holinde, M.B. Johnson and J. Speth, IKP Annual Report, p. 136 (1994).

9. F.E. Close, in *Spin and Isospin in Nuclear Interactions*, S.W. Wissink, C.D. Goodman and G.E. Walker *eds.*, Plennum Press, p. 75 (1990).

10. D.L. Prout, *et al.*, Phys. Rev. Lett. **76**, 4448 (1996).

11. K.J. Heller, in *Proc. 7th Intl. Conf. on Polarization Phenomena in Nuclear Physics*, Paris, 1990, Colloque de Phys. **51**, C6-163 (1990)

12. J. Lach, in *Flavour and Spin in Hadronic and Electromagnetic Interactions*, (Torino, 1993), eds. F. Balestra, *et al.* (Italian Physical Society, Bologna, 1993)

13. T. A. DeGrand and H.I. Miettinen, Phys. Rev. D **23**, 1227 (1981) and Phys. Rev. D **24**, 2419 (1981); T.A. DeGrand, J. Markkanen and H.I. Miettinen, Phys. Rev. D **32**, 2445 (1985); B. Andersson *et al.*, Phys. Rep. **97**, 31 (1983).

14. P. Kroll, *High-Energy Spin Phys., 8th Int'l Symp.*, ed., K. Heller, AIP Conf. **187**, V1, 48 (1989) and references therein.

15. A. Morelos *et al.*, Phys. Rev. Lett. **71**, 2172 (1993).

16. A. Dzierba, in *Proceedings of the Workshop on Future Directions in Particle and Nuclear Physics at Multi-GeV Hadron Facilities*, March, 1993 BNL-52389 (1993) 311.

THE STUDY OF THE ELEMENTARY PHOTO- AND ELECTRO-PRODUCTION OF KAONS AT JEFFERSON LAB

M.IODICE[1], E.CISBANI[1], S.FRULLANI[1], F.GARIBALDI[1], G.M.URCIUOLI[1], R.DE LEO[2], R.PERRINO[3], M.SOTONA[4]

[1] *INFN, Sezione Sanitá, Viale Regina Elena, 299,*
I-00161 Roma, Italy

[2] *University of Bari, Physics Department - INFN - Sezione di Bari,*
Via Amendola 173, I-70126 Bari, Italy

[3] *INFN, Sezione di Lecce, Via per Arnesano,*
Lecce, Italy

[4] *Nuclear Physics Institute,*
25064 Rez, Prague, Czech Republic

The subject of electromagnetic production of strangeness, covers an important part of the planned CEBAF experimental program at Jefferson Lab. In this review we will mainly focus on those experiments aiming to investigate the elementary mechanism of the associated production of kaon–hyperon pairs, on hydrogen target, induced by electron and by real photon beams. Complementary experiments, proposed for all the three experimental halls, allow to access a wide kinematical region where different theoretical approaches can be used for the interpretation of the (upcoming) data.

1   Introduction

The construction of the Continuous Electron Beam Accelerator Facility is now completed at Jefferson Lab (JLab in Virginia, USA). The accelerator delivers three simultaneous electron beams into three experimental halls with independent energy and intensity. The maximum energy now available is 4 GeV, but, due to the very good performances of the superconducting cavities, it is supposed that, with only minor changes to the accelerator, in two years a 6 GeV continuous beam will be available. The maximum current is $200\mu A$, allowing very high luminosity experiments. At Hall C, is currently taking data for experiments, while Hall A and B are under the commissioning phase and will soon start their planned physics program.

The study of strangeness production will be investigated in all the three halls where, due to the complementarity of the experimental setup [1], different reactions, as well as different kinematical regions, will be explored.

The Hall A is a 53 m diameter circular experimental area equipped with a pair of High Resolution Spectrometers allowing high resolution (10^{-4}) detection of electrons and hadrons with a maximum momentum of 4 GeV/c. These focusing twin spectrometers are 45 deg vertical bending systems composed by two quadrupoles, one dipole and one more quadrupole magnet (QQDQ) along the 25 meters flight path of the travelling particles. The momentum and angular acceptances are 10% and 7.0 msr, respectively.

The focal plane detection system has tracking capabilities, hodoscopes to provide fast trigger, and a system of two Cherenkov counters used for particle identification, one with CO_2 gas as active material and one with aerogel.

The Hall C experimental setup is similar to that of Hall A, being equipped with two magnetic spectrometers, as well. In this case, however, an "asymmetric" pair of spectrometers was chosen: one is able to detect particles up to 6 GeV/c (the High Momentum Spectrometer) with moderate resolution ($\delta p/p \sim 5 \times 10^{-3}$), the other one has a short flight path (the Short Orbit Spectrometer) being mainly conceived to detect decaying particles (such as kaons) with maximum momenta of 1.8 GeV/c.

Hence, both Hall A and C are well suited to carry out high luminosity ($\sim 10^{37} cm^{-2} s^{-1}$) coincidence ($e, e'K^+$) experiments (of associated production of $K-Hyperon$ pairs) detecting in the final state the scattered electron and the produced kaon. While in Hall C this reaction can be well investigated at low and moderate energies (due to limitations on the highest measurable kaon momentum), Hall A is more suited to detect kaons of high momenta whose longer life time compensate for the longest flight path of the HRS spectrometer with respect to SOS.

The detector assembly of the Hall B is conceptually different. The Cebaf Large Acceptance Spectrometer (CLAS) is employed to detect (at lower luminosity, $\leq 10^{34} cm^{-2} s^{-1}$) multiple particles in the final state. It is composed by a toroidal magnet optimized to track from 8 to 140 degrees charged particle of momenta between 250 MeV/c and 4 GeV/c. An electromagnetic calorimeter and a gas Cherenkov detector provide electron, photon and π^o detection for angles below 45°. The particle identification capability limit the strangeness production experiment to kaon momenta lower than 2 GeV/c.

A real photon beam is also available in Hall B where a photon tagging system operates from 20% to 95% of the bremsstrahlung endpoint energy, with about 5 MeV energy resolution.

In the following sections we will describe the theoretical framework and the models that can be adopted to investigate the strangeness production process. Then, we will outline those experiments aiming to investigate the elementary mechanism of the associated production of kaon–hyperon pairs, on nucleons, induced by electron and by real photon beams.

Due to limitations of space, the present review could not fully treat the experimental program on kaon production at Jefferson Lab. Instead of going through a quick review of all the experiments (perhaps resulting only in a list of them) our choice was to focus on few (relevant in our opinion) subjects. Thus, we apologize to all of the authors who are going to contribute to this field of physics at JLab and have not been properly honoured in this paper.

2 The $(e, e'K^+)$ electro-production reaction - The elementary process on protons.

In order to write the cross section of the $(e, e'K^+)$ reaction and discuss the physics, let us introduce the appropriate variables. We will consider the exclusive channel of the Λ Hyperon production associated to the kaon in the $e + p \rightarrow e' + K^+ + \Lambda$ reaction, with a straightforward extension to other channels (i.e. Σ Hyperon production, or semiexclusive 'X' production). The 4-momenta that are directly measured in the reaction are e, e', P, k (related to the incident electron, the scattered electron, the proton target at rest in the lab. and the knocked-out kaon, respectively) while P_Λ (Λ 4-momentum) is calculated from the conservation laws.

The following Lorentz invariant, related to the $\gamma^* + p \rightarrow K^+ + \Lambda$ virtual photo-production binary process can be defined:

$$q^2 = (e - e')^2 = -Q^2 \qquad \textit{squared 4-momentum transfer}$$

$$s = (q + P)^2 = (k + P_\Lambda)^2 = W^2 \qquad \textit{squared invariant mass of the } \gamma^*\textit{-N system}$$

$$t = (q - k)^2 = (P_\Lambda - P)^2 \qquad \textit{squared momentum transferred to the hadronic system}$$

$$u = (q - P_\Lambda)^2 = (k - P)^2$$

being s, t and u the Mandelstam variables.

Of particular note is the variable t which can be thought of as the momentum transferred to the remnant hadronic system, playing the same role in the (γ^*, K^+) virtual-photo-production reaction as that played by q^2 in the (e, e') inclusive experiments.

The electro-production cross section can be expressed in terms of the $\gamma^* + p \rightarrow K^+ + \Lambda$ cross section as:

$$\frac{d^5\sigma}{dE'_e d\Omega_e d\Omega_k} = \Gamma \frac{d\sigma(\gamma^*, k)}{d\Omega_k} \tag{1}$$

where Γ is the virtual photon flux.

In turn the virtual-photo-production cross section can be expressed in terms of 4 response functions:

$$\frac{d\sigma}{d\Omega_k} = \sigma_T + \varepsilon_L \sigma_L + \varepsilon \sigma_{TT} \cos 2\Phi + \sqrt{2\varepsilon_L(\varepsilon+1)}\sigma_{LT}\cos\Phi \qquad (2)$$

where:

$$\varepsilon = \frac{1}{1 + 2|\vec{q}|^2/Q^2 \tan^2 \vartheta_e/2} \qquad \text{and} \qquad \varepsilon_L = \frac{Q^2}{\omega^2}\varepsilon$$

are the transverse and the longitudinal photon polarization, respectively, and Φ, in formula (2), is the "out of plane" angle that is the angle between the leptonic plane (defined by the incoming and outgoing electrons) and the reaction plane (defined by the direction of the 3-momentum transfer and that of the kaon).

The pieces correspond to the cross section for transverse (σ_T), longitudinal (σ_L), transverse interference (σ_{TT}) and longitudinal-transverse interference (σ_{LT}) kaon production by virtual photons and they only depend on the variables Q^2, W (or s) and t.

In order to separate all the four pieces of the cross section, out-of-plane detection capabilities are needed to determine the Φ dependence of the cross section. In Hall B, the CLAS detector allows in principle to determine such a Φ distribution. However the separation of the response functions is a challenging goal in this case, since very good measurements accouracy, at the limit of CLAS possibilities, are needed. With two coplanar spectrometers, only when the kaon is detected along the direction of the virtual photon the interference terms vanish, and σ_L and σ_T can be separated using at least two measurements at different values of ε.

3 Theoretical models

To calculate the cross section of the process under investigation, different approaches can be adopted. In principle it could be calculated within the QCD theory making use of quark hadronization models. At CEBAF energies, however, non perturbative QCD degrees of freedom have to be taken into account. A more reliable approach in this case, is based on hadronic field theories (Quantum Hadron Dinamics) and makes use of semiphenomenological diagrammatic models where the explicit degrees of freedom are mesons and baryons and the tree level Feynman diagrams reported in figure 1 can be calculated.

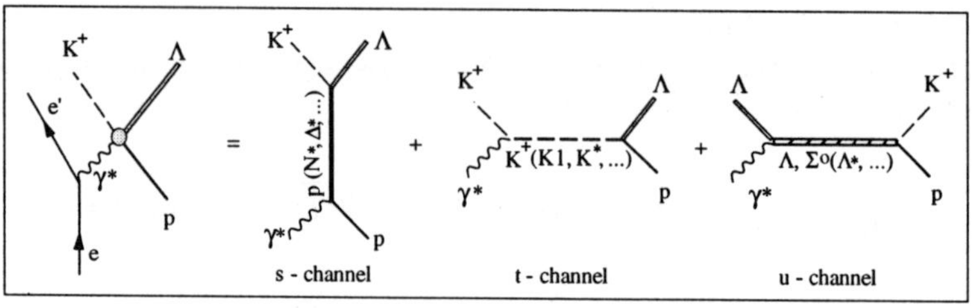

Figure 1: Tree level Feynman diagrams for the process $p(\gamma^*, K^+)\Lambda$

Contributions to the cross section from these diagrams are usually calculated in a semiphenomenological way, trying to fit the unknown parameters to the existing data. One of the main problems encountered in this kind of approach is the determination of the proper set of particles as propagators (p, K^+, Λ, Σ are Born terms). Moreover, in the fitting procedure, the values of the coupling constants (like the leading coupling constant $g_{K\Lambda N}$) have also to be established. So far, quite a large variety of models, in this framework, have been developed (see e.g. ref.[2]), but, at present, with the available photo production data (collected e.g. in ref.[3]) and electro-production data (collected e.g. in ref.[4]) it is not possible to determine unambiguously a given set of propagators and coupling constants. Due to the lack of high quality experimental data in a wide kinematical region, different models, taking into account very different set of diagrams and parameters, are able to satisfactorily reproduce most of the data. On the other hand, they give very different predictions in those regions where a comparison with the data is not possible yet. Thus, the situation has to be clarified and a consistent data set in a broad kinematical range is needed.

At CEBAF energies of 6 GeV (and maybe even more, in the next future) the transition into the hard scattering regime could be explored. Thus, the reliability of different models, based on perturbative QCD or phenomenologically inspired to QCD, can be tested.

In the region where non perturbative effects are still present, an approach based on the idea of diquark can be appropriate and promising. Recently, theoretical predictions of exclusive photo- and electro-production of kaons based on QCD semiphenomenological diquark models have become available[5]

The main ingredients of the diquark model are: $i)$ baryon treated as quark-diquark systems; $ii)$ phenomenological diquark form factors taking into account the composite nature of diquarks; $iii)$ the gluons and photons couplings to diquarks.

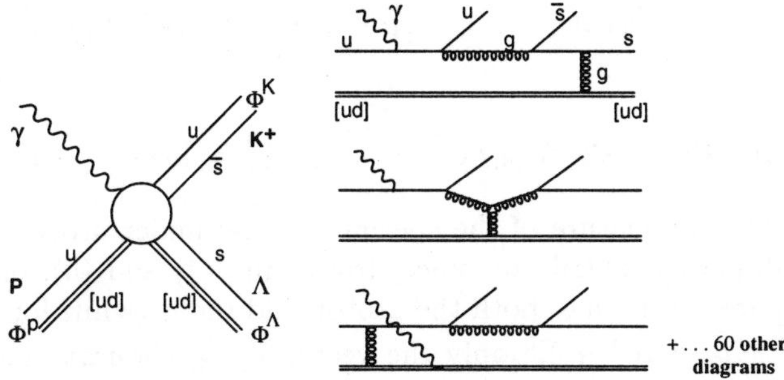

Figure 2: A few representative examples of diagrams contributing to the process $\gamma u[ud] \rightarrow u\bar{s}s[ud]$

In its ground state a diquark has positive parity and may be an axial-vector (spin 1) or a scalar (spin 0) boson. The baryon wave function can be written as

$$\Psi^B(p,\lambda) = f_S\Phi_S^B(x_1)\chi_S^B u(p,\lambda) + f_V\Phi_V^B(x_1)\chi_V^B\frac{1}{\sqrt{3}}(\gamma^\alpha + p^\alpha/m_B)\gamma_5 u(p,\lambda) \quad (3)$$

where the two terms represent configurations consisting of a quark and either a scalar (subscript S) or vector (subscript V) diquark and:

i) Φ^B are the baryon distribution amplitudes, the valence Fock-state (consisting of a quark and a diquark) wave functions integrated over the transverse momentum. The argument x_1 is the momentum fraction of the parent baryon carried by the quark ($p_q = x_1 p_B$). It is assumed (collinear approximation) that it sums up to 1 with the momentum fraction of the diquark ($p_D = x_2 p_B = (1 - x_1)p_B$)

ii) f are the $r = 0$ values of the distribution amplitudes

iii) χ are the flavour functions and $u(p,\lambda)$ are spinors with helicity λ.

Investigating the $\gamma^{(*)} + p \rightarrow K^+ + \Lambda$ reaction, the hard scattering amplitude can be calculated perturbatively taking into account all the possible tree diagrams contributing to the elementary scattering process $\gamma u D \rightarrow u\bar{s}s D$ where D is a u-d diquark. In figure 2, only few diagrams are reported as an example, together with a blob representation of the process.

The flavour functions $\chi_{S,V}^{p,\Lambda}$ for the proton and the lambda hyperon take on the form :

134

$$\chi^{p}_{S} = uS_{[u,d]} \quad , \quad \chi^{p}_{V} = [uV_{\{u,d\}} - \sqrt{2}dV_{\{u,u\}}] \tag{4}$$

$$\chi^{\Lambda}_{S} = [uS_{[d,s]} - dS_{[u,s]} - 2sS_{[u,d]}/\sqrt{6} \quad , \quad \chi^{\Lambda}_{V} = [uV_{\{d,s\}} - dV_{\{u,s\}}]/\sqrt{2} \tag{5}$$

The simplifying feature of the specific process under investigation is that only scalar diquarks contribute, since, from eqn. (4) and (5), only scalar diquarks $S_{[u,d]}$ are common to both the proton and the Λ (while for the exclusive channel $\gamma^{(*)} + p \rightarrow K^{+} + \Sigma^{0}$ only the vector $V_{\{u,d\}}$ diquark contribute). As a consequence, the number of diagrams is reduced to 63 "only". A physical consequence of that is the vanishing of the polarization of the produced Λ. Thus, the measurement of the Λ polarization, as well as the comparision of the Λ and Σ^{0} production, turns to be very important.

4 The planned experimental program

In tables 1, 2, 3 the experiment on the subject of strangeness productions, planned at JLab, are listed.

At present, the experiment 94-108, proposed within a collaboration where our group is involved, is not yet fully approved. Fully approval is conditioned under the demonstration of the feasibility of Longitudinal-Transverse separation of exp. 93-018 which has now just taken data. Nevertheless in the following, we will refer to this experiment as well, since it exploits the unique possibility at Jlab to study the elementary process of exclusive kaon electro-production through the $(e, e'K^{+})$ reaction on hydrogen target up to rather high values of the momentum transfer $(Q^{2} \leq 3(GeV/c)^{2})$ and of the center of mass energy $(W = \sqrt{s} \leq 2.2GeV)$.

4.1 Kaon electro-production experiments on proton target

The main goals of the kaon electro-production experiments $(e, e'K^{+})$ are :

- The partial or total separation of the four terms of the cross section, σ_{L}, σ_{T}, σ_{LT}, σ_{TT}; their separate measurements generally provide a very stringent test of the models. As a matter of fact:

  the longitudinal photons contributing to σ_{L} are sensitive to scalar particle exchange in the t-channel, hence sensitive to kaons as propagator particles, or, for example, to scalar diquarks. The possibility to measure, in addition, the kaon electromagnetic form factor from the t-dependence

Table 1: The elementary process for kaon electro-(photo)-production

electro-production experiments: $p(e, e'K^+)\ \Lambda, \Sigma, \ldots$

Exp n.	Hall	Title	Spokespersons
93-030	B	Measurement of the Structure Functions for Kaon Electro-production	M.Mestayer, K.H.Hicks
93-018	C	Longitudinal/Transverse Cross Section Separation in $p(e, e'K^+)\ \Lambda, \Sigma$ for $0.5 < Q^2 < 2.0(GeV/c)^2$, $W > 1.7GeV$, $t_{min} > 0.1(GeV/c)^2$	O.K.Baker
89-043	B	Measurements of the electro-production of the Λ(gnd), $\Lambda^*(1520)$ and $f_0(975)$ via K^+K^-p and $K^+\pi^-p$ Final States	L.Dennis, H.Funsten
94-108	A	Electro-production of Kaons up to $Q^2 = 3(GeV/c)^2$	O.K.Baker, C.C.Chang, S.Frullani, M.Iodice, P.Markowitz
95-003	B	Measurement of the K^0 electro-production	R.A.Magahiz

photo-production experiments: $p(\gamma, K^+)\Lambda, \Sigma^0, \ldots,\quad p(\gamma, K^0)\Sigma^+$

Exp n.	Hall	Title	Spokespersons
89-004	B	Electromagnetic Production of Hyperons	R.Schumacher
89-024	B	Radiative Decays of the Low-Lying Hyperons	G.S.Mutchler

Table 2: Kaon electro-(photo)-production on light nuclei

Exp n.	Hall	Title	Spokespersons
89-045	B	Study of the Kaon Photo-production on Deuterium	B.Meking
91-016	C	Electro-production of Kaons on Light Hypernuclei	B.Zeidman

Table 3: Kaon electro-(photo)-production on heavier nuclei

Exp n.	Hall	Title	Spokespersons
91-014	B	Quasi-Free Strangeness Production in Nuclei	C.Hyde-Wright
89-009	C	Investigation of the Spin Dependence of the ΛN Effective Interaction in P Shell	R.Chrien, E.Hungerford, L.G.Tang
94-107	A	High Resolution 1p Shell Hypernuclear Spectroscopy (on ^7Li, ^9Be, ^{12}C, ^{16}O)	S.Frullani, F.Garibaldi, J.LeRose, P.Markowitz, T.Saito
95-002	C	Direct Measurement of the Lifetime of Heavy Hypernuclei at CEBAF	L.G.Tang, A.Margarian

of the σ_L term, in the same way as what has been done in the past for the pion form factor, will be discussed later.

The longitudinal-transverse interference term is very sensitive to the different available models and could also give information on the magnitude of the transverse momentum of the quarks[6].

- The measurement of the Λ/Σ production ratio. Since all the experiments have exclusive character, from the missing energy spectra the two reactions $e + p \to e' + K^+ + \Lambda$ and $e + p \to e' + K^+ + \Sigma^0$ can be disentangled. The different contribution of scalar and vector diquarks to the two reactions can therefore be investigated.

- The study of the t-dependence of the full cross section up to large values of $|t|$ and Q^2. In this region, the transition from a semiphenomenological description in terms of mesons and baryons to a hard scattering regime in which the theoretical approach based on quarks or diquarks description of the elementary process is applicable.

Most of these goals constitute the expectations of experiment 94-108[7]. This experiment will place severe constraints on the models used to reproduce the data. In particular, important parameters such as the $g_{k\Lambda N}$ coupling constant used in semi-phenomenological models based on hadron dynamics formalism, can be much better established. Additionally, the transition to a more fundamental description of the reaction in terms of quarks can be identified. In common with exp. 93-018, in a complementary (highest) region of Q^2, is the goal to study the problem of the determination of the kaon electromagnetic form factor from the t-dependence of the longitudinal cross section.

The kaon form factor

In the space-like region, the only existing measurements for the kaon form factor come from $k - e$ scattering[8] and are limited to values of 4-momentum transfer $\leq 0.1(GeV/c)^2$. The possibility to determine the kaon form factor through the electro-production reaction, relies on the fact that the t-channel diagram of fig.1, in which the exchanged meson is the kaon (kaon pole diagram), dominates all the other diagrams in the chosen kinematic. If this is true, an extrapolation procedure has still to be carried out, as the probed kaon is not on mass shell, having t, its 4-momentum squared, a negative value.

The method chosen to maximize the t-channel contribution, while minimizing the s- and u-channel contributions is to examine the longitudinal response as a function of t, and to extrapolate to $t = m_k^2$, similar to what has been done in the case of the pion[9]. In such a way the kaon form factor comes from the following extrapolation:

$$F_k^2(Q^2) = F^2(Q^2,t)|_{t\to m_k^2} = \frac{(t - m_k^2)^2 \sigma_L}{N(t)}|_{t\to m_k^2} \tag{6}$$

where $N(t)$ is a known function of t.

Theoretical guidance would be necessary in the extrapolation to exploit model-independent constraints and analytical properties of the amplitudes. Without such theoretical indications and in the absence of the experimental data showing how much the kaon pole diagram contributes to the longitudinal cross section, it is very difficult and risky to face an analysis of uncertainties in the extraction of the kaon form factor. However, in order to understand in some detail the problem, we have estimated the uncertainties from a linear fit, extrapolated to the kaon pole, of the function $(t - m_k^2)^2 \sigma_L/N(t) = F^2(Q^2,t)$. Such a function has been calculated on the basis of particular models and the projected statistical errors on the extracted longitudinal terms have also been taken into account. In this procedure, we also made a self-consistency test by checking that the obtained values of the form factor were consistent with those used by the model itself[7]. This exercise was made in the framework of the "wjc4" model[10]. The results obtained for $Q^2 F_k(Q^2)$ have been reported in figure 3 as projected measurements at CEBAF. Also shown are other simpler

138

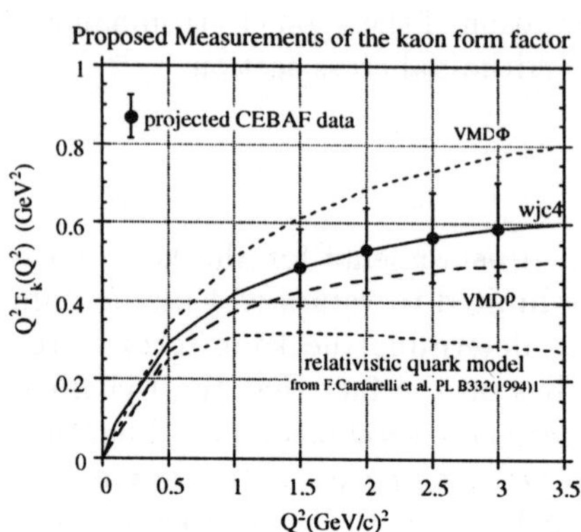

Figure 3: Measurements of the kaon for-m factor. Shown are the four points of the proposed experiment together with the projected experimental uncertainties. For the theoretical curves see the text.

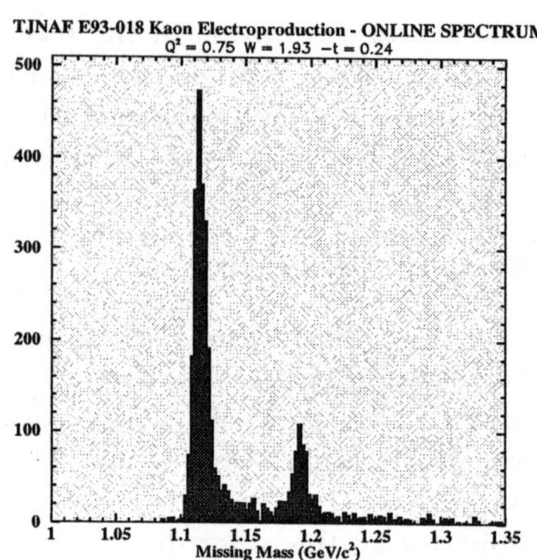

Figure 4: ON-LINE (preliminary) missing energy spectrum of the $p(e, e'K^+)$ 93-018 experiment The two reaction channels $p(e, e'K^+)\Lambda$ (first peak) and $p(e, e'K^+)\Sigma^0$ (second peak) are clearly distinguishable. The beam energy was $4.045 GeV$.

Vector Mesons Dominance Model predictions which consider the Φ and ρ-mesons and predictions from a model based on a more fundamental approach[11].

Initial studies of kaon electro-production have been completed at CEBAF. The measurements included angular and momentum transfer dependences of the $(e, e'K^+)$ reaction in hydrogen, deuterium, and carbon nuclei. A very preliminary (almost on-line) example of how the missing energy spectrum looks like, with the clear identification of the two channels $K^+-\Lambda$ and $K^+-\Sigma^0$, is reported in figure 4.

4.2 Kaon photo-production experiments on proton target

Important complementary information for the studies of the elementary electromagnetic production of kaons come from the photo-production experiments planned for the Hall B with the use of the CLAS detector.

The main goals of these experiments are :

i) investigation of the three reaction channels:

$$\gamma + p \rightarrow K^+ + \Lambda \ , \quad \gamma + p \rightarrow K^+ + \Sigma^0 \ , \quad \gamma + p \rightarrow K^0 + \Sigma^+$$

ii) Measurement of the polarization variables of the hyperon.

Emphasis is especially given to the Λ polarization observables (e.g. exp. 89-004).

The weak hyperon decay is a self analysing process for the measurement of the hyperon polarization. As an example, in the weak decay $\Lambda \to p\pi^-$, parity violation permits a mixture of S and P waves in the final state. This results in a spatial asymmetry in the decay distribution:

$$\frac{dN}{d\Omega} = \frac{1}{4\pi}(1 + \alpha\Pi\cos\psi) \tag{7}$$

where ψ is the angle between the proton momentum and the Λ's polarization axis, and α is the decay asymmetry parameter (0.642 for Λ decay). When the polarization vector $\vec{\Pi}$ is explicited in a given system of reference, it can be then determined from the angular distribution of the decay products (see e.g. ref. [12])

The measurement of hyperon polarization adds more selectivity against models. In the hard scattering regime, both "pure quark" and diquark models predict the Λ polarization to vanish.

Although not strictly related to the photo-production process, It is worth while mentioning the studies of the radiative hyperon decays (exp. 89-024), although not strictly related to the photo-production process. The measurements of the electromagnetic branching ratio of low lying states of Λ and Σ will also be carried out taking advantage of the multiparticle detection capability of CLAS, where kaons, as well as photons, pions and protons from the decay chain

$$\gamma + p \to K^+ \quad Y^*$$
$$\quad\quad\quad\quad \hookrightarrow \gamma \quad \Lambda$$
$$\quad\quad\quad\quad\quad\quad \hookrightarrow \pi^- p$$

can be detected.

These measurements represent an important test of the internal structure of the hyperons (test of quark models; see e.g. table 2 in ref. [13] and references therein).

To conclude, we only want to emphasize that Jefferson Lab will soon provide a large, self consistent set of data on the $(e, e'K^+)$ and (γ, K^+) processes, contributing in a substantial way to improve our understanding of compositeness.

References

1. CONCEPTUAL DESIGN REPORT - CEBAF - April 13, 1990.
2. J.Adam, J.Mares, O.Richter, M.Sotona, and, S.Frullani, Czech.Jou.Phys. **42**, 1167 (1992).
3. R.A.Adelseck and B.Saghai, *Phys. Rev.* C **42**, 108 (1990).
4. C.J.Bebek et al., *Phys. Rev.* D **15**, 594 (1977).
5. P.Kroll, M.Schurmann, K.Passek, W.Schweiger, *preprint UNIGRAZ-UTP 15-04-96*;
   W.Schweiger contribution to this Workshop
6. R.N.Cahn, *Phys. Lett.* B **78**, 269 (1978);
   E.L.Berger,*Z. Phys.* C **4**, 289 (1980);
   A.Brandenburg, V.V.Khoze, D.Müller, SLAC-PUB-6688 (1994).
7. O.K.Baker, C.C.Chang, S.Frullani, M.Iodice, P.Markowitz, CEBAF Experimental proposal PR-94-108 (1994)
8. S.R.Amendolia et al., *Phys. Lett.* B **178**, 435 (1986).
9. R.C.E. Devenish, D.H.Lyth, *Phys. Rev.* D **5**, 47 (1972);
   C.J.Bebek et al. *Phys. Rev.* D **17**, 1693 (1978) and erlier experiments, see references therein.
10. R.A.Williams, Chueng-Ryong Ji and C.R.Cotanch, *Phys. Rev.* C **46**, 1617 (1992).
11. F.Cardarelli et al., *Phys. Lett.* B **332**, 1 (1994).
12. A.D. Panagiotou, *Int. J. Mod. Phys.* A **5**, 1197 (1990)
13. R.A.Schumacher, *Nucl. Phys.* A **585**, 63c (1995)

THE PHYSICS at ELFE

J.M. Laget

Commissariat à l'Energie Atomique
Service de Physique Nucléaire, Centre d'Etudes de Saclay
F91191, Gif-sur-Yvette CEDEX, France

The ELFE project aims at a better understanding of the quark-gluon structure of hadronic matter in the confinement regime. The battle horse is the study of exclusive reactions at large momentum transfers. Active discussions are on going in Europe to define and choose the corresponding facility.

1   Introduction

Hadronic Physics is a branch of physics which developed out of Nuclear Physics and Particle Physics. It aims at understanding the structure and the interactions of extended hadronic systems in terms of their constituents. The relevant constituents depend on the scale at which those systems are studied. At a scale of 1 fm, or larger, nuclei are made of baryons which interact by exchanging mesons. Several decades of works at energies up to 1 GeV have put this description on firm grounds. In contrast, it is known that at a scale much smaller than 0.1 fm the relevant degrees of freedom are the quarks which interact perturbatively by exchanging gluons. This has been beautifully shown by experiments performed at high energies. However the link between these two descriptions is still missing and little is known about the quark-gluon structure of hadrons, in the confinement regime. The central issue in Hadronic Physics is:

"How are hadrons and nuclei built of quarks and gluons?"

The answer to this question lies in the non-perturbative regime of Quantum Chromo-dynamics (QCD). One has to know how current quarks dress themselves into constituent quarks, which in turn form the observed baryons and mesons. This is a formidable challenge, which requires different and complementary approaches.

One of the most promising ways is to use *exclusive* reactions, induced on nucleon and nuclei by electrons and photons at high momentum transfer (in the range 10–30 GeV^2), to select the simplest configurations of quarks (three valence current quarks of a baryon, a current quark-antiquark pair of a meson, ...). The corresponding hard mechanisms can be disentangled from the soft mechanisms which confine the quarks in the hadron ground or excited state. The study of many channels (elastic and inelastic nucleon form factors, Compton scattering, meson electroproduction,..) will lead to a better understanding of this simplest valence quark component of the wave function. In addition, the study of the space-time evolution of such configurations, in vacuum or in nuclear matter, will tell us how current quark dress themselves and build the various hadrons.

To day, none of the existing facilities allows to undertake and achieve such a program. Since the momentum transfer is large, cross-sections are small: a high luminosity is mandatory. Since the reactions are exclusive, many particles are detected in coincidence: a continuous beam is mandatory. Last but not least, spin observables will be a must in being more exclusive and determining the various transition amplitudes. Those are the characteristics of the ELFE Project [1], which is currently under discussion in Europe. It has been endorsed by NuPECC (Nuclear Physics European Collaboration Committee), which found compelling the physics case and recommended to investigate the best ways to make it a reality. The past couple of years has been devoted to optimize the design of possible schemes for the machine, to finalize the experimental set-up, and to enlarge the community potentially interested. These works have been summarized last September in Saint-Malo: the interested reader will found more details, and comprehensive reviews, in the proceedings [2].

2 The ELFE Project

ELFE is the acronym of Electron Laboratory for Europe. This facility is intended to operate a CW electron accelerator in the 15–30 GeV range and to strengthen collaborations (primarily on a european basis, but open to world-wide participations), which will use the corresponding beams.

A first design [1] could be achieved with the technology available to day, by recirculating three times the electron beam in a supra-conducting linac. The design assumes a nominal accelerating gradient of 10 MV/m, a conservative figure consistent with the present state of the art of the technology of electromagnetic hyper-frequencies in supra-conducting cavities. Rapid progresses in this field open up possibilities to cut down the cost of this solution. The design of the recirculating arcs is made in such a way to afford the bending a 30 GeV beam, without catastrophic energy losses by synchrotron radiation. A second design [2] uses the HERA ring as a stretcher to convert the pulsed beam of the TESLA injector into a continuous beam. It has the advantage to start from an existing facility. Both solutions appear to be technically feasible and the decision depends on the European scientific policy.

The nominal intensity of the extracted beam will be large enough, in the range 10–50 μA, to make possible experiments with high luminosities, in the range 10^{35} (Large acceptance detectors) to 10^{38} $cm^{-2}s^{-1}$ (well shielded magnetic spectrometers). The large duty factor will make easier and even possible coincidence experiments.

3 The Physics Case

The main physical goal of this project is to provide access to a whole class of such coincidence experiments which cannot be performed with existing facilities: the study of exclusive reactions at high momentum transfer. In those reactions, both the initial and the final state are fully determined. The high momentum transfer insures that the minimum number of participant quarks are present, at the same time, in the small interaction volume. This is the way of selecting the simplest component in the Fock expansion of the hadron wave functions: three valence current quarks in the proton wave function, for instance. It is this configuration which will be first computed when solving QCD on lattice. It can be also deduced from sum rules. The study of many exclusive channels (baryon and meson form factors, real and virtual Compton scattering, meson electroproduction, etc...) will provide many ways of probing these simple quark configurations, in order to eventually reach a mapping of the corresponding wave function.

The other originality of the project is the use of the nucleus as a "microdetector" or as a filter. Two extreme cases are worth to be considered. The first is known as "color transparency". The idea is to select one of these simplest configurations of a nucleon (or a hadron) in a nuclei and to see how it evolves toward its asymptotic wave function. The study of the interaction of the outgoing hadron with the nuclear medium, as a function of the size of the nucleus, will give us informations on the corresponding evolution. Recent studies do not show any significant effects: It is very likely that the values of Q^2 are too low to observe color transparency in the quasi-free kinematics channels. An alternative way is to study reactions induced by electrons in few body system: Exclusive reactions allow to adjust the formation length of the hadron to the distance between nucleons. The kinematics should be chosen such that the interactions of the emerging hadron with a second nucleon are maximal. This maximum occurs when the produced hadron propagates on-shell. A clear signal for color transparency would be the suppression of final state interactions when the momentum transfer increases. The onset of color transparency will indicate the range of momentum transfer where hard scattering dominate the reaction amplitude: it may well depend on each channel.

The second way is the study of hadronization in nuclei. When a virtual or a real photon hits a quark and ejects it from a nucleon, this quark first propagates freely on a certain distance before dressing itself with $q\bar{q}$ pairs, as a constituent quark, and eventually resulting in a jet of hadrons. The characteristic distance over which each of these steps occurs is comparable to the size of a nucleus. Varying the size of the nucleus from which a quark is ejected is therefore a unique way to access the various steps of hadronization.

Finally, the production of heavy flavors (strangeness and charm) will open up a new way to probe hadronic matter. As in the study of all extended complex systems, the creation of an impurity (a strange or charmed quark, for instance), and the study of its propagation, will provide us with new tools. Production of vector mesons is particularly interesting. Since $s\bar{s}$ and $c\bar{c}$ pairs are not normal building blocks of nuclear matter, the simplest mechanism in the exclusive production of ϕ or J/Ψ mesons is the exchange of two gluons. It is a way to model the Pomeron whose the structure is under active debate at HERA. It provides us not only with a way to prepare a quark-antiquark pair which will eventually evolve in the nuclear medium, but also with a way to reach quark correlations in the ground state of hadronic matter or exotic configurations, as hidden color components.

In all these cases, the determination of the various spin observables will enlarge our ability to achieve a deeper understanding of hadronic matter at short distances. I refer to my talk [3] at the SPIN94 Conference for a more comprehensive review on this topic.

4   Conclusion

ELFE will offer the unique opportunity to study *exclusive* reactions in a wide range of energy and momentum transfer. This will eventually lead to an accurate mapping of the simplest components of the Fock wave function of hadrons. In addition, ELFE will enlarge considerably the kinematical domains already accessible, in the sector of semi-inclusive reactions. In both case spin observables will be a unique tool. None of the present facilities provides us at the same time all the conditions to successfully complete the full physics program. However, present facilities should be used in order to prepare the ELFE program. Two ways are open. Following works already done at SLAC and CERN, HERMES will offer us a way to start the study of semi-inclusive reactions. This is the approach from the top, where one starts in the right energy range and then increase the duty factor and the luminosity. It allows to go from semi-inclusive to exclusive reactions. Following works already done at low energy facilities, of which the duty factor has already been increased, CEBAF will offer us a significant increase of the available energy and transfer. This is the approach from the bottom, where one first increases the duty factor and then increases the energy. It allows to go from the meson sector to the constituent quark sector and eventually to the current quark sector.

References

1. J. Arvieux and E. De Sanctis, eds., "The ELFE Project", *Conference Proceedings of the Italian Physical Society* **44** (1993).
2. N. D'Hose *et al.*, eds., "Prospects of Quark and Hadron Physics with Electromagnetic Probes", Saint-Malo (Sept. 23-27, 1996). *Nucl. Phys. A (in press)*
3. J.M. Laget, in Proceedings of the SPIN94 Conference, Bloomington (September 1994). Eds. S. Vigdor *et al.*, *AIP Conference Proceedings* **339** (1995).

DIQUARK MODEL OF EXOTIC MESONS

D.B. LICHTENBERG AND RENATO RONCAGLIA

Physics Department, Indiana University, Bloomington, IN 47405, USA

E. PREDAZZI

Dipartimento di Fisica Teorica, Università di Torino and
INFN, Sezione di Torino, I-10125, Torino, Italy

The masses of a number of exotic mesons are obtained in a diquark model. Most of the exotics have calculated masses well above the threshold for decay into two mesons, and are likely to have widths too large to make them observable. A possible exception is a $ud\bar{b}\bar{b}$ exotic, but it will be a long time before such a state can be observed even if it is stable against strong decay.

1   Introduction

A normal meson has the quantum numbers of a possible bound state of a quark and an antiquark: so-called normal quantum numbers. A meson which does not have normal quantum numbers is said to have exotic quantum numbers, and is by definition exotic. Mesons with exotic quantum numbers ought to exist because QCD does not obviously forbid them, but there is not yet definitive experimental evidence for the existence of any such meson.

A meson may have normal quantum numbers and still be exotic if its internal structure differs from that of a normal meson. Although there are candidates for such exotics, none has yet been positively identified. The problem of how to distinguish between a normal and an exotic meson with the same normal quantum numbers is a difficult one and remains unsolved, although progress has been made.

Even the structure of a normal meson is not simple, as such a meson contains, in addition to a valence quark and antiquark, gluons and quark-antiquark pairs of the sea. It is convenient to simplify by considering a model in which hadrons are made of constituent quarks, antiquarks, and gluons. A constituent quark is composed of a current quark plus a cloud of gluons and quark-antiquark pairs.

In a model with only constituent quarks and constituent gluons, a normal meson is composed of a quark and antiquark. Among possible exotic mesons there might be glueballs (composed of gluons only), hybrids (composed of a quark, antiquark, and gluons), meson "molecules" (composed of two normal mesons), and diquark-antidiquark states. The evidence for the existence of exotic mesons is slowly accumulating, but is not yet definitive. [1] We use the notation "four-quark" state to refer to any exotic meson containing two quarks and two antiquarks.

An unambiguous signature for an exotic meson would be the discovery of a meson with quantum numbers which are impossible for a quark-antiquark pair. Unfortunately for the prospect of easy discovery of exotics, those with the lowest masses are likely to have normal quantum numbers, and furthermore, are likely to mix with ordinary quark-antiquark mesons.

2   Rough Estimate of Exotic Meson Masses

Let us crudely estimate the constituent mass of a light u or d quark and also the constituent mass of a gluon. We begin with the mass m_q of a light u or d quark, neglecting their mass difference. One way we can estimate the mass is from the proton magnetic moment, assuming that the quarks have Dirac moments. This procedure gives

$$m_q = m_p/\mu = 336 \text{ MeV},\tag{2}$$

where m_p is the proton mass and μ is the dimensionless ratio of the proton magnetic moment to one nuclear magneton. Another way to estimate m_q is to take 1/2 the spin-averaged mass of the ρ and π mesons. (We are neglecting the meson binding energy in the absence of spin-dependent forces). This gives

$$m_q = (3m_\rho + m_\pi)/8 = 306 \text{ MeV}.\tag{3}$$

Still a third way is to take 1/3 the appropriately spin-averaged [2] mass of the nucleon and Δ, which gives

$$m_q = (N + \Delta)/6 = 362 \text{ MeV}.\tag{4}$$

There is no fundamental reason for these methods to yield the same mass, but to a rough approximation they do. The average is 334 MeV; rounded to the nearest 10 MeV, it is

$$m_q = 330 \text{ MeV}.\tag{5}$$

We interpret the constituent mass of a valence quark as its current mass plus the extra inertia of the sea of quark-antiquark pairs and gluons surrounding it. When the quark moves adiabatically, it drags this sea with it. We can safely neglect the current masses of the u and d, which are just a few MeV. Therefore, we can interpret the light quark constituent mass of about 330 MeV as in effect the mass of the part of the sea that is dragged along as the quark moves adiabatically.

We now conjecture that the mass of the relevant part of the sea is proportional to the square of the strong charge (or colour) of the quark. If so, the relevant quantity is the Casimir operator of SU(3), namely, F^2. This has the value $F_q^2 = 4/3$ for a quark.

This little exercise enables us to estimate the mass of a constituent gluon. A gluon has a larger colour-charge than a quark; for a gluon, we have $F_g^2 = 3$. Therefore, we estimate that the constituent gluon mass is

$$m_g = (F_g^2/F_q^2)m_q = (9/4)334 \text{ MeV} = 750 \text{ MeV}, \tag{6}$$

again rounding to the nearest 10 MeV.

We are now able to obtain rough estimates of the masses of exotic mesons containing light quarks and/or glue, neglecting hadron binding energies.

We denote the mass of a meson by M, with subscripts denoting its constituents. A four-quark state has an estimated mass

$$M_{qqqq} = 4m_q = 1320 \text{ MeV}, \tag{7}$$

a hybrid has an estimated mass

$$M_{qqg} = 2m_q + m_g = 1410 \text{ MeV}, \tag{8}$$

and a glueball made of two constituent gluons has an estimated mass

$$M_{gg} = 2m_g = 1500 \text{ MeV}. \tag{9}$$

All these masses are in the same ballpark. Furthermore, excited quark-antiquark (two-quark) mesons also have masses in this region. Consequently, if quantum numbers permit, observed mesons with masses in the region, say, between 1300 and 1600 MeV are likely to be mixtures of two-quark, four-quark, hybrid, and glueball mesons. This mixing ought to occur even though most ground-state two-quark mesons are largely unmixed.

We can also estimate the masses of exotics containing other quark flavours. The mass differences between different flavoured quarks are roughly (in MeV)

$$m_s - m_q = 170 - 200, \quad m_c - m_s = 1160 - 1200, \quad m_b - m_c = 3340 - 3400. \tag{10}$$

Therefore, replacing a light quark in an exotic by a heavier one will raise the mass of the exotic approximately by the appropriate quark mass difference. Actually, this approximation overestimates the mass of the heavier exotic because the binding energy, which we have neglected so far, increases with mass.

3 A Diquark Model

In order to go beyond rough estimates of exotic masses and to calculate exotic decay rates and branching fractions, we need to consider a specific model. We concentrate on a model in which an exotic meson is composed of a diquark and an antidiquark. Work has been done with a number of other four-quark models, of which we mention the papers of Sylvestre-Brac and Semay[3] and Pepin et al.[4] Other references are contained in these papers and in a review of diquarks.[5]

In our scheme, an exotic hadron is composed of quark clusters,[6] and we confine ourselves to the case in which the clusters are diquarks or antidiquarks. Now the $\mathbf{F}_1 \cdot \mathbf{F}_2$ factor in the lowest-order QCD potential between two quarks is positive for a sextet diquark but negative for a triplet (actually antitriplet) diquark, leading to a repulsive interaction between two quarks at small separations for the former and an attractive interaction for the latter. Therefore, we expect that colour-sextet diquarks will lie higher in energy than the colour-triplet diquarks. While we neglect the former, on general grounds we expect that their effect could slightly lower the ground state energy compared to our finding. We do not, however, have a simple recipe to propose in order to give a quantitative estimate of this effect.

We now consider a detailed model which allows us to obtain the masses of exotic mesons in terms of the masses of mesons and baryons. Many of the meson and baryon masses are known either from experiment or from predictions for unknown hadrons based on the systematics of known hadrons.[7,8] We can use this information as input data to obtain specific predictions for the masses of exotics. Our model has the advantage that we do not need to assume an explicit Hamiltonian to describe the interaction, but there is the accompanying disadvantage that the model does not allow us to calculate decay rates. We can, however, make some qualitative remarks about decays.

According to QCD, in first approximation, the force between two coloured particles depends only on their colour configuration and not on their mass or spin. This fact gives rise to an approximate supersymmetry between a diquark and antiquark. A discussion of this supersymmetry, with references, is contained in the diquark review.[5] We neglect spin at the outset of our treatment, but subsequently take it into account.

The mass of a composite particle is the sum of the masses of the constituents plus an interaction energy. In general, the interaction energy depends on the constituent masses even if the force does not. In our approximation, the force between a quark and an antiquark in an overall colour-singlet state is the same as the force between an antidiquark and a diquark in an overall colour singlet. Consequently, we can equate the mass of an exotic meson composed of a diquark and antidiquark to the mass of a meson composed of a fictitious quark and antiquark having the same masses as the diquark and antidiquark. Two obstacles must be overcome in order to carry out this procedure: we must obtain estimates of the masses of diquarks, and we must obtain the interaction energy of two fictitious quarks having the same masses as the diquarks.

We first consider the problem of obtaining the interaction energy of a bound quark and antiquark with any masses. The interaction energy turns out to be a fairly smooth function of the reduced mass of the two particles. [7] We assign reasonable values of the masses to the different flavoured quarks. To be specific, we use the same quark masses as in ref. 8, which are (in MeV):

$$m_q = 300, \quad m_s = 475, \quad m_c = 1640, \quad m_b = 4985. \tag{11}$$

These masses are a little below those given in our rough estimates in Eqs. (5) and (10). This is not important, as we have found [7,8] that we may vary the input quark masses appreciably without significantly changing the ouput hadron masses, as we can make compensating changes in the interaction energy. Our approach is much more sensitive to the differences of quark input masses than to their absolute values.

If we use the experimental values M_{12} of meson masses [1] as input, we can calculate the interaction energies E_{12} from the simple formula:

$$E_{12} = M_{12} - m_1 - m_2. \tag{12}$$

In order to eliminate the effects of spin as much as possible in this procedure, we use spin-averaged meson and baryon masses according to the prescription of Ref. 2. This procedure cannot always be carried out in terms of known hadron masses. Where necessary, we approximately remove the effect of the colour-magnetic interaction by a semi-empirical mass formula. [9]

We can then plot the values of E_{12} against the reduced mass μ_{12}. We obtain the value of E_{12} for any μ_{12} by interpolation or extrapolation, or, in other words, from the mass given by a smooth curve fitting the known points. Thus, if we are given the masses of any two diquarks, we compute the reduced mass, find the relevant interaction energy from the curve for mesons, add the two diquark masses, and thereby obtain the mass of the exotic composed of the two diquarks.

4 Diquark Masses

We next turn to the problem of estimating the diquark masses. Consider the spin average of the masses of ground-state baryons with a given quark content.

The individual baryon masses are known either from experiment [1] or from estimates based on the systematics of known baryons. [7,8]. For a baryon like the Ω, which contains the quarks sss, spin averaging in terms of known baryons cannot be done. In such cases we use the semi-empirical formula. [9]

We assume the baryon is composed of a diquark and a quark. If the quarks have different flavours, then we assume that the two heaviest quarks form the diquark. (We neglect the mass difference between the d and u quarks.) We then guess a trial value for the diquark mass; it is convenient to take the sum of the masses of the two quarks in the diquark. We then find the reduced mass of the diquark and the third (or spectator) quark, use the meson curve to obtain the interaction energy, and so obtain a prediction for the baryon mass, which, in general, will not be the same as the input baryon mass. However, we can then use a revised diquark mass and repeat the process by iteration, stopping when we have found an input diquark mass that gives the input value of the baryon mass. The difference in mass between a diquark and the sum of the masses of the quarks it contains is the triplet interaction energy E_{12}^t. We expect that the energy E_{12}^t is larger than E_{12} because the interaction at small distances contains the factor $\mathbf{F}_1 \cdot \mathbf{F}_2$, which is smaller in magnitude for two quarks in a colour-triplet (actually, antitriplet) state than in a colour-singlet state. We find numerically that our expectation is satisfied.

We must overcome still another difficulty. If we use the above iteration procedure to calculate the mass of, say, the ss diquark, the result depends mildly on whether the spectator quark in the baryon is q (u or d) or s. We resolve the ambiguity by averaging the two values.

The above procedure gives us diquark masses that are suitable for use in calculating the masses of exotic mesons except that so far we have not included the effects of spin. We therefore have to correct the diquark masses for spin-dependent forces. We do so under the assumption that these forces arise from the colour-magnetic interaction of perturbative QCD or from a generalization of this interaction. [2,7,8] Then, any pair of quarks in a baryon is subject to essentially the same spin-dependent force as the same two quarks in an exotic meson, provided the two quarks belong to a single diquark.

The magnitude of the colour-magnetic energy arising from two quarks in a baryon has already been estimated [2,7,8] and found to be only weakly dependent on the mass of the spectator quark. This dependence is such that the magnitude of the colour-magnetic energy increases slightly as the mass of the third quark increases. We approximate the colour-magnetic energy by taking an average of the energy obtained with the q, s, and c as spectator. (We do not include the b as spectator because the result is very similar to the c with a larger error and because we do not want to give too much importance to the heavy quarks.) A different prescription will change our results by only a few MeV at most. We can now obtain all the necessary colour-magnetic energies from the results in Refs. 2,7, and 8.

The above procedure lets us correct the masses of the diquarks for the effects of spin. However, in general, in an exotic meson there are still other spin-dependent forces arising between the quarks in the diquark and the quarks in the antidiquark. However, if either the diquark or antidiquark has spin zero, the net effects of these spin-dependent inter-diquark forces vanish. Although we have a method that applies to exotics containing two spin-one diquarks, here we restrict ourselves for simplicity to exotics with at least one diquark of spin zero. These latter exotics have the lowest masses in any case.

5   Results and Discussion

As we stated in the previous section, we obtain diquark masses in terms of input baryon masses and estimates of the colour-magnetic interaction energies between pairs of quarks in baryons. Strictly speaking, the calculated diquark masses depend slightly on the mass of the spectator quark in the baryon. The differences, however, are usually less than 20 MeV. In those cases in which we obtain more than one value of the diquark mass, we give the average value. These average values of the diquark masses are shown in Table 1.

Using the diquark masses in Table 1, we can compute the reduced mass of a system of a diquark and antidiquark. We then obtain the interaction energy for the reduced mass, using the meson curve as input. This procedure gives us the exotic meson mass. We have already taken into account the colour-magnetic interaction in computing the diquark masses. Because we restrict ourselves to those cases in which at least one diquark has spin zero, no further spin-dependent forces enter the problem. A sample of the exotic meson masses that we have calculated is shown in Table 2. We also give in Table 2 the lightest mesons into which the exotic can decay strongly, provided the energy permits. The sum of the masses of the decay products is shown in the last column of Table 2.

The calculated exotic masses given in Table 2 have errors of up to 30 MeV associated with the fact that we have chosen the fixed diquark masses of Table 1, whereas actually, in our model, the diquark masses depend on their environment. If future experiments should yield masses which differ from those in Table 2 by much more than 30 MeV, we will have to discard our model in its present form.

We see from Table 2 that only the exotic $qq\bar{b}\bar{b}$ lies below the threshold of the lightest mesons into which it can decay strongly. The exotic $qs\bar{b}\bar{b}$ lies only a little above threshold. It may be observable as a narrow resonance or even a bound state, as our model is probably not accurate enough to distinguish these possibilities in this case. However, exotic mesons with two b quarks are unlikely to be seen for quite some time.

Table 1. Masses M_0 and M_1 of spin-zero and spin-one diquarks obtained from input baryon masses and colour-magnetic interaction energies in baryons. The method is described in more detail in Sec. 3. The symbol q stands for u or d, and masses are rounded to the nearest 5 MeV. Errors of up to 20 MeV arise because, in our model, the calculated diquark masses depend on the spectator quark, and we have averaged over this dependence.

Quark content	M_0 (MeV)	M_1 (MeV)
qq	595	800
qs	835	975
ss	–	1150
qc	2100	2150
sc	2250	2295
cc	–	3415
qb	5465	5485
sb	5630	5650
cb	6735	6750
bb	–	10075

The exotic masses we have calculated depend on the input masses of mesons and baryons. Because some of these input masses are not known from experiment but are estimated, we have an additional source of error. In particular, the estimation of the masses of baryons containing two heavy quarks involves a considerable extrapolation from known data.[8] This fact adds to the uncertainty in our calculated masses of exotics containing two heavy quarks.

All exotics with at most one heavy quark have masses well above the masses of the decay products; in particular, this is true for the exotic masses exhibited in Table 2. Therefore, these exotics will decay rapidly into two mesons. Our method does not allow us to calculate the decay widths (because we do not have a Hamiltonian or spatial wave functions). However, we observe that the colour wave functions of the exotics are not orthogonal to the colour wave functions of the two mesons into which they can decay. This means that there is nothing to prevent the exotics from "falling apart" into their decay products.

Table 2. Predicted masses M_E of exotic mesons obtained from the diquark masses in Table 1. These are all ground-state mesons with the given spin and quark content, and the parities are all positive. The symbol q stands for u or d, and masses are rounded to the nearest 10 MeV. The two numbers in column 2 are the spin of the diquark and antidiquark respectively; the spin of the exotic is just the sum of these spins. The next-to-last column gives the lowest-mass mesons into which the exotic can decay strongly if energetically permitted, and the last column gives the threshold energy E_t of the decay. Errors of up to 30 MeV in the exotic masses arise because of errors in the diquark masses in addition to errors resulting from defects in our model.

Quark content	Diquark spins	M_E (MeV)	Decay products	E_t (MeV)
$qq\bar{q}\bar{q}$	0, 0	1180	$\pi\pi$	280
$qq\bar{q}\bar{q}$	0, 1	1370	$\pi\rho$	910
$qq\bar{q}\bar{s}$	0, 0	1400	πK	630
$qq\bar{q}\bar{s}$	1, 0	1580	πK^*	1030
$qq\bar{s}\bar{s}$	0, 1	1700	KK^*	1390
$qs\bar{q}\bar{s}$	0, 0	1610	$K\bar{K}$	990
$qs\bar{q}\bar{s}$	0, 1	1740	$K^*\bar{K},\, K\bar{K}^*$	1390
$qq\bar{q}\bar{c}$	0, 0	2620	$\pi\bar{D}$	2010
$qq\bar{q}\bar{c}$	0, 1	2660	$\pi\bar{D}^*$	2150
$qq\bar{q}\bar{c}$	1, 0	2770	$\pi\bar{D}^*$	2150
$qq\bar{q}\bar{b}$	0, 0	5950	πB	5420
$qq\bar{q}\bar{b}$	0, 1	5970	πB^*	5460
$qq\bar{s}\bar{c}$	0, 0	2760	$K\bar{D}$	2360
$qs\bar{q}\bar{c}$	0, 0	2800	$\pi\bar{D}_s$	2110
$qq\bar{s}\bar{b}$	0, 0	6120	KB	5780
$qq\bar{c}\bar{c}$	0, 1	3910	$\bar{D}\bar{D}^*$	3880
$qc\bar{q}\bar{c}$	0, 0	3920	$\pi\eta_c$	3120
$qq\bar{c}\bar{b}$	0, 0	7220	$\bar{D}B$	7150
$qc\bar{q}\bar{b}$	0, 0	7180	πB_c	6390
$qs\bar{c}\bar{b}$	0, 0	7380	$\bar{D}B_s$	7240
$qq\bar{b}\bar{b}$	0, 1	10550	BB^*	10600
$qb\bar{q}\bar{b}$	0, 0	10380	$\pi\eta_b$	9540
$qs\bar{b}\bar{b}$	0, 1	10710	BB_s^*	10700

Under these circumstances, we guess that the exotics will have widths too large to make these mesons observable.

It is possible that the spin-dependent forces arise, not from the colour-magnetic interaction but from pseudoscalar meson exchange.[10,11] If so, some exotic mesons with two c quarks are predicted[4] to have masses well below those given in our Table 2 and to be stable against strong decay. Future experiments, by their observation or non-observation of weakly decaying exotics containing two c quarks, will rule out either our model or the one of Ref. 4.

Acknowledgments

We should like to thank Ted Barnes, Michael Pennington, and Jean-Marc Richard for valuable discussions. This work was supported in part by the U.S. Department of Energy, the Italian Institute for Nuclear Physics (INFN), and the Ministry of Universities, Research, Science and Technology (MURST) of Italy.

References

1. Particle Data Group: R. M. Barnett et al., *Phys. Rev.* D **54**, 1 (1996)
2. M. Anselmino, D.B. Lichtenberg, and E. Predazzi, Z. Phys. C. **48**, 605 (1990).
3. B. Silvestre-Brac and C. Semay, Z. Phys. C **59**, 457 (1993).
4. S. Pepin, Fl. Stancu, M. Genovese, and J.-M. Richard, nucl-th/9608058 and ph-9609348 (unpublished).
5. M. Anselmino, E. Predazzi, S. Ekelin, S. Fredriksson, and D.B. Lichtenberg, Rev. Mod. Phys. **65**, 1199 (1993).
6. D.B. Lichtenberg, E. Predazzi, D.H. Weingarten, and J.G. Wills, *Phys. Rev.* D **18**, 2569 (1978).
7. R. Roncaglia, A.R. Dzierba, D.B. Lichtenberg, and E. Predazzi, *Phys. Rev.* D **51**, 1248 (1995).
8. R. Roncaglia, D.B. Lichtenberg, and E. Predazzi, *Phys. Rev.* D **52**, 1722 (1995).
9. Yong Wang and D.B. Lichtenberg, PRD **42**, 2404 (1990).
10. L.Ya. Glozman and D.O. Riska, Phys. Rep. **208**, 263 (1996).
11. L.Ya. Glozman, Z. Papp, and W. Plessas, PLB **381**, 311 (1996).

DIQUARK MODEL OF DIBARYONS

D. B. LICHTENBERG and RENATO RONCAGLIA

Physics Department, Indiana University, Bloomington, IN 47405, USA

ENRICO PREDAZZI

*Dipartimento di Fisica Teorica, Università di Torino and INFN, Sezione di Torino,
I-10125, Torino, Italy*

A diquark model previously formulated to describe exotic mesons is extended to
dibaryons. In the model, dibaryons containing only light quarks are unbound. The
H dibaryon, consisting of *uuddss* quarks, is unstable by about 90 MeV or more,
and should decay strongly into two Λ baryons. A charmed dibaryon H_c, composed
of *uuddsc*, is unstable by about 60 MeV or more and should decay strongly into
$\Lambda + \Lambda_c$. On the other hand, we find that a bottom dibaryon H_b, made of *uuddsb*,
may be just bound by about 10 MeV with respect to $\Lambda + \Lambda_b$. If so, it should decay
weakly into various final states with a charmed hadron. A possible two-body decay
would be into $\Sigma + \Lambda_c$.

1 Introduction

The deuteron is the only known dibaryon. However, many authors have proposed that there might exist others, some of them not well-separated resonant
states of two nucleons but rather six-quark states in other configurations. A
recent review[1] contains some references to the subject.

From time to time experiments have indicated that dibaryons other than
the deuteron might exist, but none of these experiments has been decisive. One
work reporting a possible dibaryon signal in a system containing only light u
and d quarks appeared very recently,[2] another paper reporting evidence for a
dibaryon resonance with strangeness -1 appeared several years earlier.[3]

Here, we concentrate on the H dibaryon, composed of *uuddss* quarks,
which Jaffe[4] proposed might be stable against decay into two Λ baryons. We
also briefly discuss the H_c, in which one of the s quarks is replaced by a c
quark, and the H_b, in which the c is replaced by a b. Thus far, the H dibaryon
has not been observed, and theorists have disagreed about whether it is bound
or not. Oka[5] has given a recent review of the theory of the H. The H_c and
H_b, which have received much less attention, have also not been observed.

In general, we are interested in the lowest-mass dibaryon states with a given quark content, as these are most likely to be observed. We can envision two different types of stable or nearly stable dibaryons. The first, like the deuteron, can be considered as a kind of "molecule" composed of two largely non-overlapping colour-singlet baryons, and the second is composed of six quarks, no three of which form a colour singlet. Thus far, the only known example of the first kind of dibaryon is the deuteron, and there is no definitely known example of the second kind. In this work, we consider only the second kind of dibaryon.

One reason to consider a model for dibaryons is that, so far, lattice gauge calculations have not been adequate to decide on the nature of the H, one calculation finding that the H is unbound,[6] and another finding the opposite.[7]

2 The Model

In our model, a dibaryon is composed of three diquarks, each of which is a colour antitriplet. In the H, H_c, and H_b each diquark has spin zero. Two of the diquarks couple to form a colour-triplet quadriquark, which, in turn couples to the third diquark to form a colour singlet. We shall often refer to a diquark as a colour triplet, although, of course, like an antiquark, it is a colour antitriplet.

There are two reasons why we confine ourselves to colour-triplet diquarks. (1) At small interquark separations, the force between two quarks arising from QCD is attractive in a triplet state, while it is repulsive in a sextet state. This fact makes it plausible that the mass of a colour-sextet diquark is considerably larger than that of a colour-triplet diquark, leading to a larger dibaryon mass, and consequently to a more unstable dibaryon that will be harder to observe. (2) With plausible assumptions motivated by QCD, we are able to obtain the masses of colour-triplet diquarks and the masses of dibaryons just from the properties of mesons and baryons, without the need for an explicit Hamiltonian. We have already applied these ideas to possible exotic mesons, each of which is composed of a diquark and antidiquark.[8]

Our diquarks are constructed so that the two quarks in each diquark satisfy the Pauli principle. In the limit that diquarks are pointlike, or at least much smaller than the typical separation between different diquarks, we can safely ignore the Pauli principle for quarks in different diquarks. This is analogous to the case in nuclear physics, in which the quark degrees of freedom in nucleons are generally ignored.

In nuclei, when two nucleons are close together, an effective repulsive potential exists between them that arises in part because of the Pauli principle for quarks. Similarly, when two diquarks in a dibaryon are close together, the Pauli principle between quarks of the same flavour in different diquarks leads to a modification of the effective potential between the diquarks. This modification should cause a short-range repulsion. We neglect this repulsion, and thereby underestimate the mass of a dibaryon containing three diquarks if each of two different diquarks contains a quark of the same flavour. To lessen this effect, we restrict ourselves to dibaryons containing three different diquarks and at most two quarks of any flavour. Furthermore, we restrict each diquark to have two different flavoured quarks.

The restriction that each diquark has two different quarks enables us to consider diquarks with spin zero. A spin-zero diquark has a lower mass than a spin-one diquark with the same quark content because of the colour-magnetic interaction. Consequently, a dibaryon made of spin-zero diquarks is likely to have a lower mass than a dibaryon made of spin-one diquarks, making the former more easily observable.

The masses of the diquarks have already been calculated (see Table 1 of Ref. [8]) in terms of properties of mesons and baryons, and we simply take over the results. We then combine two diquarks into a colour-triplet quadriquark. We can obtain the mass of the quadriquark by adding the appropriate interaction energy E_{12}^t, where the superscript t denotes the fact that the quadriquark is a colour triplet. The interaction energy can be computed for any reduced mass by interpolation or extrapolation by the methods of Ref. [8]. Once we have the mass of the quadriquark, we can obtain the reduced mass of it and the remaining diquark and so obtain the appropriate interaction energy E_{12}. Of all possible ways to choose the quadriquark from two out of three diquarks, we pick the way that yields the lowest dibaryon mass, as we are looking for a lower limit. We now have all the ingredients to obtain lower limits on the masses of dibaryons from the diquark masses of Ref. [8] and the interaction energies E_{12}^t and E_{12}.

3   Results and Discussion

We obtain that the mass of the H dibaryon is

$$M(H) \geq 2320 \text{ MeV}, \tag{1}$$

rounded to the nearest 10 MeV. This mass is about 90 MeV above the mass of two Λ baryons, and so we obtain that the H can decay strongly into two Λ's. Because in our model there is nothing to inhibit the decay, the decay width may be too large to make the state readily observable. Similarly, we obtain that the mass of the H_c dibaryon is

$$M(H_c) \geq 3460 \text{ MeV},\tag{2}$$

a value 60 MeV above threshold for decay into $\Lambda + \Lambda_c$. Again, the decay width might be too large to allow the state to be observable. Within our model, we estimate errors on the lower limits in (1) and (2) to be about 30 MeV, owing to the methods used to calculate the diquark masses. Lastly, we obtain that the mass of the H_b is

$$M(H_b) \geq 6730 \text{ MeV},\tag{3}$$

It is interesting that this last value is 10 ± 20 MeV *below* the threshold for decay into $\Lambda + \Lambda_b$. Because our method allows us only to calculate a lower limit on the mass of the H_b, we cannot say definitely whether it is bound. However, this dibaryon is well worth searching for. If it is bound, it should decay weakly into the two-body final state $\Sigma + \Lambda_c$ as well as into other hadrons including a charmed hadron.

We estimate that dibaryons containing diquarks with only u and d quarks will be even more unstable than the H. We conclude that, according to our model in which dibaryons are composed of colour-triplet diquarks, dibaryons containing at most one c quark and lighter quarks will be unstable against strong decay and hard to observe, but a dibaryon containing one b quark might be stable against strong decay.

Our model of a dibaryon is that it is composed of three diquarks, each of which is a colour triplet. Within our model, we have made approximations that lower the calculated dibaryon mass compared to its mass in an exact model calculation. This means that our calculated masses should be regarded as lower limits. Consequently, if a weakly decaying H or H_c dibaryon is observed, it must have a structure quite different from the one we have assumed.

Acknowledgments

Part of this work was done while one of us (E.P.) visited Indiana University. This work was supported in part by the U.S. Department of Energy, the Italian Institute for Nuclear Physics (INFN), and the Ministry of Universities, Research, Science and Technology (MURST) of Italy.

References

1. M. Anselmino, E. Predazzi, S. Ekelin, S. Fredriksson, and D.B. Lichtenberg, Rev. Mod. Phys **65**, 1199 (1993).

2. W. Brodowski et al., Z. Phys. A **355** (1996) 5.
3. H. Pickarz, in *Intersections Between Particle and Nuclear Physics*, AIP Conf. Proc. 243, editor, W.T.H. Van Oers, AIP New York (1992).
4. R.L. Jaffe, Phys. Rev. Lett. 38 (1977) 195.
5. M. Oka, in Intern. Symposium on Exotic Atoms and Nuclei, June 7–10, 1995, Hakono, Japan. Proceddings to be published in Hyperfine Interactions, R. Hayano, ed.
6. P.B. Mackenzie and H.B. Thacker, Phys. Rev. Lett. **55** (1985) 2539.
7. Y.T. Iwasaki, T. Yoshie, and Y. Tsuboi, Phys. Rev. Lett. **60** (1988) 1371.
8. D.B. Lichtenberg, R. Roncaglia, and E. Predazzi, Diquark model of exotic mesons, preceding talk given at the Diquark III workshop (Torino, Oct. 28–30, 1996).

WHICH SCALAR MESON IS THE GLUE-STATE ?

M. BOGLIONE

Dipartimento di Fisica Teorica, Universita' di Torino,
I.N.F.N, Sezione di Torino, Via P.Giuria 1, I-10125 Torino, Italy
and
Centre for Particle Theory, University of Durham,
Durham DH1 3LE, U.K.

Preliminary results of a work in collaboration with M.R. Pennington are presented. Extending a scheme introduced by Tornqvist, we investigate a dynamical model in which the spectrum of scalar mesons can be derived, with the aim of locating the lightest glue-state. Adding hadronic interaction contributions to the bare propagator, to 'dress' the bare quark-model $q\bar{q}$ states, we are able to write the amplitudes and the phase shifts in the approximation in which scalar resonances decay only into two pseudoscalar channels. The fit of these quantities to experimental data gives a satisfactory understanding of how hadronic interactions modify the underlying 'bare' spectrum. In particular, we examine the case in which a glue-state is introduced into the model.

1 Introduction

Gluons carry colour charge which means they interact together. Consequently it is possible for them to cluster and form objects which are colourless overall. These would be just like conventional hadrons, but with constituents that are massless gauge bosons. These are known as *glueballs*.

Interest in glueballs has increased since new candidates for gluonic states have emerged from experimental results [1-5] especially in the $1.5 - 2.0$ GeV energy region. Moreover, Lattice QCD calculations (in quenched approximation) have recently suggested the presence of light scalar glue-states in this same energy region [7]. From the experimental point of view, it is now clear that there are too many confirmed scalar (0^{++}) mesons to form one $q\bar{q}$ meson nonet. The problem is particularly pronounced for the I=0 sector, since at least four f_0's now appear in the particle data listing. As a consequence, we can infer that some of them have to be extra states.

The quark model gives a reasonably good description of the vector and tensor meson spectra and properties, but its predictions for the scalar sector are very disappointing. To understand how and why scalars are so different from vectors and tensors, we consider a simple model in which all bare meson states belong to ideally mixed quark multiplets. We call $n\bar{n}$ the nonstrange light state and suppose that substituting a strange quark for a light one increases the mass of the state by $\Delta m_s \simeq 100$ MeV, as illustrated in Fig. 1.

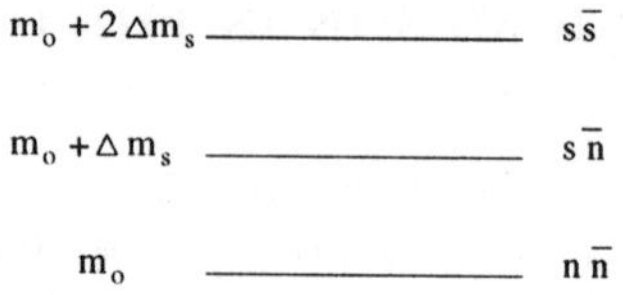

$$m_0 + 2\,\Delta m_s \;\underline{\hspace{6cm}}\; s\bar{s}$$

$$m_0 + \Delta m_s \;\underline{\hspace{6cm}}\; s\bar{n}$$

$$m_0 \;\underline{\hspace{6cm}}\; n\bar{n}$$

Figure 1: Spectrum of the bare $q\bar{q}$ states

The bare propagator for each of these bound states will be of the form [6]

$$P = \frac{1}{\mathcal{M}_0^2 - s}\,, \tag{1}$$

with a pole on the real axis, corresponding to a non decaying state, for example

$$|\phi\rangle_0 = |s\bar{s}\rangle$$

for the vector $I = 0$ state, and

$$|f_0'\rangle_0 = |s\bar{s}\rangle$$

for the scalar $I = 0$ state. If we now assume that the experimentally observed hadrons are obtained from the bare states ($n\bar{n}$, $s\bar{n}$, $s\bar{s}$, ...) by dressing them with hadronic interactions, the propagator becomes

$$P(s) = \frac{1}{\mathcal{M}^2(s) - s - i\,\mathcal{M}(s)\,\Gamma(s)}\,, \tag{2}$$

and the pole moves into the complex s-plane. The corresponding state can be decomposed as

$$|\phi\rangle = \sqrt{1 - \epsilon^2}\,|s\bar{s}\rangle + \epsilon_1|K\overline{K}\rangle + \epsilon_2|\rho\pi\rangle + ... \tag{3}$$

where calculation would give $\epsilon^2 = \epsilon_1^2 + \epsilon_2^2 + ... \ll 1$. The hadronic loop contributions allow the bare states ($s\bar{s}$ in this example) to communicate with all hadronic channels permitted by quantum numbers, and this enables the ϕ meson to decay. A similar picture works for the tensors.

For scalars the situation is different because the dominant decays are just into two pseudoscalars, the couplings are bigger and they couple strongly to more than one channel, creating overlapping and interfering resonance structures. Furthermore, being S-waves, the opening at thresholds produces a more dramatic s-dependence in the propagator. As a consequence, the $f_0(980)$, for example, turns out to be predominantly a $|K\overline{K}\rangle$ state, and not an $|s\bar{s}\rangle$ one:

$$|f_0(980)\rangle = \sqrt{1 - \epsilon^2}\,|K\overline{K}\rangle + \epsilon_1|s\bar{s}\rangle + ... \;. \tag{4}$$

This does not mean that the $f_0(980)$ is a $K\overline{K}$ molecule, because the *seeds* of the model are conventional $q\overline{q}$ states and the binding forces are not due to inter-hadron interactions alone. The crucial point here is the fact that hadronic interactions allow the $q\overline{q}$ bare states to communicate with all possible hadronic channels and these channels to communicate with each other, giving rise to a *mixing* which means, for example, that the $f_0(980)$ spend most of its time in a $K\overline{K}$ state and not in an $s\overline{s}$ one.

2 A closer look at the model

Let us now examine the model in more detail starting, for simplicity, from the case in which just one resonance is produced. If we define a vacuum polarization function $\Pi(s)$ which takes into account all the possible two pseudoscalar loop contributions to the propagator $P(s)$, we can easily write its imaginary part [8]:

$$\mathrm{Im}\Pi(s) = -\sum_i G_i^2(s) = -\sum_i g_i^2 \, \frac{k_i(s)}{\sqrt{s}} \, (s - s_{A,i}) \, F_i^2(s) \, \theta(s - s_{th,i}) \qquad (5)$$

where the index i runs over the pseudoscalar channels, the g_i's are the $SU(3)$ flavour couplings, the k_i's are the c.m. momenta and $(s - s_{A,i})$ are the Adler zeros. $F_i(s)$ are the form factors, which take into account the fact that the interaction is not pointlike but has a spatial extension. These are parametrized by

$$F_i = \exp\left(\frac{-k_i^2(s)}{2\,k_0^2}\right), \qquad (6)$$

where the momentum k_0 is inversely proportional to the range of the interaction. The real part of the vacuum polarization function can be found from the dispersion relation

$$\mathrm{Re}\,\Pi(s) = \frac{1}{\pi} P \int_{s_{th,1}}^{\infty} ds' \, \frac{\mathrm{Im}\Pi(s')}{s' - s} \qquad (7)$$

No subtraction is needed, since the form factors decrease fast enough when $|s| \to \infty$. At this point we are able to write the propagator as

$$P(s) = \frac{1}{m_0^2 + \Pi(s) - s}, \qquad (8)$$

and the contribution to the $i \to j$ amplitude as

$$R_{ij}(s) = \frac{G_i(s)\,G_j(s)}{m_0^2 + \Pi(s) - s}. \qquad (9)$$

Notice that $R_{ij}(s)$, having its numerator and denominator related to each other by the same G_i couplings, is not the most general amplitude satisfying unitarity. This would look like:

$$A_{ij}(s) \; = \; R_{ij}(s) \, e^{\,2i\,\alpha(s)} + \sin\alpha(s) \, e^{\,i\,\alpha(s)} \tag{10}$$

where $\alpha(s)$ is an unknown function of s real along the right hand cut, and the second term in Eq. (8) accounts for the background contribution. To avoid an increasing number of parameters in the model, we will consider just the approximate amplitude R_{ij} of Eq. (9), having checked that a constant value of α of about $15°$ in Eq. (10) would typically work just as well.

Fig. 2 shows the behaviour of the real and imaginary part of the running complex mass function $m^2(s) = m_0^2 + \Pi(s)$, for the $I = 1/2$, $K_0^*(1430)$, and the $I = 1$, $a_0(980)$. Here the parameters of the model (the mass of the bare $n\overline{n}$ state, the form factor cut-off k_0, an overall coupling γ and the position of the Adler zero) are determined by fitting the amplitudes and the phase shifts to the experimental data from LASS [9]. As we can see, the opening of each threshold gives an extra contribution (square root cusp) to the imaginary part of the vacuum polarization function. This is reflected in the shape of its real part, which presents a very strong s-dependence at every threshold. The intersection of $\operatorname{Re} m^2(s)$ with the curve s represents the square of the Breit Wigner mass, and the value of $\operatorname{Im}\Pi(s)$ at this point is related to the Breit Wigner width by

$$\Gamma_{BW} = \frac{-\operatorname{Im}\Pi(m_{BW})}{m_{BW}} \; . \tag{11}$$

Notice how the negative contribution of $\operatorname{Re}\Pi(s)$ shifts down the actual mass of the resonance with respect to the value of the corresponding bare mass. This effect is particularly pronounced for the $a_0(980)$ because the mass of the dressed bound state happens to coincide with the first of two thresholds, the $K\overline{K}$ and the $\pi\eta'$ ones, which are very close to each other and have similar couplings: despite the bare mass m_0 being fixed at 1420 MeV, the Breit Wigner mass of the a_0 is found to be as low as 987 MeV.

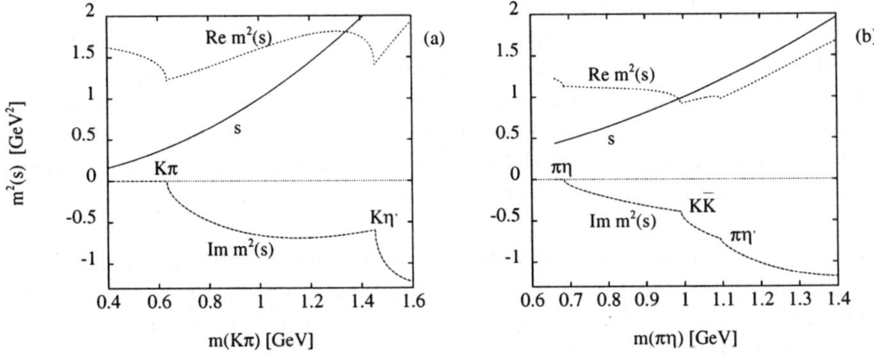

Figure 2: Real and imaginary parts of the complex running mass functions for the (a) $K_0^*(1430)$, and (b) $a_0(980)$, illustrating the effect of thresholds.

To analyze the more controversial $I = 0$ scalar sector, we need to examine the case in which more than one resonance is produced. Now, the polarization function has to incorporate the features (couplings, masses, form factors, ...) of each of the intermediate particles created in the process:

$$\text{Im}\Pi_{\alpha\beta}(s) = -\sum_i G_{\alpha,i}(s)\,G_{i,\beta}(s)\,, \tag{12}$$

$$\text{Re}\,\Pi_{\alpha\beta}(s) = \frac{1}{\pi}P\int_{s_{th,1}}^{\infty} ds' \,\frac{\text{Im}\Pi_{\alpha\beta}(s')}{s'-s}\,, \tag{13}$$

where the indices α and β run over the N different resonances. As a consequence, the propagator

$$P_{\alpha\beta}(s) = \frac{1}{(m_0^2 - s)\delta_{\alpha\beta} + \Pi_{\alpha\beta}(s)} \tag{14}$$

becomes an $N \times N$ complex matrix, and the couplings are N-dimensional column vectors. The amplitudes are determined using a diagonalization procedure and can be written as

$$R_{ij}(s) = \sum_\alpha \frac{G'_{\alpha,i}(s)\,G'_{j,\alpha}(s)}{m_{\alpha,diag}^2(s) - s} \tag{15}$$

where the diagonalized masses $m_{\alpha,diag}(s)$ are admixtures of all the bare masses and the vacuum polarization matrix elements. They are, like the new couplings

$G'_{\alpha,i}(s)$, complex and s-dependent.

What is the interpretation of this more complicated picture ? The physical observed hadrons are the states we obtain after diagonalization, an admixture of the *seed* states of the model. The mixing among them is embodied in the diagonalization procedure, which allow all the channels to communicate with each other and to create new physical states with masses, couplings and widths different from the ones of the primitive states.

We now want to consider explicitly the case in which not only the two conventional f_0 and f'_0 are present in the $I = 0$ sector, but also a third state arises, thanks to the presence of a glue gg seed. Since, in principle, we do not know the mass of the bare glue-state, m_{gg} is assumed to be an extra parameter of the model. To avoid a further increase in the number of parameters we will also assume that the gg state cannot mix with the $q\bar{q}$ states at the bare level: all the mixing will occur via hadron interactions. The bare couplings for the glueball to the two pseudoscalar channels are the ones of an $SU(3)$ flavour singlet. Fig. 3 presents the results we obtain if we choose a gg bare mass of 1.8 GeV and readjust the other parameters to fit the data from $\pi\pi$ scattering [10,11], in such a way that the agreement with the LASS data on $K_0^*(1439)$ [9] remains satisfactory. The two lower states have the same features that would be obtained if only two resonances were considered [8]: a very broad state with a mass of about 1 GeV and a narrow one with a mass of about 1.2 GeV. The higher glue-state almost decouples from the lowest channels, but shows relatively strong couplings not only to the heaviest $\eta'\eta'$, but also to the $\eta\eta'$ channel, to which it had a zero underlying bare coupling. Again, the Breit Wigner mass is considerably lower then the input one: for these values of the parameters our glueball mass is about 1.58 GeV. Notice how the masses of all the scalar $I = 0$ mesons are related to the positions of the two pseudoscalar thresholds. The amplitude presents two dips: the former corresponds to the presence of a very narrow resonance sitting on top of a very broad one, the second is related to the third heavier resonance, in agreement with the experimental data from Crystal Ball [1], GAMS [13], and the analysis from Bugg et al. [14] and Anisovich et al. [15]

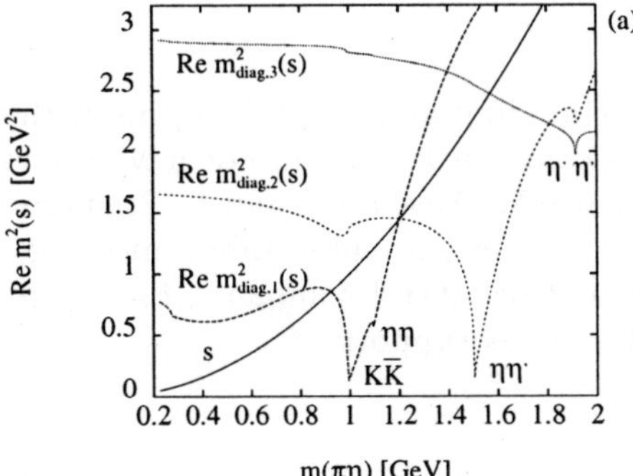
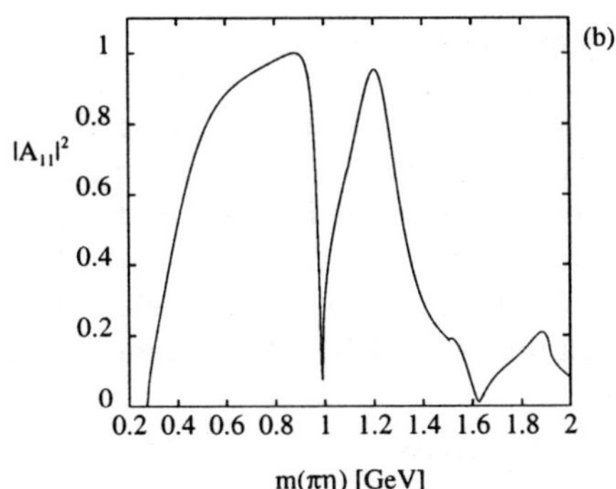

Figure 3: (a) Real parts of the diagonalized masses of the three $I = 0$ scalar resonances created in $\pi\pi$ scattering. (b) Modulus of the amplitude corresponding to the same process

3 Conclusions

In this talk I have presented some preliminary results of a work performed in collaboration with M.R.Pennington, which is still in progress. In particular, a proper fit of the more recent experimental data up to about 1.8 GeV [12,13] has still to be completed.

The model we are using requires approximations, to allow the calculation to be performed. The amplitudes are assumed to be pole dominated and so the unitarization chosen is not the most general one. Furthermore, only decays into two pseudoscalars are considered, whereas other thresholds like multipion, vector-vector or axial-pseudoscalar ones are neglected.

Despite the assumptions, this scheme has many great advantages: first of all it gives a *dynamical* description of the particularly complex mechanism which leads to the creation of scalar resonances, naturally taking into account the mixing amongst different states, and explaining why the scalars differ from vector and tensor mesons. Moreover, it is simple and with very few parameters.

Consequently, we believe this model is not a machine from which one can blindly extract numbers from a fitting program, but an efficient schematic way to approach the *dynamics* of the non-perturbative hadronic world: a world in which so much remains to be understood.

Acknowledgments

I am very grateful to M.R. Pennington for guiding and supervising me with great expertise and infinite patience. This contribution would not have been possible without his help and support. I thank M. Anselmino and E. Predazzi for having invited me to Turin for this interesting and entertaining meeting, and D. Lichtenberg for useful discussions and delightful company. Finally, I acknowledge I.N.F.N, Sezione di Torino, for travel support.

References

1. C. Amsler *et al.*, *Phys. Lett.* B **342**, 433 (1995), *Phys. Lett.* B **353**, 571 (1995), *Phys. Lett.* B **355**, 425 (1995).
2. D. Alde *et al.*, *Phys. Lett.* B **201**, 160 (1988).
3. F. Binon *et al.*, *Nuovo Cimento* **80**, 363 (1984).
4. F. Antinori *et al.*, *Phys. Lett.* B **353**, 589 (1995).
5. D. Bugg *et al.*, *Phys. Lett.* B **353**, 378 (1995).
6. M.R. Pennington, Proceedings of 6th International Conference on Hadron Spectroscopy (HADRON 95), Manchester (10-14 Jul 1995), eds. M.C. Birse, G.D. Lafferty and J.A. McGovern (World Scientific, Singapore, 1996), p. 3.
7. J. Sexton *et al.*, *Phys. Rev. Lett.* **75**, 4563 (1995).
8. N.A. Tornqvist, *Z. Phys.* C **68**, 647 (1995).
9. D. Aston *et al.*, *Nucl. Phys.* B **296**, 493 (1988)
10. W. Ochs, Ph.D. thesis, Munich University, 1973 (unpublished); B. Hyams *et al.*, *Nucl. Phys.* B **134**, 1973 ().
11. N.M. Cason *et al.*, *Phys. Rev.* D **28**, 1586 (1986)
12. By R. Kaminski, L. Lesniak, K. Rybicki, INP-1730-PH, Jun 1996, hep-ph/9606362
13. A.A. Kondashov *et al.*, *Phys. Atom. Nucl.* **59**, 1624 (1996).
14. D.V. Bugg *et al.*, *Nucl. Phys.* B **471**, 59 (1996)
15. V.V. Anisovich, A.V. Sarantsev, *Phys. Lett.* B **382**, 429 (1996), V.V. Anisovich, *Phys. Lett.* B **364**, 195 (1995)

Study of Multiquarks Systems in a Chiral Quark Model

M. GENOVESE [a] and J.-M. RICHARD
Institut des Sciences Nucléaires
Université Joseph Fourier–IN2P3–CNRS,
53, avenue des Martyrs, F-38026 Grenoble Cedex, France

S. PEPIN and Fl. STANCU
Université de Liège
Institut de Physique, B.5, Sart Tilman,
B-4000 Liège 1, Belgium

We discuss the stability of multiquark systems within the recent model of Glozman
et al. where the chromomagnetic hyperfine interaction is replaced by pseudoscalar-
meson exchange contributions. In this model the (u, d) diquark $S = 0$ $I = 0$ is
strongly bound (more than for the chromomagnetic interaction) and this leads to
a bound heavy tetraquark $QQ\bar{q}\bar{q}$ where the light quarks are in the $S = 0$ $I = 0$
configuration. We study the stability of other multiquark systems as well.

1 Introduction

The interest for the possible existence of multiquark hadrons has been raised
twenty years ago by Jaffe, who suggested that states of two quarks - two
antiquarks [1] and of six quarks [2] could be bound.

In the following years this problem has been studied within a large variety of models. Some earlier studies in MIT bag indicated the presence of a
dense spectrum of tetraquark states in the light sector [1] (and more generally of
multiquarks [3]). Later on, tetraquark systems have been examined in potential
models [4,5,6,7] and flux tube models [8]. Weinstein and Isgur showed [4] that there
are only a few weakly bound states of resonant meson–meson structure in the
light (u, d, s) sector. Usually the ground state of these systems lies closely
above the lowest $(q\bar{q}) + (q\bar{q})$ threshold. The other extreme result, as compared
to MIT bag model, was obtained by Carlson and Pandharipande [8] in their
flux-tube model with quarks of equal masses, where no bound state was found.

Altogether the theoretical predictions about the existence of these bound
states are still unclear.

As far as the experimental situation is concerned, there are some candidates for non-$q\bar{q}$ states, but data are not yet conclusive. Furthermore, even
if a resonance would be clearly identified as an exotic (not $q\bar{q}$ or qqq) state,
nevertheless a careful phenomenological study would be necessary in order to

---

[a]Supported by the EU Program ERBFMBICT 950427

understand its internal structure among many different possibilities (multi-quarks, hybrids, glueballs,...).

The description of these candidates and of the phenomenological properties which permit a distinction between different exotics is beyond the purposes of this proceeding and we refer to Ref.s [10,11] and the last issue of Review of Particle Properties [12] for further details.

Anyway, a well-established theoretical result is that systems with a larger mass difference among their components are more easily bound [9,5]. For example, for a system of two heavy quarks and two light antiquarks $QQ\bar{q}\bar{q}$ ($Q = c$ or b, $q = u$, d or s) stability can be achieved without spin–spin interaction, provided the mass ratio $m(Q)/m(q)$ is larger [5] than about 15, which means that Q must be a b-quark.

Recently, the interest for multiquark systems containing charm quarks has grown, considering that new experiments are being planned at Fermilab and CERN, to search for new hadrons and in particular for doubly charmed tetra-quarks [13,14,15]

In this context, we have carried out [16] a study of the $QQ\bar{q}\bar{q}$ system in the framework of the chiral quark model of Glozman *et al.* [18,19]. This model is somehow quite "extreme" because it includes meson-exchange forces between quarks and entirely neglects the chromomagnetic interaction. However it permits a very good description of the baryon spectrum, it is thus worth testing it in further predictions.

Considering that in this model light-quark mesons are "quasiparticles" with a spectrum which must be assumed and cannot be evaluated directly (this is of course rather an unpleasant feature of the model), multiquark systems represent an obvious testing ground for it. Other possible tests could come from a careful analysis of the spectrum of charmed baryons [17].

2 The Glozman model

Before presenting our results about multiquarks, let us briefly consider the Glozman *et al.* model and compare it with other potential models used in hadron spectroscopy.

In a general Hamiltonian, which would approximate the low energy limit of QCD, one can introduce both a chromomagnetic interaction and a meson-exchange contribution, obtaining an explicit form as

$$H = \sum_i \frac{\vec{p}_i^{\,2}}{2m_i} - \frac{3}{16} \sum_{i<j} \tilde{\lambda}_i^c \cdot \tilde{\lambda}_j^c \, V_{\mathrm{conf}}(r_{ij})$$

$$-\sum_{i<j} \tilde{\lambda}_i^c \cdot \tilde{\lambda}_j^c \, \vec{\sigma}_i \cdot \vec{\sigma}_j \, V_g(r_{ij}) - \sum_{i<j} \tilde{\lambda}_i^F \cdot \tilde{\lambda}_j^F \, \vec{\sigma}_i \cdot \vec{\sigma}_j \, V_F(r_{ij}), \tag{1}$$

where m_i is the constituent mass of the quark located at $\vec{r}_i$; $r_{ij} = |\vec{r}_j - \vec{r}_i|$ denotes the interquark distance; $\vec{\sigma}_i$, $\tilde{\lambda}_i^c$, $\tilde{\lambda}_i^F$ are the spin, colour and flavour operators, respectively. Spin-orbit and tensor components may supplement the above spin-spin forces for studying orbital excitations (they give no contribution for $L = 0$ systems). The potential in (1) has three parts containing the confining, the chromomagnetic and the meson-exchange contribution.

Usually, the confining term V_{conf} is assumed to include a Coulomb plus a linear term,

$$V_{\text{conf}} = -\frac{a}{r} + br + c. \tag{2}$$

In the following, we shall either use the very weak linear potential of Glozman et al. [19] corresponding to

$$(C_1) \qquad a = c = 0, \qquad \text{and} \qquad b = 0.01839 \, \text{GeV}^2, \tag{3}$$

or the more conventional choice

$$(C_2) \qquad a = 0.5203, \qquad b = 0.1857 \, \text{GeV}^2, \qquad c = -0.9135 \, \text{GeV}, \tag{4}$$

which has already been applied to the study of tetraquarks by Silvestre-Brac and Semay [6].

The term $\sum_{i<j} \tilde{\lambda}_i^c \cdot \tilde{\lambda}_j^c \, \vec{\sigma}_i \cdot \vec{\sigma}_j \, V_g(r_{ij})$ is the chromomagnetic analogue of the Breit–Fermi term of QED. In the interaction between a quark and an antiquark (as e.g. in a meson) one finds $\tilde{\lambda}_1^c \cdot \tilde{\lambda}_2^c = -16/3$. Then a positive V_g shifts each vector meson above its pseudoscalar partner, for instance $D^* > D$ in the charm sector. For baryons, where $\tilde{\lambda}_1^c \cdot \tilde{\lambda}_2^c = -8/3$ for each quark pair, such a positive V_g pushes the spin $3/2$ ground states up, and the spin $1/2$ down, for instance $\Delta > N$. For the radial shape, as an example, we mention

$$V_g = \frac{a}{m_i m_j d^2} \frac{\exp -r/d}{r}, \tag{5}$$

which was used in Ref. [6], with the same value of a as in Eq. (4) and $d = 0.454 \, \text{GeV}^{-1}$.

Finally, the last term of H corresponds to meson exchange, and an explicit sum over F is understood. If the system contains light quarks only (as in Ref.s [18,19]), the sum over F runs from 0 to 8, i.e. over the members of the $J^{PC} = 0^{-+}$ nonet, which represent the Goldstone bosons of the spontaneusly broken $SU(3)_A$ symmetry $(1-3 \to \pi, 4-7 \to K, 8 \to \eta$ and $0 \to \eta')$. If a heavy flavour is incorporated, a phenomenological extension from SU(3)$_F$ to SU(4)$_F$ would further extend the sum to F = 9 − 12 corresponding to a D-exchange, F = 13 − 14 to a D_s-exchange and F = 15 to an η_c-exchange (of course in this case the interpretation as Goldstone bosons is not really possible). Similar terms should then be introduced if one includes the beauty sector as well. The radial form of $V_F \neq 0$ is derived from the usual pion-exchange potential which contains a long-range part and a short-range one

$$\sum_{i<j} \vec{\tau}_i \cdot \vec{\tau}_j \, \vec{\sigma}_i \cdot \vec{\sigma}_j \frac{g^2}{4\pi} \frac{1}{4m^2} \left[\mu^2 \frac{\exp(-\mu r_{ij})}{r_{ij}} - 4\pi\delta^{(3)}(r_{ij}) \right], \tag{6}$$

where μ is the pion mass.

When constructing NN forces from meson exchanges, one usually disregards the short-range term in Eq. (6), for it is hidden by the hard core, and anyhow the potential in that region is parameterized empirically. For example, when Törnqvist[21], Manohar and Wise[22] or Ericson and Karl[23] considered pion exchange in multiquark states, they used the Yukawa term $\exp(-\mu r)/r$ acting between two well-separated quark clusters. Weber et al. [24], in their model with hyperfine plus pion-exchange interaction, studied both the cases with or without delta-term, showing that good results could be obtained without it in baryon spectroscopy. Thus the relevance of an ad-hoc regularized delta-term [18,19] in the study of baryon spectroscopy is somehow surprising. Nevertheless, this ansatz permits to give a good description of the baryon spectrum and, in particular, allows one to solve the problem of the ordering of the lowest parity-odd and parity-even states of N, Λ and Σ resonances, which did not find a solution in conventional chromomagnetic models[20].

Incidentally, this result is not completely unexpected, in fact Buchman et al. had shown, already some years ago, that this term is essential for obtaining a good description of magnetic moments[26].

For the sake of completeness we report here the explicit form of the regularized delta-term[19]

$$V_\mu = \Theta(r - r_0)\mu^2 \frac{\exp(-\mu r)}{r} - \frac{4\epsilon^3}{\sqrt{\pi}} \exp(-\epsilon^2(r - r_0)^2), \tag{7}$$

where $r_0 = 2.18$ GeV^{-1}, $\epsilon = 0.573$ GeV, and $\mu = 0.139$ GeV for π, 0.547 GeV for η and 0.958 GeV for η'. Furthermore, the Yukawa-type part is cut off for $r \leq r_0$. This smearing should account for the fact that both the pseudoscalar mesons and the constituent quarks have a finite size and that boson fields cannot be described by a linear equation near their source.

The explicit form of the Hamiltonian which extends the results of Ref. [18] from SU(3) to SU(4) is given in Ref. [16]. It includes also the exchange of D, D_s

and η_c mesons. However, considering that little $u\bar{u}$ or $d\bar{d}$ mixing is expected in η_c, this contribution can be neglected when considering systems with none or one charm quark. Moreover when the meson mass μ reaches values of a few GeV as for D or η_c the two terms in Eq. (6) basically cancel each other and one recovers practically the SU(3) form [18]. This is in agreement with Ref. [25] where it has been explicitly shown that the dominant contribution to the Σ_c and Σ_c^* masses is due to meson exchange between light quarks and the contribution of the matrix elements with the D (D_s) and D^* (D_s^*) quantum numbers (which are evaluated phenomenologically, fitting the mass difference $\Sigma_c - \Lambda_c$) play a minor rôle. In the following numerical calculations we will neglect the exchange of heavy mesons.

Finally one has to fix quark masses. The light quarks ones are fixed to 0.34 GeV according to Ref.s [19,6]. The heavy quark masses $m_Q = m_c$ and m_b are adjusted to reproduce the experimental average mass $\overline{M} = (M + 3M^*)/4$ between the 0^- and 1^- $M = D$ or B mesons by a variational calculation, where a trial wave function of type $\phi \propto \exp(-\alpha r^2/2)$ is used (with α as a variational parameter). It has been checked that the error never exceeds a few MeV with respect to the exact value. The variational approximation is retained for consistency with the treatment of 3-, 4- and 6-body systems discussed below. This leads to $m_c = 1.35$ GeV and $m_b = 4.66$ GeV for the potential C_1 and $m_c = 1.87$ GeV and $m_b = 5.259$ GeV for C_2.

Before proceeding further, let us briefly discuss the calculation of baryons masses in the model of Glozman et $al.$ The explicit form of the Hamiltonian integrated in the spin–flavour space is :

$$H = H_0 + \frac{g^2}{48\pi m^2} \begin{cases} 15V_\pi - V_\eta - 2\left(g_0/g\right)^2 V_{\eta'} & \text{for} \quad N \\ 3V_\pi + V_\eta + 2\left(g_0/g\right)^2 V_{\eta'} & \text{for} \quad \Delta \end{cases} \tag{8}$$

with

$$H_0 = 3m + \sum_i \frac{\vec{p}_i^{\,2}}{2m} + \frac{b}{2} \sum_{i<j} r_{ij}, \tag{9}$$

where $g^2/4\pi = 0.67$ (which leads to the usual strength $g_{\pi NN}/4\pi \simeq 14$ for the Yukawa tail of the nucleon–nucleon potential) and $(g_0/g)^2 = 1.8$. We have performed variational estimates with a wave function $\phi \propto \exp(-\alpha(\rho^2 + \lambda^2)/2)$, where $\vec{\rho} = \vec{r}_2 - \vec{r}_3$, $\vec{\lambda} = (2\vec{r}_1 - \vec{r}_2 - \vec{r}_3)/\sqrt{3}$: our results agree with the more elaborated Faddeev calculations of Ref. [19].

When the meson–exchange terms are switched off, the N and Δ ground states are degenerate at 1.63 GeV. Introducing the coupling the nucleon mass drops stronger leading to a reasonable splitting (≈ 0.3 GeV). It is interesting to notice that in this model one has an attraction both for the $S = 0$, $I = 0$

and also, albeit smaller, for the $S = 1$, $I = 1$ diquarks. Then both the nucleon and Δ masses decrease when the interaction is turned on. The diquarks $S = 0$, $I = 1$ and $S = 1$, $I = 0$ on the contrary are repulsive configurations. The situation is thus different from the chromomagnetic case where one has a smaller attraction for the $S = 0$, $I = 0$ diquark and repulsion for the $S = 1$, $I = 1$ one.

We have also calculated the ground state of cqq baryons using a trial wave function $\phi \propto \exp(-(\alpha \rho^2 + \beta \lambda^2)/2)$ and found $\Lambda_c = 2.32$ GeV and $\Sigma_c = \Sigma_c^* = 2.48$ GeV, close to the experimental values and consistent with the findings of Ref.[25], although the Hamiltonian, its treatment, and the input parameters are somewhat different there.

3 Evaluation of multiquarks masses

Due to arguments at the beginning of this contribution, here we discuss tetra-quarks containing heavy flavours, i.e. $QQ\bar{q}\bar{q}$, studying the most favourable configuration: $\bar{3}3$, $S = 1$, $I = 0$. This means that QQ is in a $\bar{3}$ colour state and $\bar{q}\bar{q}$ in a 3 colour state. The mixing with $6\bar{6}$ is neglected because one expects this to play a negligible rôle in deeply-bound heavy systems[5]. Then the Pauli principle requires $S_{12} = 1$ for QQ, and $S_{34} = 0$, $I_{34} = 0$ for $\bar{q}\bar{q}$ (which we have seen to be the diquark with the largest binding), if the relative angular momenta are zero for both subsystems. This gives a state of total spin $S = 1$ and isospin $I = 0$.

The tetraquark Hamiltonian integrated in the colour–spin–flavour space, and incorporating the approximations discussed in the former section, reduces to

$$H = 2(m + m_Q) + \frac{\vec{\mathrm{p}}_x^2}{m_Q} + \frac{\vec{\mathrm{p}}_y^2}{m} + \frac{m + m_Q}{2 m m_Q}\vec{\mathrm{p}}_z^2 + \sum_{i<j} V_{ij}, \tag{10}$$

where

$$V_{12} = \frac{1}{2}\left(-\frac{a}{r_{12}} + b\, r_{12} + c\right),$$

$$V_{ij} = \frac{1}{4}\left(-\frac{a}{r_{ij}} + b\, r_{ij} + c\right), \qquad i = 1 \text{ or } 2,\ j = 3 \text{ or } 4, \tag{11}$$

$$V_{34} = \frac{1}{2}\left(-\frac{a}{r_{34}} + b\, r_{34} + c\right) + 9 V_\pi - V_\eta - 2 V_{\eta'}.$$

The momenta $\vec{\mathrm{p}}_x$, etc., are conjugate to the relative distances $\vec{\mathrm{x}} = \vec{\mathrm{r}}_1 - \vec{\mathrm{r}}_2$, $\vec{\mathrm{y}} = \vec{\mathrm{r}}_3 - \vec{\mathrm{r}}_4$, and $\vec{\mathrm{z}} = (\vec{\mathrm{r}}_1 + \vec{\mathrm{r}}_2 - \vec{\mathrm{r}}_3 - \vec{\mathrm{r}}_4)/\sqrt{2}$. The wave function is parameterized as

$$\psi \propto \exp[-(\alpha x^2 + \beta y^2 + \gamma z^2)/2], \tag{12}$$

and the minimization with respect to α, β and γ shows that both the $cc\bar{q}\bar{q}$ and $bb\bar{q}\bar{q}$ systems are bound whatever is the potential, (C_1) or (C_2), provided meson exchange is incorporated. This is in contradistinction to previous studies based on conventional models where the flavour-independent confining potential is supplemented by one gluon exchange. For example the authors of Ref.[6] found that the $cc\bar{q}\bar{q}$ state is about 20MeV above threshold (while the $bb\bar{q}\bar{q}$ is bound of 135 MeV). A similar situation is also obtained in Ref. [27], where studying the four quarks states in a diquark model only the $bb\bar{q}\bar{q}$ is found to be bound (by 50MeV, while $cc\bar{q}\bar{q}$ is 30 MeV above the threshold).

Quantitatively in our model we find that using the confining potential C_1 plus meson exchange the double charmed tetraquark is bound by 185 MeV and the double bottom one by 226 MeV. Using the potential C_2 the results are nearly two times larger (332 and 497 MeV respectively) and also much more different from each other. The reason is that (C_2) contains a Coulomb part which binds more, heavier is the system, leading thus to a larger separation among levels as well. This is related to the fact that the potential (C_2) has been fitted to reproduce the J/Ψ and the Υ meson masses (anyway it also gives overall good results both for other mesons and baryons) while, by construction [18,19], the potential (C_1) was designed and fitted to light baryons only. Altogether, our results shows that the prediction of binding for the $cc\bar{q}\bar{q}$ is substantially independent on the choice of the confining potential.

Considering this result one could rise the question if in the model of Glozman *et al.* a proliferation of multiquark systems appears. We have therefore tried to investigate $QQqqqq$ and q^6 systems as well.

As a general procedure, for a given multiquark system, one searches for the spin-isospin wave functions corresponding to a colour singlet, then selects the most favourable configuration. The contribution of global spin-flavour-averaged interaction is reduced to the calculation of the matrix elements of the two body operator $(\vec{\sigma}_i\cdot\vec{\sigma}_j)(\vec{\tau}_i\cdot\vec{\tau}_j)$; this is accomplished by using Clebsh–Gordan coefficients of the permutation group according to Ref.s [28,29].

Let us begin with $QQqqqq$. In this case the most favourable configuration is the one where the light quark subsystem has $S = 1$, $I = 0$, which leads to the spin-flavour-averaged interaction

$$\langle V \rangle = 10V_\pi - 2/3V_\eta - 4/3(g_0/g)^2 V_{\eta'}. \tag{13}$$

In the numerical calculation we have used the variational Gaussian wave function

$$\Psi \propto \exp[-(\alpha x^2 + \beta y^2 + \gamma(u^2 + v^2 + w^2))] \tag{14}$$

where appear the Jacobi variables $\vec{x} = \vec{r}_1 - \vec{r}_2$, $\vec{y} = \vec{R}_q - \vec{R}_Q$, in terms of $\vec{R}_q = (\vec{r}_1 + \vec{r}_2)/2$ and $\vec{R}_Q = (\vec{r}_3 + \vec{r}_4 + \vec{r}_5 + \vec{r}_6)/4$, $\vec{u} = (\vec{r}_3 + \vec{r}_4 - \vec{r}_5 - \vec{r}_6)/\sqrt{2}$,

176

$\vec{v} = (\vec{r}_3 - \vec{r}_4 + \vec{r}_5 - \vec{r}_6)/\sqrt{2}$, $\vec{w} = (\vec{r}_3 - \vec{r}_4 - \vec{r}_5 + \vec{r}_6)/\sqrt{2}$, where indices $1, 2$ refer to the heavy quarks and $3, 4, 5, 6$ to the light ones.

The result of the numerical calculation is that this potential is largely insufficient to bind both the $ccqqqq$ (e.g. of 500 MeV with the potential C_1) and the $bbqqqq$ systems. Incidentally, at this workshop results about heavy hexaquarks were presented also by Lichtenberg et al. [30], in their diquark model. Albeit they do not treat directly the case $ccqqqq$, they find that the system $csqqqq$ is unbound, while the one containing a beauty quark instead of the charm is bound.

For the sake of completeness we have also made a study of the q^6 system. The most favourable configuration is $S = 1$, $I = 0$, leading to

$$\langle V \rangle = 11V_\pi - 5/3V_\eta - 10/3(g_0/g)^2 V_{\eta'} \tag{15}$$

Our result shows that, also in this case, the system is largely insufficiently bound (by almost 1.5GeV with the potential C_1) for being under the two baryons threshold.

4   Conclusions

In conclusion, considering the success of the Glozman et al. model in describing the baryon spectrum, we have searched for new possible tests of this model. Unluckily the model does not permit an analysis of mesons (light-quark mesons at least), which are interpreted as "quasiparticles" related to the breaking of flavour $SU(3)$. Further tests of it would then involve multiquarks systems. We have thus studied four and six-quark systems involving two heavy quarks, showing that the Glozman et al. model leads to predictions for the $cc\bar{q}\bar{q}$ which differ from the ones of more conventional models. In fact this system is found to be strongly bound. The search of such a resonance in forthcoming experiments will therefore be a good test for understanding the dynamics which produces the hadron spectrum.

We have also found that the $bb\bar{q}\bar{q}$ state is bound, but this result is obtained also in other models and the possibility to observe this resonance is relegated to a more remote future.

Finally we have found that the six quark system with two heavy quarks is unbound, also this result does not differ from the one of other more conventional models. However, together with the outcome that also the q^6 system is unbound, it is a useful indication that in the framework of the model under investigation there is no proliferation of multiquark bound states.

Acknowledgements

We thank D.B. Lichtenberg for enlighting discussions and the organizers for the stimulating atmosphere of this workshop.

1. R.L. Jaffe, *Phys. Rev.* D 15, 267 (1977); *Phys. Rev.* D 17, 1444 (1978).
2. R.L. Jaffe, *Phys. Rev. Lett.* 38, 195 (1977).
3. A. T. M. Aerts *et al.*, *Phys. Rev.* D 17, 260 (1977).
4. J. Weinstein and N. Isgur, *Phys. Rev.* D 27, 588 (1983), *Phys. Rev.* D 41, 2236 (1990).
5. S.Zouzou, B. Silvestre-Brac, C. Gignoux and J.-M. Richard, *Z. Phys.* C 30, 457 (1986).
6. B. Silvestre Brac and C. Semay, *Z. Phys.* C 59, 457 (1993)); *Z. Phys.* C 61, 271 (1994).
7. D.M. Brink and Fl. Stancu, *Phys. Rev.* D 49, 4665 (1994).
8. J. Carlson and V.R. Pandharipande, *Phys. Rev.* D 43, 1652 (1991).
9. J.P. Ader, J.-M. Richard and P. Taxil, *Phys. Rev.* D 25, 2370 (1982).
10. G. Karl, *Int. J. of Mod. Phys.* E1, 491 (1992); *Nucl. Phys.* A 558, 113c (1993); N.A. Törnqvist, Proc. of "Int. Europh. Conf. on High Energy Phys.", Brussels (Belgium), edit. J. Lemonne et al., 84 (1995); G. Landsberg, *Phys. of Atom. Nucl.* 57, 42 (1994).
11. M. Genovese, *Nuovo Cimento* A 107, 1249 (1994) and references therein.
12. Particle Data Group, *Phys. Rev.* D 54, 1 (1996).
13. M.A. Moinester, *Z. Phys.* A 355, 349 (1996).
14. D.M. Kaplan, Proc. Int. Workshop "Production and Decay of Hyperons, Charm and Beauty Hadrons", Strasbourg (France) September 5–8, 1995.
15. COMPASS Collaboration (G. Baum et al.), CERN–SPSLC–96-14, March 1996.
16. S. Pepin, Fl. Stancu, M. Genovese and J.-M.Richard, hep-ph 9609348, ISN-96.99, to be published in *Phys. Lett.* B.
17. S. Pepin, Fl. Stancu, M. Genovese, J.-M.Richard and S. Zouzou, work in progress.
18. L.Ya. Glozman and D.O. Riska, *Phys. Rep.* 268, 263 (1996).
19. L.Ya. Glozman, Z. Papp and W. Plessas, *Phys. Lett.* B 381, 311 (1996).
20. L. A. Copley, N. Isgur and G. Karl, *Phys. Rev.* D 20, 768 (1979); S. Capstick and N. Isgur, *Phys. Rev.* D 34, 2809 (1986); C. S. Kalman and D. Pfeffer, *Phys. Rev.* D 27, 1648 (1983).
21. N. Törnqvist, *Phys. Rev. Lett.* 67, 556 (1991); *Z. Phys.* C 61, 525 (1994).

22. A.V. Manohar and M.B. Wise, *Nucl. Phys.* B 399, 17 (1993).

23. T.E.O. Ericson and G. Karl, *Phys. Lett.* B 309, 426 (1993).

24. M. Weyrauch and H.J. Weber, *Phys. Lett.* B 171, 13 (1986); H.J. Weber and H.T. Williams, *Phys. Lett.* B 205, 118 (1988).

25. L.Ya. Glozman and D.O. Riska, *Nucl. Phys.* A 603, 326 (1996).

26. A. Buchman *et al.*, *Nucl. Phys.* A 569, 661 (1994).

27. D. B. Lichtenberg, R. Roncaglia and E. Predazzi, IUHET-344, published in these proceedings.

28. Fl. Stancu, Group Theory in Subnuclear Physics, Oxford University Press, 1996, chapter 4;

29. S. Pepin and Fl. Stancu Preprint ULG-PNT-96-1-J.

30. D. B. Lichtenberg, R. Roncaglia and E. Predazzi, DFTT 65/96, published in these proceedings.

HEAVY BARYONS IN A QUARK-DIQUARK-PICTURE

D. EBERT

Institut für Physik, Humboldt–Universität zu Berlin,
Invalidenstr.110, D-10115 Berlin, Germany

R. N. FAUSTOV, V. O. GALKIN

Russian Academy of Sciences, Scientific Council for Cybernetics,
Vavilov Street 40, Moscow 117333, Russia

The mass spectrum of baryons with two heavy quarks is calculated within a quark-diquark picture. The quasipotentials for interactions of two quarks and of a quark with a scalar and axial vector diquark are evaluated. The bound state masses of baryons with $J^P = \frac{1}{2}^+, \frac{3}{2}^+$ are computed.

1 Introduction

In this talk we present new results for the mass spectrum of heavy baryons obtained on the basis of a quark-diquark-picture. To calculate the meson and baryon mass spectra QCD sum rules [1], potential models [2,3,4,5], lattice QCD [6], heavy quark effective theory (HQET) [7,8], the method of vacuum correlators [9] and some other approaches are widely used. Presently at the LHC, B-factories and the Tevatron with high luminosity, several experiments have been proposed, in which a detailed study of heavy baryons containing two heavy quarks can be performed. The possible quark composition of such baryons looks as follows: (ccq, cbq, bbq), where c and b are heavy quarks (Q) and q denotes a light u, d and s quark. In this connection theoretical predictions for heavy baryon masses acquire important significance. The properties of heavy hadrons (containing b and/or c quarks) are essentially different from those of light hadrons composed from u, d and s quarks. This difference originates from the fact that the heavy quark mass strongly exceeds the scale of QCD interaction, $\Lambda_{QCD} \approx 300 - 400$ MeV, namely $m_{c,b} \gg \Lambda_{QCD}$. Baryons, containing two heavy quarks (QQq), may be considered as a localized source (QQ) of the colour field, in which the light quark moves. Interaction forces between heavy quarks then lead to the formation of a two-particle bound state of a QQ-diquark, the scale of which is determined by the quantity $1/m_Q$ which is small compared to the QCD scale $1/\Lambda_{QCD}$. Thereby it seems justified to treat the heavy diquark as a pointlike object with definite colour, spin and mass and to ignore the configuration with a qQ-diquark which size is of order $1/m_q \sim 1/\Lambda_{QCD}$. The baryon formation then occurs as a result of a

180

diquark interaction with the light quark. Using a quark-diquark-picture we have calculated the mass spectrum of heavy baryons on the basis of a local Schrödinger-like quasipotential equation [10]:

$$\left(\frac{b^2(M)}{2\mu_R} - \frac{\mathbf{p}^2}{2\mu_R} \right) \psi_M(\mathbf{p}) = \int \frac{d^3q}{(2\pi)^3} V(\mathbf{p}, \mathbf{q}, M) \psi_M(\mathbf{q}), \tag{1}$$

where the relativistic reduced mass is

$$\mu_R = \frac{E_1 E_2}{E_1 + E_2} = \frac{M^4 - (m_1^2 - m_2^2)^2}{4M^3}, \tag{2}$$

$$E_1 = \frac{M^2 - m_2^2 + m_1^2}{2M}, \quad E_2 = \frac{M^2 - m_1^2 + m_2^2}{2M}, \quad E_1 + E_2 = M,$$

and the center of mass system relative momentum squared on the mass shell reads

$$b^2(M) = \frac{[M^2 - (m_1 + m_2)^2][M^2 - (m_1 - m_2)^2]}{4M^2}, \tag{3}$$

with $m_{1,2}$ the masses of the constituent particles. In refs. [11,12] the operator of quark-antiquark interaction has been constructed and the mass spectra and decay rates of mesons have been investigated. The relativistic quasipotential quark model gives in the meson sector results which nicely agree with experimental data. Relativistic effects play an important role in describing properties of quark bound states and may be consistently taken into account within the quasipotential method.

2 Diquarks

Diquarks represent by themselves two-particle clusters, which are produced inside the three-quark systems as a result of quark interactions in which the spin-dependent forces play an important role [13,14]. There arise attractive forces between two quarks if the quarks are in the antisymmetric colour state. The interaction potential at large distances in the quark-quark system is similar to that in the quark-antiquark system since the baryon size is practically the same as the meson size. In the ground state the diquarks are two-particle bound states of quarks in an antisymmetric colour state with zero angular momentum and definite flavour and spin. In the case of identical quarks the diquark has spin $S = 1$, whereas for quarks of different flavours the state with spin $S = 0$ is also possible.

Now we construct the interaction quasipotential for two heavy quarks Q, which enters in eq. (1). The interaction operator in the QQ system contains a perturbative part, which is defined at small distances in the quasipotential approach by the one-gluon exchange amplitude and can be evaluated within QCD. Moreover there is a nonperturbative term in the operator $V(\mathbf{p}, \mathbf{q}, M)$, which cannot be obtained consistently within QCD and leads to the quark confinement at large distances.

The quasipotential $V(\mathbf{p}, \mathbf{q}, M)$ of eq. (1) is constructed with the help of the off-mass-shell scattering amplitude, projected onto the positive energy states. Then the perturbative part of the quasipotential can be represented in the form:

$$V_{QQ}^{pert}(\mathbf{p}, \mathbf{q}) = \bar{u}_1(\mathbf{p})\bar{u}_2(-\mathbf{p}) \frac{\frac{2}{3}\alpha_s}{4\sqrt{\epsilon_1(\mathbf{p})\epsilon_1(\mathbf{q})\epsilon_2(\mathbf{p})\epsilon_2(\mathbf{q})}} D_{\mu\nu}(k)\gamma_1^\mu \gamma_2^\nu u_1(\mathbf{q})u_2(-\mathbf{q}),$$

(4)

where $D_{\mu\nu}(k)$ is the gluon propagator conveniently taken in the Coulomb gauge, $\mathbf{k} = \mathbf{p} - \mathbf{q}$, and $u(\mathbf{p})$ is the Dirac spinor.

In the approximation of one-gluon exchange the quark-quark potential in a baryon is equal to one half of the quark-antiquark potential in a meson. Calculations based on the baryon Wilson loop area law in QCD [15] show that this relation holds even true for nonperturbative interactions. Thus, while constructing the long-range nonperturbative part of the quark interaction operator we used the known "1/2 rule", i. e. $V_{qq} = \frac{1}{2}V_{q\bar{q}}$ [5] and chose this part of the quasipotential in the standard linearly rising in r form. As a result in this approximation, neglecting corrections $O(v^2/c^2)$, the complete interaction operator in the quark-quark system is chosen in the form:

$$V_{QQ}(r) = -\frac{2}{3}\frac{\alpha_s}{r} + \frac{1}{2}(Ar + B) + \frac{4\pi\alpha_s}{9m_1 m_2}(\sigma_1\sigma_2)\delta(\mathbf{r}), \qquad (5)$$

where the parameters $A = 0.18$ GeV^2 and $B = -0.3$ GeV had been fixed previously [11,12] in calculating the meson mass spectrum. The numerical solution of the quasipotential equation (1) with interaction operator (5) yields the mass spectrum of the s-wave axial vector and scalar diquarks. The results are presented [a] in Table 1. Here the usual values of quark masses (the same as for mesons) were used [11,12]:

$$m_u = m_d = 0.33 \text{ GeV}, \ m_s = 0.5 \text{ GeV}, \ m_c = 1.55 \text{ GeV}, \ m_b = 4.88 \text{ GeV}.$$

3 Interaction of a quark with a scalar diquark

Now to calculate the baryon mass spectrum in the quark-diquark approximation it is necessary to construct the quark-diquark interaction operator. In

---

[a]The masses of light diquarks are presented for completeness where $q = u, d$.

Table 1: Masses of s-wave scalar (S) and axial vector (AV) diquarks (in GeV).

Diquark	qq	qs	qc	qb	ss	sc	sb	cc	cb	bb
M_S	0.85	1.04	2.08	5.39	-	2.27	5.57	-	6.52	-
M_{AV}	1.02	1.15	2.12	5.41	1.32	2.29	5.58	3.26	6.52	9.79

the case of a scalar diquark the perturbative part of the quasipotential is described by the Feynman diagram in Fig. 1. The corresponding expression for

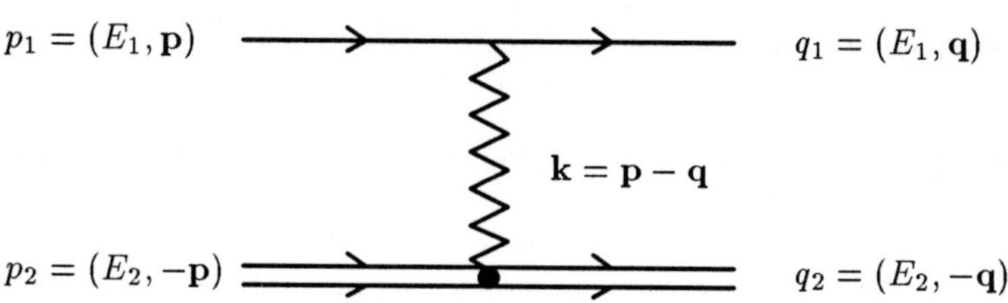

Figure 1: Feynman diagram of the one-gluon interaction in the quark-diquark system.

the interaction operator has the form:

$$V_{q+SD}^{pert}(\mathbf{p}, \mathbf{q}) = \frac{\frac{4}{3}g_s^2}{4\sqrt{\epsilon_1(\mathbf{p})\epsilon_1(\mathbf{q})\epsilon_2(\mathbf{p})\epsilon_2(\mathbf{q})}}(p_2 + q_2)^\mu D_{\mu\nu}(\mathbf{k})\bar{u}(\mathbf{q})\gamma^\nu u(\mathbf{p}), \quad (6)$$

where the interaction vertex of a scalar diquark with a gluon is given by the factor $ig_s T_{ij}^a (p_2 + q_2)^\mu$. Using the standard representation of the Dirac spinors, we get for V_{q+SD}^{pert} in configuration space up to terms $O(v^2/c^2)$ the following expression:

$$\begin{aligned}
V_{q+SD}^{pert}(r) &= -\frac{4\alpha_s}{3r} - \frac{4\alpha_s}{3m_1 m_2}\mathbf{p}\frac{1}{r}\mathbf{p} + \frac{2\pi\alpha_s}{3m_1^2}\delta(\mathbf{r}) \\
&+ \frac{2\alpha_s}{3r^3}\frac{\mathbf{L}^2}{m_1 m_2} + \frac{4\alpha_s}{3m_1 m_2}\left(\frac{m_2}{2m_1} + 1\right)\frac{\mathbf{L}\mathbf{S}_1}{r^3},
\end{aligned} \quad (7)$$

where $\mathbf{S}_1$ is a quark spin operator.

Then we add to this expression the confining potential

$$V_{q+SD}^{nonpert}(r) = V_{q+SD}^{conf(S)}(r) + V_{q+SD}^{conf(V)}(r), \quad (8)$$

where $V_{q+SD}^{conf(S)}$ and $V_{q+SD}^{conf(V)}$ are Lorentz scalar and vector confining potentials. We also assume that the long-range vector interaction vertex with quark
contains the Pauli term as in the case of mesons [11,12]:

$$\Gamma_\mu(\mathbf{k}) = \gamma_\mu + \frac{i\kappa}{2m}\sigma_{\mu\nu}k^\nu, \tag{9}$$

κ is the Pauli interaction constant.

Carrying out the similar calculations as for the perturbative part we get
for the nonperturbative part up to terms $O(v^2/c^2)$ the expression

$$V_{q+SD}^{conf(S)}(r) = V^S(r) - \frac{1}{2m_1^2}\mathbf{p}V^S(r)\mathbf{p} + \frac{1}{8m_1^2}\Delta V^S(r), \tag{10}$$

$$\begin{aligned}
V_{q+SD}^{conf(V)}(r) = {} & V^V(r) - \frac{1}{4m_1}\left(\frac{1}{2\mu} + \frac{1}{2m_2} - \frac{1+\kappa}{m_1}\right)\Delta V^V(r) \\
& + \frac{1}{m_1 m_2}\mathbf{p}V^V(r)\mathbf{p} + \frac{1}{m_1}\left(\frac{1+\kappa}{\mu} - \frac{1}{2m_1}\right)\frac{V'^V(r)}{r}\mathbf{L}\mathbf{S}_1,
\end{aligned} \tag{11}$$

where m_2 is the scalar diquark mass, m_1 is the light quark mass, μ is their
reduced mass and the prime denotes differentiation with respect to r. The
scalar and vector confining potentials reduce in the nonrelativistic limit to

$$V^S(r) = \varepsilon(Ar + B), \qquad V^V(r) = (1 - \varepsilon)(Ar + B), \tag{12}$$

reproducing $V_{nonrel}^{conf} = V^S + V^V = Ar + B$, where ε is the mixing coefficient.
We use the values $\kappa = -1$ and $\varepsilon = -1$ found from the analysis of meson-
s [11,12], and we take $B = -0.39$ GeV. The complete quasipotential, which is
used to calculate the mass spectrum of baryons with $J^P = \frac{1}{2}^+$ is the sum of
perturbative and nonperturbative parts:

$$V_{q+SD}^{tot} = V_{q+SD}^{pert} + V_{q+SD}^{nonpert}. \tag{13}$$

The numerical results for the mass spectrum of the ground states of baryons
with $J^P = \frac{1}{2}^+$ on the basis of eqs. (1) and (13) are given in Table 2.

4   Interaction of a quark with an axial vector diquark

As in the previous section, the perturbative part of the interaction operator for
spin 1/2 and spin 1 particles is given by the diagram in Fig. 1. It includes the
matrix element of the chromomagnetic current operator between states of the
axial vector diquark. This matrix element is defined by three chromomagnetic

184

form factors. We neglect the chromoquadrupole moment of the axial vector particle. As a result the one-gluon exchange quasipotential takes the form:

$$V_{q+AD}^{pert}(\mathbf{p}, \mathbf{q}) = \frac{\frac{4}{3}g_s^2}{4\sqrt{\epsilon_1(\mathbf{p})\epsilon_1(\mathbf{q})\epsilon_2(\mathbf{p})\epsilon_2(\mathbf{q})}} D_{\mu\nu}(\mathbf{k})\bar{u}(\mathbf{q})\gamma^\nu u(\mathbf{p}) \tag{14}$$

$$\times\left[G_1(k^2)(p_2+q_2)^\mu + G_2(k^2)[\frac{1}{m_2}W^\mu(\mathbf{p}_2)(\mathbf{S}_2\boldsymbol{\Delta}_2) - \frac{1}{m_2}(\mathbf{S}_2\boldsymbol{\Delta}_2)W^\mu(\mathbf{p}_2)]\right],$$

$$\boldsymbol{\Delta}_2 = -\mathbf{q} + \frac{\mathbf{p}}{m_2}\left(E_2 - \frac{\mathbf{q}\mathbf{p}}{E_2+m_2}\right),$$

where $\mathbf{S}_2$ is the spin operator of a vector particle, W^μ is the four vector of the relativistic spin (the Pauli-Lubansky vector). The chromomagnetic form factor of the axial vector particle $G_2(k^2)$ at $k^2 = 0$ defines its total chromomagnetic moment: $G_2(0) = \mu_V = 1 + \zeta$. Setting here the anomalous chromomagnetic moment of the axial vector diquark $\zeta = 1$ and $G_1(0) = 1$, we obtain the following expression for V_{q+AD}^{pert} in configuration space up to terms $O(v^2/c^2)$:

$$\begin{aligned}
V_{q+AD}^{pert}(r) &= -\frac{4\alpha_s}{3r} - \frac{4\alpha_s}{3m_1m_2}\mathbf{p}\frac{1}{r}\mathbf{p} + \frac{2\pi\alpha_s}{3m_1^2}\delta(\mathbf{r}) + \frac{2\alpha_s}{3r^3}\frac{\mathbf{L}^2}{m_1m_2} \\
&+ \frac{4\alpha_s}{3m_1m_2}\left(\frac{m_2}{2m_1}+1\right)\frac{\mathbf{LS}_1}{r^3} + \frac{4\alpha_s}{3m_1m_2}\left(\frac{m_1}{2m_2}+1\right)\frac{\mathbf{LS}_2}{r^3} \\
&+ \frac{32\pi\alpha_s}{9m_1m_2}(\mathbf{S}_1\mathbf{S}_2)\delta(\mathbf{r}) - \frac{4\alpha_s}{3m_1m_2r^3}[(\mathbf{S}_1\mathbf{S}_2) - \frac{3}{r^2}(\mathbf{S}_1\mathbf{r})(\mathbf{S}_2\mathbf{r})].
\end{aligned} \tag{15}$$

We assume that the nonperturbative part of the interaction potential for a quark and an axial vector diquark consists of scalar and vector confining terms. It is taken in the form

$$V_{q+AD}^{nonpert}(r) = V_{q+AD}^{conf(S)}(r) + V_{q+AD}^{conf(V)}(r), \tag{16}$$

where

$$\begin{aligned}
V_{q+AD}^{conf(S)}(r) &= V^S(r) - \frac{1}{2m_1^2}\mathbf{p}V^S(r)\mathbf{p} + \frac{1}{8m_1^2}\Delta V^S(r) \\
&\quad - \frac{1}{2m_1^2}\frac{V'^S(r)}{r}\mathbf{LS}_1 - \frac{1}{2m_2^2}\frac{V'^S(r)}{r}\mathbf{LS}_2,
\end{aligned} \tag{17}$$

$$\begin{aligned}
V_{q+AD}^{conf(V)}(r) &= V^V(r) - \frac{1}{4m_1}\left(\frac{1}{2\mu}+\frac{1}{2m_2}-\frac{1+\kappa}{m_1}\right)\Delta V^V(r) \\
&\quad + \frac{1}{m_1m_2}\mathbf{p}V^V(r)\mathbf{p} + \frac{1}{m_1}\left(\frac{1+\kappa}{\mu}-\frac{1}{2m_1}\right)\frac{V'^V(r)}{r}\mathbf{LS}_1
\end{aligned} \tag{18}$$

$$-\frac{1}{2m_2^2}\frac{V'^V(r)}{r}\mathbf{LS}_2 + \frac{2(1+\kappa)}{3m_1m_2}\Delta V^V(r)\mathbf{S}_1\mathbf{S}_2$$
$$-\frac{1+\kappa}{3m_1m_2}\left(\frac{V'^V(r)}{r} - V''^V(r)\right)\left(\mathbf{S}_1\mathbf{S}_2 - \frac{3}{r^2}(\mathbf{S}_1\mathbf{r})(\mathbf{S}_2\mathbf{r})\right).$$

Note that for $\kappa = -1$ the contributions of the vector confining potential to spin-spin and tensor interactions vanish as in the case of mesons.

The resulting total potential has been used for calculating the mass spectrum of baryons with $J^P = \frac{1}{2}^+, \frac{3}{2}^+$ on the basis of the local quasipotential equation (1). Corresponding numerical results are presented in Table 2.

Table 2: Masses of baryons (in GeV) containing two heavy quarks. $[QQ]$ denotes the diquark subsystem with the antisymmetric spin wave function, $\{QQ\}$ denotes the diquark subsystem with the symmetric spin wave function ($q = u, d$).

Notation	Quark content	J^P	M_B (our results)	M_B [16]	M_B [8]
Ξ_{cc}	$\{cc\}q$	$\frac{1}{2}^+$	3.66	3.66	3.61
Ξ_{cc}^*	$\{cc\}q$	$\frac{3}{2}^+$	3.81	3.74	3.68
Ω_{cc}	$\{cc\}s$	$\frac{1}{2}^+$	3.76	3.74	3.71
Ω_{cc}^*	$\{cc\}s$	$\frac{3}{2}^+$	3.89	3.82	3.76
Ξ_{bb}	$\{bb\}q$	$\frac{1}{2}^+$	10.23	10.34	-
Ξ_{bb}^*	$\{bb\}q$	$\frac{3}{2}^+$	10.28	10.37	-
Ω_{bb}	$\{bb\}s$	$\frac{1}{2}^+$	10.32	10.37	-
Ω_{bb}^*	$\{bb\}s$	$\frac{3}{2}^+$	10.36	10.40	-
Ξ_{cb}	$\{cb\}q$	$\frac{1}{2}^+$	6.95	7.04	-
Ξ_{cb}'	$[cb]q$	$\frac{1}{2}^+$	7.00	6.99	-
Ξ_{cb}^*	$\{cb\}q$	$\frac{3}{2}^+$	7.02	7.06	-
Ω_{cb}	$\{cb\}s$	$\frac{1}{2}^+$	7.05	7.09	-
Ω_{cb}'	$[cb]s$	$\frac{1}{2}^+$	7.09	7.06	-
Ω_{cb}^*	$\{cb\}s$	$\frac{3}{2}^+$	7.11	7.12	-

5   Summary

In this talk we have discussed the mass spectrum of baryons with two heavy quarks calculated in the quark-diquark approximation on the basis of a local quasipotential equation. As it follows from Table 2, our results are in good agreement with calculations made within other approaches. One can see from Table 2 that our predictions are especially close to those obtained in [16]. The approach used in [16,17] is based on some semiempirical regularities of the baryon mass spectrum following from experimental data and the constituent quark model. The authors of [16,17] found a successful parametrization of the interpolating curves which describe the known meson and baryon masses, and give predictions for some yet unobserved mesons and baryons. Our results for the B_c meson mass spectrum [12] are also rather close to those of [17]. The quoted authors use in particular the Feynman-Hellmann theorem in order to obtain the monotonic decrease of the hadron mass spectrum with respect to quark masses. Here they use a simplified assumption about the interquark potential, namely they neglect the specific form of the Breit-Fermi potential, but retain the information coming from its derivative with respect to masses. This leads, for the vector mesons (and for spin 3/2 baryons) to a smooth behaviour in the reduced mass, from which inequalities between quark masses and hadron masses follow. Deviations from these inequalities are found in some constituent quark models (including ours) in which widely adopted quark masses are used. From our point of view there is nothing serious in these contradictions and the recovery of the appropriate spin-independent (flavour-dependent) part of quark potential may eliminate them. This is indirectly suggested by the spectacular agreement between these two different approaches which leads us to conclude that our model correctly describes some important features of the baryon mass spectrum.

There exists yet another type of baryons with one heavy and two light quarks for which the quark-diquark approximation gives reasonable results. Thus, in ref. [18] Qqq baryons were treated by considering the interaction of the diquark system with the third quark on the basis of additional quark exchange forces [19,20,21]. The resulting effective lagrangian, which incorporates heavy quark and chiral symmetry, describes interactions of heavy baryons with Goldstone bosons in the low energy region.

Acknowledgments

We are grateful to A. Martynenko and V. Saleev for their contribution to the results presented here and to T. Feldmann and S. Gerasimov for discussions. The work of D. E. was supported in part by *Deutsche Forschungsgemeinschaft* under contract Eb 139/1-2. He also thanks the organizers for financial support and kind hospitality. R. N. F. and V. O. G. were supported in part by *Russian Foundation for Fundamental Research* under Grant No. 96-02-17171.

References

1. M.A. Shifman, A.I. Vainshtein and V.I. Zakharov, *Nucl. Phys.* B **147**, 385 (1979).
2. S. Godfrey and N. Isgur, *Phys. Rev.* D **32**, 189 (1985).
3. E. Bagan *et al.*, Prepr. CERN-TH.7141/94.
4. V.V. Kiselev, A.K. Likhoded and A.V. Tkabladze, *Yad. Fiz.* **58**, 1045 (1995).
5. J.-M. Richard, *Phys. Rep.* **212**, 1 (1992).
6. C. Alexandrou *et al.*, *Phys. Lett.* B **337**, 340 (1994); UKQCD Collaboration, (K.C. Bowler *et al.*), hep-lat/9601022.
7. J. Savage and M.B. Wise, *Phys. Lett.* B **248**, 177 (1990).
8. J.G. Körner, M. Krämer and D. Pirjol, *Prog. Part. Nucl. Phys.* **33**, 787 (1994).
9. M. Fabre de la Ripelle and Yu.A. Simonov, *Ann. Phys. (N.Y.)* **212**, 235 (1991).
10. A.P. Martynenko and R.N. Faustov, *Teor. Mat. Fiz.* **64**, 179 (1985).
11. V.O. Galkin and R.N. Faustov, *Yad. Fiz.* **44**, 1575 (1986); V.O. Galkin, A.Yu. Mishurov and R.N. Faustov, *Yad. Fiz.* **51**, 1101 (1990); R.N. Faustov and V.O. Galkin, *Z. Phys.* C **66**, 119 (1995); *Phys. Rev.* D **52**, 5131 (1995); R.N. Faustov, V.O. Galkin and A.Yu. Mishurov, *Phys. Lett.* B **356**, 516 (1995); *Phys. Rev.* D **53**, 1391 (1996); *ibid* 6302.
12. V.O. Galkin, A.Yu. Mishurov and R.N. Faustov, *Yad. Fiz.* **55**, 2175 (1992).
13. S. Fleck, B. Silvestre-Brac and J.-M. Richard, *Phys. Rev.* D **38**, 1519 (1988).
14. M. Anselmino *et al.*, *Rev. Mod. Phys.* **65**, 1199 (1993).
15. J.M. Cornwall, Prepr. UCLA/96/TEP/15 (May 1996), hep-th/9605116.
16. R. Roncaglia, D.B. Lichtenberg and E. Predazzi, *Phys. Rev.* D **52**, 1722 (1995).
17. R. Roncaglia, A. Dzierba, D.B. Lichtenberg and E. Predazzi, *Phys. Rev.* D **51**, 1248 (1995).
18. D. Ebert, T. Feldmann, C. Kettner and H. Reinhardt, *Z. Phys.* C **71**, 329 (1996); Prepr. DESY-96-010 (1996).
19. R.T. Cahill, C.D. Roberts and J. Praschifka, *Aust. J. Phys.* **42**, 129 (1989).
20. H. Reinhardt, *Phys. Lett.* B **244**, 316 (1990).
21. D. Ebert and L. Kaschluhn, *Phys. Lett.* B **297**, 367 (1992).

HIGHER TWIST EFFECTS IN DIS: AN OUTLINE OF DIQUARK CONTRIBUTIONS

M. Anselmino

Dipartimento di Fisica Teorica dell'Università and INFN, Sezione di Torino
Via Pietro Giuria 1, 10125, Torino, Italy

F. Caruso

Centro Brasileiro de Pesquisas Físicas/CNPq
Rua Dr. Xavier Sigaud 150, 22290-180, Rio de Janeiro, Brazil, and
Instituto de Física da Universidade do Estado do Rio de Janeiro
R. São Francisco Xavier 524, 20550-013, Rio de Janeiro, Brazil

A quark–diquark picture of the nucleon, previously introduced in the description of several exclusive and inclusive processes at intermediate Q^2 values, is found to accurately model the higher-twist data on the unpolarized proton structure function $F_2^p(x, Q^2)$. The main results of the model are summarized. The emerging set of parameters is consistent with the diquark properties suggested by other experimental and theoretical analyses. Higher-twist corrections to the Bjorken and Gottfried sum rule are also estimated in the framework of the same quark-diquark model. The resulting corrections to both sum rules turn out to be negligible.

1   Introduction

Recent data from Deep Inelastic Scattering (DIS) experiments at CERN and SLAC have provided precise information on the unpolarized proton structure function $F_2^p(x, Q^2)$ and have allowed a quantitative estimate of higher-twist terms.[1,2]

Higher-twist contributions to DIS are expected to originate from quark and gluon correlations, and we assume here that they can be described by a quark–diquark picture of the nucleon. Diquarks originate from QCD colour forces and model correlations between two quarks inside the nucleon which, at moderate Q^2 values, interact collectively and behave as bound states.

The diquark model is supported by a number of phenomenological successes in the description of many physical phenomena, such as: the behaviour of the DIS structure functions at Bjorken $x \to 1$, the inclusive large p_T production of deuterons, protons and other baryons in pp interactions, several exclusive processes and decays.[3,4] There is little doubt that two quark correlations are present inside nucleons and that they might play a significant rôle in processes at moderate Q^2 values, precisely the region where the higher-twist effects have been observed [Q^2 up to $\simeq 10 - 20 \ (\text{GeV}/c)^2$].

2 Diquark contributions to $\mathbf{F}_2$

The diquark contributions to DIS have been studied in several papers,[5,6] mainly taking into account only spin 0, scalar diquarks, in simplified versions of the quark-diquark model of the nucleon. A most general calculation of spin 0 and spin 1 diquark contributions to DIS was performed in Ref. 7, allowing for a vector diquark anomalous magnetic moment and for scalar–vector and vector–scalar diquark transitions; however, at that time, no attempt was made of fitting the experimental data due to the lack of detailed information on the higher-twist contributions alone. This was performed more recently.[8]

Let us assume the nucleon to be a quark–diquark state where the diquark mimics the correlations between two quarks which, at moderate momentum transfer, interact collectively and behave as a bound state. When probing the nucleon with the virtual photon in DIS we have then three kinds of contributions: the scattering off the single quark, the elastic scattering off the diquark, and the inelastic diquark contribution, that is the scattering off one of the quarks inside the diquark. Eventually, at large enough Q^2 values, the elastic diquark contribution, weighted by form factors, vanishes and one recovers the usual pure quark results.

The expression of the structure functions F, in the quark-diquark parton model, is then given in general by:

$$
\begin{aligned}
F(x, Q^2) \;=\; & \sum_q F^{(q)} + \sum_S F^{(S)} + \sum_V F^{(V)} \\
+ \; & \sum_{q_S} F^{(q_S)} + \sum_{q_V} F^{(q_V)} + \sum_{S,V} F^{(S-V)} + \sum_{S,V} F^{(V-S)},
\end{aligned}
\tag{1}
$$

where (q) denotes the single quark contribution, (S) and (V) respectively the scalar and vector diquark ones, and (q_S) $[(q_V)]$ the contribution of the quark inside the scalar [vector] diquark. We have also allowed for elastic diquark contributions with a scalar–vector $(S-V)$ or vector–scalar $(V-S)$ transition.

Let us consider here the unpolarized structure function $F_2(x, Q^2)$ only. It is straightforward to extract from the full contribution of the quark–diquark model to F_2 all terms which are proportional to the diquark form factors.[8] These terms vanish at large Q^2 values, but at moderate Q^2 values they give non negligible contributions. These are the terms which we assume to model higher-twist effects in DIS. Explicitly – following the notations of Refs. 7, 8 – they are given by

$$
F_2^{HT} \;=\; \sum_S e_S^2 \, S(x) \, x \, D_S^2 + \sum_V \frac{1}{3} e_V^2 \, V(x) \, x \left\{ \left[\left(1 + \frac{Q^2}{2m_N^2 x^2} \right) D_1 + \right. \right.
$$

$$-\ \frac{Q^2}{2m_N^2 x^2}D_2 + Q^2\left(1 + \frac{Q^2}{4m_N^2 x^2}\right)D_3\Big]^2 + 2\Big[D_1^2 + \frac{Q^2}{4m_N^2 x^2}D_2^2\Big]\Big\}$$

$$+\ \frac{1}{4}\sum_S e_S^2\, S(x)\, xQ^2\, D_T^2 + \frac{1}{12}\sum_V e_S^2\, V(x)\, xQ^2\, D_T^2 \tag{2}$$

$$-\ \sum_{q_S} e_{q_S}^2\, x\, q_s(x, Q^2)D_S^2 - \sum_{q_V} e_{q_V}^2\, x\, q_v(x, Q^2)D_V^2\ .$$

Notice that, depending on the actual behaviour of the form factors, one not only obtains contributions proportional to $1/Q^2$, but also to higher powers of $1/Q^2$, which might be significant in the lowest Q^2 range of the experimental data.[1,2] Notice also that Eq. (2) has both positive and negative contributions, so that it might yield positive or negative higher-twist values, according to the different x or Q^2 ranges.

The valence quark content of the proton in the quark-diquark model[9] is given by the flavour and spin wave function:

$$|p, S_z = \pm 1/2\rangle\ =\ \pm\frac{1}{3}\Big\{\sin\Omega\big[\sqrt{2}V_{(ud)}^{\pm 1}u^{\mp} - 2V_{(uu)}^{\pm 1}d^{\mp} \tag{3}$$

$$+\ \sqrt{2}V_{(uu)}^{0}d^{\pm} - V_{(ud)}^{0}u^{\pm}\big] \mp 3\cos\Omega\, S_{(ud)}u^{\pm}\Big\},$$

where $V_{(ud)}^{\pm 1}$ stands for a (ud) vector diquark with the third component of the spin $S_z = \pm 1$, $u^{\mp}$ is a u quark with $S_z = \mp 1/2$ and so on. The vector and scalar diquark components have different weights, so that the probabilities of finding a vector or scalar diquark in the proton are $\sin^2\Omega$ and $\cos^2\Omega$ respectively. From Eqs. (2, 3) and following the notation of Ref. 8 we finally get, for a proton

$$(F_2^{HT})_p\ =\ \frac{1}{9}\cos^2\Omega\ \Big[f_S(x) - [4f_{u_S}(x, Q^2) + f_{d_S}(x, Q^2)]\Big]\, x\, D_S^2$$

$$+\ \frac{1}{81}\sin^2\Omega\,[f_{V_{(ud)}}(x) + 32f_{V_{(uu)}}(x)]\, x\Big\{\Big[\Big(1 + \frac{Q^2}{2m_N^2 x^2}\Big)D_1$$

$$-\ \frac{Q^2}{2m_N^2 x^2}D_2 + Q^2\Big(1 + \frac{Q^2}{4m_N^2 x^2}\Big)D_3\Big]^2 + 2\Big[D_1^2 + \frac{Q^2}{4m_N^2 x^2}D_2^2\Big]\Big\}$$

$$-\ \frac{1}{27}\sin^2\Omega\,[16f_{u_{V_{(uu)}}}(x, Q^2) + 5f_{u_{V_{(ud)}}}(x, Q^2)]\, x\, D_V^2$$

$$+\ \frac{1}{36}\Big[\cos^2\Omega\, f_S(x) + \frac{1}{9}\sin^2\Omega\, f_{V_{(ud)}}(x)\Big]\, xQ^2\, D_T^2\ . \tag{4}$$

We have adopted the following parametrization of the different distribution functions:

$$P(x) = \langle P\rangle\, f_P(x) = \langle P\rangle\, N_P\, x^{\alpha_P}(1 - x)^{\beta_P} \tag{5}$$

for each kind of parton $P = S, V, q_s, q_v$; the N_P's are the proper normalization constants[8] such that $\int_0^1 dx\, f_P(x) = 1$; $\langle P \rangle$ gives the average number of valence partons P inside the proton according to Eq. (3).

Concerning the diquark form factors we have chosen the most simple expressions which agree both with the asymptotic perturbative QCD predictions[10] and the pointlike, $Q^2 \to 0$, limits:

$$
\begin{aligned}
D_S &= \frac{Q_S^2}{Q_S^2 + Q^2} \\[2mm]
D_1 &= \left(\frac{Q_V^2}{Q_V^2 + Q^2} \right)^2 \\[2mm]
D_2 &= (1 + \kappa) D_1 \\[2mm]
D_3 &= \frac{Q^2}{m_N^4} D_1^2 \\[2mm]
D_V &= D_1 \\[2mm]
D_T &= \frac{\sqrt{Q^2}}{m_N} D_1
\end{aligned}
\tag{6}
$$

where κ is the vector diquark anomalous magnetic moment.

We have then used Eq. (4), with the parametrizations given in Eqs. (5) and (6), as a model to fit the experimental data on the higher-twist contributions to F_2^{p}.[1,11] We have neglected the small effect due to the perturbative QCD Q^2 evolution of the single quark distribution functions f_{u_S}, f_{d_S}, $f_{u_{V_{(ud)}}}$ and $f_{u_{V_{(uu)}}}$.

The results of our best fits are indeed very good and are given in Ref. 8.

The corresponding values of the free parameters turn out to be:

$$
\cos^2 \Omega = 0.81 \qquad Q_S^2 = 2.02 \ (\text{GeV}/c)^2 \qquad Q_V^2 = 1.21 \ (\text{GeV}/c)^2
$$

$$
\alpha_S = 2.13 \qquad \beta_S = 18.51 \qquad \alpha_V = 7.93 \qquad \beta_V = 3.32
\tag{7}
$$

$$
\beta_{u_S} = 5.13 \qquad \beta_{d_S} = 5.13 \qquad \beta_{u_{V_{(ud)}}} = 8.41 \qquad \beta_{u_{V_{(uu)}}} = 8.41
$$

The values of α_{u_S}, α_{d_S}, $\alpha_{u_{V_{(ud)}}}$ and $\alpha_{u_{V_{(uu)}}}$ have been fixed to be -0.5 in agreement with the expected small x behaviour of the valence quark distributions. The value of κ which allows the best fits is $\kappa = 0$.

It is important to stress that the diquark features, as emerging from the values of the best fit parameters, Eqs. (7), are consistent with the expectations from many other studies and applications of quark-diquark models of the nucleon.[3] Scalar diquarks turn out to be more abundant ($\cos^2 \Omega = 0.81$) and more pointlike ($Q_S^2 > Q_V^2$) than vector ones. Also, the average momentum fraction, x, carried by a scalar diquark, proportional to its average mass, is smaller, as expected, than the average x of a vector diquark.

The deep inelastic scattering of neutrinos on unpolarized nucleons has also been considered in the framework of the same (and most general) quark–diquark model.[12] The resulting scaling violations have been discussed, but a detailed study of the higher-twist terms and their possible evaluation are at present impossible, due to lack of experimental data.

3   Diquark contribution to Bjorken sum rule

Several recent measurements of the polarized structure functions $g_1^p(x)$ and $g_1^n(x)$ both for protons and neutrons [13] have allowed the first tests of the fundamental Bjorken sum rule: [14]

$$\int_0^1 dx \, [g_1^p(x) - g_1^n(x)] = S_{Bj} = \frac{a_3}{6} E_{NS}(Q^2) \,, \tag{8}$$

where $a_3 = 1.2573 \pm 0.0028$ is the neutron β-decay axial coupling. The above equation holds on ground of isospin invariance, at leading twist in the Operator Product Expansion; the coefficient function E_{NS} has been computed up to third order [15] in the strong coupling constant α_s:

$$E_{NS}(Q^2) = 1 - \frac{\alpha_s}{\pi} - 3.58 \left(\frac{\alpha_s}{\pi}\right)^2 - 20.22 \left(\frac{\alpha_s}{\pi}\right)^3 \,, \tag{9}$$

where a number of three active flavours is assumed.

Data are available at $Q^2 = 2$ and 4.6 $(\text{GeV}/c)^2$: at such values of Q^2 a comparison of Bjorken sum rule with experiment should also take into account possible higher-twist or target mass corrections, not included in Eq. (8):

$$\int_0^1 dx \, [g_1^p(x) - g_1^n(x)] = S_{Bj}(Q^2) + \delta S_{HT}^{Bj} + \delta S_T^{Bj} \,. \tag{10}$$

Some evaluations of target mass [16–18] and higher-twist corrections can be found in the literature,[13] either in the framework of the Drell-Hearn-Gerasimov sum rule [19] or of the QCD sum rules [16,20]. Some of these corrections may be sizeable at $Q^2 \simeq 2$ $(\text{GeV/c})^2$, but the final comparison with data shows no significant violation of the Bjorken sum rule. Let us notice that reliable estimates of the higher-twist contributions are of fundamental importance in view of the attempts to exploit the Bjorken sum rule (8) in order to extract the value of the strong coupling constant α_s.

It is then natural to evaluate the higher-twist corrections to S_{Bj} in the framework of the quark-diquark model of the nucleon.[21] We would like to stress that the diquark approach, being a physically motivated and self consistent model, rather than an expansion in powers of $1/Q^2$, has the unique feature of taking into account a full set of higher-twist corrections and not only the leading ones. Confident in its physical content, we have adopted the same diquark model, with the same parameters, as given in Eq. (7): therefore the results presented here are parameter free predictions.

The elastic diquark contributions have been studied in Ref. [7] and, for the polarized structure function $g_1(x, Q^2)$, they read

$$g_1^{(V)}(x, Q^2) = \frac{1}{4} e_V^2 \, \Delta V \left[\left(2 + \frac{Q^2}{2m_N^2 x^2} \right) (D_1 D_2 + \frac{1}{2} Q^2 D_2 D_3) - \frac{Q^2}{4m_N^2 x^2} D_2^2 \right]$$

$$(11)$$

where ΔV denotes the difference between the number density of vector diquarks with spin parallel and antiparallel to the nucleon spin.

Following the notations of Ref. [7] the higher-twist diquark contributions to g_1 for protons and neutrons are given by:

$$[g_1^{HT}]_p = \frac{1}{4} \left(\frac{16}{9} \Delta V_{uu} + \frac{1}{9} \Delta V_{ud} \right) \left[\left(2 + \frac{Q^2}{2m_N^2 x^2} \right) (D_1 D_2 + \frac{1}{2} Q^2 D_2 D_3) \right.$$
$$\left. - \frac{Q^2}{4m_N^2 x^2} D_2^2 \right] - \frac{1}{2} \left(\frac{4}{9} \Delta u_{V_{uu}} + \frac{4}{9} \Delta u_{V_{ud}} + \frac{1}{9} \Delta d_{V_{ud}} \right) D_V^2 , \qquad (12)$$

$$[g_1^{HT}]_n = \frac{1}{4} \left(\frac{1}{9} \Delta V_{uu} + \frac{1}{9} \Delta V_{ud} \right) \left[\left(2 + \frac{Q^2}{2m_N^2 x^2} \right) (D_1 D_2 + \frac{1}{2} Q^2 D_2 D_3) \right.$$
$$\left. - \frac{Q^2}{4m_N^2 x^2} D_2^2 \right] - \frac{1}{2} \left(\frac{1}{9} \Delta u_{V_{uu}} + \frac{4}{9} \Delta u_{V_{ud}} + \frac{1}{9} \Delta d_{V_{ud}} \right) D_V^2 . \qquad (13)$$

The negative contributions come from the scattering off the quarks inside the diquarks, weighted by a factor $(1 - D_V^2)$, with D_V given in Eqs. (6, 7). In Eq. (13) we have already used isospin relationships and all distribution functions refer to the proton; Δq is the usual helicity density carried by quark q and the suffices refer to the type of diquark it comes from.

Subtracting Eq. (13) from (12) yields

$$[g_1^{HT}]_{p-n} = \frac{1}{3} \Delta V_{uu} \left[\left(2 + \frac{Q^2}{2m_N^2 x^2} \right) (D_1 D_2 + \frac{1}{2} Q^2 D_2 D_3) - \frac{Q^2}{4m_N^2 x^2} D_2^2 \right]$$
$$- \frac{1}{6} \Delta u_{V_{uu}} D_V^2 , \qquad (14)$$

where, according to Eq. (3),

$$\Delta V_{uu}(x) = \frac{4}{9}\sin^2\Omega\, f_{V_{uu}}(x)$$

$$\Delta u_{V_{uu}}(x) = \frac{8}{9}\sin^2\Omega\, f_{u_{V_{uu}}}(x) \tag{15}$$

and the f_P distributions are the same as defined in Eq. (5).

From Eqs. (14), (15) and (6), upon integration over x, we obtain

$$\left[\delta S_{HT}^{Bj}\right]_{diquarks} = \frac{\sin^2\Omega}{27}\left(D_1^2 + Q^2 D_1 D_3\right)$$

$$\times\ \left\{4 + \frac{Q^2}{m_N^2}\int_0^1 \frac{dx}{x^2}\, f_{V_{uu}}(x)\right\}, \tag{16}$$

which, via Eqs. (5) and (7), yields the numerical estimates at the Q^2 values [in $(\mathrm{GeV}/c)^2$] for which data are available:

$$\left[\delta S_{HT}^{Bj}\left(Q^2 = 2.0\right)\right]_{diquarks} \simeq\ 0.2\times 10^{-2}$$

$$\left[\delta S_{HT}^{Bj}\left(Q^2 = 4.6\right)\right]_{diquarks} \simeq\ 0.5\times 10^{-3}. \tag{17}$$

The prediction of the model in the full range $1 \le Q^2 \le 10\ (\mathrm{GeV}/c)^2$ can be found in Ref. 21.

We notice that the higher-twist contributions to the Bjorken sum rule, evaluated in the framework of the quark-diquark model of the nucleon, turn out to be quite small. They are small, and opposite in sign, even when compared with the perturbative QCD higher order corrections of Eqs. (8) and (9), namely with $a_3[E_{NS}(Q^2) - 1]/6$.

$$\frac{\left[\delta S_{HT}^{Bj}\left(Q^2\right)\right]_{diquarks}}{a_3[1 - E_{NS}(Q^2)]/6} \simeq 7\%\ (Q^2 = 2)\quad \text{and}\quad 2\%\ (Q^2 = 4.6). \tag{18}$$

4 Diquark contribution to Gottfried sum rule

Similarly, one can compute, within the same model and using the same set of parameters, higher-twist contributions [22] to Gottfried sum rule:

$$S_G \equiv \int_0^1 \frac{dx}{x}[F_2^p(x, Q^2) - F_2^n(x, Q^2)] = \frac{1}{3} + \frac{2}{3}\int_0^1 dx[\bar{u}(x, Q^2) - \bar{d}(x, Q^2)]. \tag{19}$$

Infact, in order to extract a reliable value of $\int dx\, [\bar{u}(x, Q^2) - \bar{d}(x, Q^2)]$ from Eq. (19) one should take into account all possible corrections, both perturbative QCD ones and higher-twist ones. The former are known to be rather small, less than 1%, and we give here an estimate of the latter; they also will turn out to be negligibly small, so that Eq. (19) can be safely used to extract precise information on the isospin asymmetry of the nucleon sea.

In the quark-diquark model of the nucleon one obtains

$$\int_0^1 \frac{dx}{x}[F_2^p - F_2^n]_{q-D} = \frac{1}{3} + \frac{2}{3}\int_0^1 dx\,[\bar{u}(x) - \bar{d}(x)]$$

$$- \frac{4}{9}\sin^2\Omega\, D_1^2 + \frac{8}{27}\sin^2\Omega \int_0^1 dx\, f_{V_{uu}}(x) \qquad (20)$$

$$\times \left\{\left[D_1 + Q^2\left(1 + \frac{Q^2}{4m_N^2 x^2}\right)D_3\right]^2 + 2\left(1 + \frac{Q^2}{4m_N^2 x^2}\right)D_1^2\right\},$$

so that

$$[\delta S_{HT}^G]_{diquarks} = \frac{4}{9}\sin^2\Omega\left[D_1^2 + \frac{4}{3}Q^2 D_1 D_3 + \frac{2}{3}Q^4 D_3^2\right]$$

$$+ \frac{8}{27}\sin^2\Omega\, \frac{Q^2}{2m_N^2}[D_1^2 + Q^2 D_1 D_3 + Q^4 D_3^2]\int_0^1 \frac{dx}{x^2} f_{V_{uu}}(x)$$

$$+ \frac{8}{27}\sin^2\Omega\, \frac{Q^8}{16m_N^4}D_3^2 \int_0^1 \frac{dx}{x^4} f_{V_{uu}}(x). \qquad (21)$$

Eqs. (21) and (5)-(7) give, at $Q^2 = 4$ (GeV/c)^2,

$$[\delta S_G]_{diquarks}(Q^2 = 4) \simeq 0.6 \times 10^{-2}. \qquad (22)$$

5 Comments and conclusions

Higher-twist diquark contributions, both to Bjorken and Gottfried sum rules (Eqs. (17) and 22) respectively) turn out to be very small. This is due to some intrinsic features of the quark-diquark model of the nucleon: the strong $SU(6)$ violation ($\sin^2\Omega = 0.19$) favouring scalar diquarks (notice that only vector diquarks contribute to Eqs. (16) and (21)); the mass scale of the vector diquark form factor which turns out to be small, $Q_V^2 = 1.21$ (GeV/c)^2, corresponding to a large size, and the vector diquark x distribution which is found to be peaked at $x \simeq 0.7$, suggesting that vector diquarks consist of two almost uncorrelated quarks.

This conclusion is confirmed by computing also the diquark contribution to the first moment of $g_1^p(x)$ alone, *i.e.*, $\Gamma_1^p = \int_0^1 dx\, g_1^p(x)$. A sizeable diquark contribution to Γ_1^p would indicate that the tiny results obtained above for the Bjorken sum, Eqs. (17), are only due to a strong cancellation between the proton and the neutron contributions. However, an explicit calculation shows that diquark contributions to Γ_1^p are comparable to those found for the Bjorken sum rule, rejecting this possibility.

We have summarized recent results on the evaluation of higher-twist contributions in DIS in terms of a quark–diquark model for the nucleon.[23] Such a work has consistently been performed within the same, physically motivated, phenomenological framework, within which it is possible to evaluate higher-twist effects (and not only the leading ones) in electron – and neutrino – initiated deep inelastic processes. In particular, we find that the diquark contributions to the Bjorken and Gottfried sum rules turn out to be quite small. This clarifies and makes more reliable both the use of the Bjorken sum rule in the experimental measurement of the strong coupling constant and of the Gottfried sum rule in the determination of the isospin asymmetry of the quark sea.

Acknowledgments

F.C. is grateful to the CNPq of Brazil for financial support.

References

1. M. Virchaux and A. Milsztajn, *Phys. Lett.* B **274**, 221 (1992).
2. S. Choi, T. Hatsuda, Y. Koike and Su H. Lee, *Phys. Lett.* B **312**, 351 (1993).
3. For a comprehensive review on Diquarks see, *e.g.*, M. Anselmino, S. Ekelin, S. Fredriksson, D.B. Lichtenberg and E. Predazzi, *Rev. Mod. Phys.* **65**, 1199 (1993).
4. F. Caruso, *Il ruolo dei diquarks come costituenti barionici nella trattazione di processi esclusivi ad energie intermedie*, PhD. Thesis, Università di Torino (1989).
5. S. Fredriksson, M. Jändel and T. Larsson, *Z. Phys.* C **14**, 35 (1982); **19**, 53 (1983); S. Ekelin and S. Fredriksson, *Phys. Lett.* B **162**, 373 (1985).
6. J.J. Dugne and P. Tavernier, *Z. Phys.* C **59**, 333 (1993); for a more detailed and general treatment see P. Tavernier, *Essai de détermination des fonctions de structure des nucleons dans l'hipothèse de l'existence de*

diquarks, Thèse de Doctorat (PCCF T 9207), Universitè Blaise Pascal (1992).

7. M. Anselmino, F. Caruso, E. Leader and J. Soares, *Z. Phys.* C **48**, 689 (1990). For further details see J. Soares, *Sobre a contribuição genérica dos diquarks escalares e pseudo–vetoriais às funções de estrutura do nucleon*, PhD. Thesis, CBPF (1995).

8. M. Anselmino, F. Caruso, J.R.T. de Mello Neto, A. Penna Firme and J. Soares, *Z. Phys.* C **71**, 625 (1996); A. Penna Firme, *Da contribuição dos diquarks à descrição de efeitos de "higher-twist" observados no espalhamento profundamente inelástico*, Thesis, CBPF (1994).

9. M.I. Pavkovič, *Phys. Rev.* D **13**, 2128 (1976).

10. A.I. Vainshtein and V.I. Zakharov, *Phys. Lett.* B **72**, 368 (1978).

11. A. Milsztajn, *private communication*.

12. M. Anselmino, V. Barone, F. Caruso, E. Cheb–Terrab and P.C.R. Quintairos, *preprint* CBPF–NF–047/95, to appear in *Intern. Jour. Mod. Phys. C*. See also P.C.R. Quintairos, *"Sobre a contribuição dos diquarks ao espalhamento profundamente inelástico neutrino–nucleon"*, Thesis, CBPF (1995).

13. For a complete list of references and a comprehensive review paper on the subject see, *e.g.*, M. Anselmino, A. Efremov and E. Leader, *Phys. Rep.* **261**, 1 (1995).

14. J.D. Bjorken, *Phys. Rev.* D **148**, 1467 (1966).

15. S.A. Larin and J.A.M. Vermaseren, *Phys. Lett.* B **259**, 345 (1991); S.A. Larin, *Phys. Lett.* B **334**, 192 (1994).

16. I.I. Balitskii, V.M. Braun and A.V. Kolesnichenko, *Phys. Lett.* B **242**, 245 (1990); **318**, 648 (1993).

17. J. Ellis and M. Karliner, *Phys. Lett.* B **313**, 131 (1993); **341**, 397 (1995).

18. H. Kawamura and T. Uematsu, *Prog. Theor. Phys. Suppl.* **120**, 225 (1995); *Phys. Lett.* B **343**, 346 (1995).

19. V.D. Burkert and B.L. Ioffe, *JETP* **78**, 619 (1994).

20. E. Stein, P. Gornicki, L. Mankiewicz and A. Schaefer, *Phys. Lett.* B **353**, 107 (1995).

21. M. Anselmino, F. Caruso and E. Levin, *Phys. Lett.* B **358**, 109 (1995).

22. M. Anselmino and F. Caruso, submitted for publication to *Anais da Academia Brasileira de Física*

23. See also F. Caruso, "Diquarks ad higher-twist effects: recent results", *in* Aguilera–Navarro, V.C., Galetti, D., Pimentel, B. and Tomio, L. (Eds.), *Topics in Theoretical Physics: Festschrift for Paulo Leal Ferreira*, São Paulo, IFT, 1995, pp. 66-77.

THE PROTON SPIN STRUCTURE IN A LIGHT-CONE QUARK-SPECTATOR-DIQUARK MODEL

BO-QIANG MA

Institute of High Energy Physics, Academia Sinica
P. O. Box 918(4), Beijing 100039, China

It is shown that the proton spin problem raised by the Ellis-Jaffe sum rule violation does not in conflict with the SU(6) quark model, provided that the relativistic effect from the quark transversal motions, the flavor asymmetry between the u and d valence quarks, and the intrinsic quark-antiquark pairs generated by the non-perturbative meson-baryon fluctuations of the nucleon sea are taken into account. The meson-baryon fluctuations of the nucleon sea provide a consistent picture for connecting the proton spin structure with a number of other anomalies observed in the proton's structure, such as the flavor asymmetry of the nucleon sea implied by the violation of the Gottfried sum rule and the discrepancy between two different measurements of the strange quark distributions in the nucleon sea.

1 The Proton Spin Puzzle and the Wigner Rotation

1.1 The Violation of the Ellis-Jaffe Sum Rule and the Proton Spin "Crisis"

Parton sum rules and similar relations played important roles in the establishment of the quark-parton picture for nucleons in deep inelastic scattering. Thus any violation of a parton sum rule may of essential importance to reveal possible new content concerning our understanding of the underlying quark-gluon structure of hadrons.

Form the SU(6) quark model one would expect that the spin of the proton is fully provided by the valence quark spins. Therefore the observation of the Ellis-Jaffe sum rule violation received attention by its implication that the sum of the quark helicities is much smaller than the proton helicity. The EMC result of a much smaller integrated spin-dependent structure function data than that expected from the Ellis-Jaffe sum rule triggered the proton "spin crisis", i.e., the intriguing question of how the spin of the proton is distributed among its quark spin, gluon spin and orbital angular momentum. It is commonly taken for granted that the EMC result implies that there must be some contribution due to gluon polarization or orbital angular momentum to the proton spin. For example, in the gluonic and the strange sea explanations of the EJSR breaking, the proton spin carried by the spin of quarks was estimated to be of about 70% in the former and of about 30% in the latter[1]. It will be reported here based on previous works[2-5], however, that the the proton spin problem raised by the Ellis-Jaffe sum rule violation does not in conflict with the SU(6) quark model in which the spin of the proton, when viewed in its rest reference frame, is provided by the vector sum of the quark spins, provided that the relativistic effect from the quark transversal motions[2,3], the flavor asymmetry between the u and d valence quarks[4], and the intrinsic quark-antiquark pairs generated by the non-perturbative meson-baryon fluctuations of the nucleon sea[5] are taken into account.

1.2 The Relativistic Effect due to Quark Transversal Motions

As it is known, spin is essentially a relativistic notion associated with the space-time symmetry of Poincaré. The conventional 3-vector spin $\mathbf{s}$ of a moving particle with finite mass m and 4-momentum p_μ can be defined by transforming its Pauli-Lubánski 4-vector $\omega_\mu = 1/2 J^{\rho\sigma} P^\nu \epsilon_{\nu\rho\sigma\mu}$ to its rest frame via a non-rotation Lorentz boost $L(p)$ which satisfies $L(p)p = (m, \mathbf{0})$, by $(0, \mathbf{s}) = L(p)\omega/m$. Under an arbitrary Lorentz transformation, a particle state with spin $\mathbf{s}$ and 4-momentum p_μ will transform to the state with spin $\mathbf{s}'$ and 4-momentum p'_μ,

$$\mathbf{s}' = R_\omega(\Lambda, p)\mathbf{s}, \qquad p' = \Lambda p, \tag{1}$$

where $R_\omega(\Lambda, p) = L(p')\Lambda L^{-1}(p)$ is a pure rotation known as Wigner rotation. When a composite system is transformed from one frame to another one, the spin of each constituent will undergo a Wigner rotation. These spin rotations are not necessarily the same since the constituents have different internal motion. In consequence, the sum of the constituent's spin is not Lorentz invariant.

The key points for understanding the proton spin puzzle lie in the facts that the vector sum of the constituent spins for a composite system is not Lorentz invariant by taking into account the relativistic effect of Wigner rotation, and that it is in the infinite momentum frame the small EMC result was interpreted as an indication that quarks carry a small amount of the total spin of the proton. From the first fact we know that the vector spin structure of hadrons could be quite different in different frames from relativistic viewpoint. We thus can naturally understand the proton "spin crisis" because there is no need to require that the sum of the quark spins be equal to the spin of the proton in the infinite momentum frame, even if the vector sum of the quark spins equals to the proton spin in the rest frame.

The effect due to the Wigner rotation can be best understood from the light-cone spin structure of the pion. It has been shown[6,7,8] that there are higher helicity ($\lambda_1 + \lambda_2 = \pm 1$) components in the light-cone spin space wavefunction for the pion besides the usual helicity ($\lambda_1 + \lambda_2 = 0$) components[6,7]. Therefore the light-cone wavefunction for the lowest valence state of pion can

200

be expressed as

$$|\psi_{q\bar{q}}^{\pi}> = \psi(x,\vec{k}_\perp,\uparrow,\downarrow)|\uparrow\downarrow> +\psi(x,\vec{k}_\perp,\downarrow,\uparrow)|\downarrow\uparrow>$$
$$+\psi(x,\vec{k}_\perp,\uparrow,\uparrow)|\uparrow\uparrow> +\psi(x,\vec{k}_\perp,\downarrow,\downarrow)|\downarrow\downarrow>, \tag{2}$$

It is interesting to be noticed that the light-cone wave function (2) is the correct pion spin wave function since it is an eigenstate of the total spin operator $(\hat{S}^F)^2$ in the light-cone formalism[8].

It is thus necessary to clarify what is meant by the quantity Δq defined by $\Delta q \cdot S_\mu = <P,S|\bar{q}\gamma_\mu\gamma_5 q|P,S>$, where S_μ is the proton polarization vector. Δq can be calculated from $\Delta q = <P,S|\bar{q}\gamma^+\gamma_5 q|P,S>$ since the instantaneous fermion lines do not contribute to the $+$ component. One can easily prove, by expressing the quark wave functions in terms of light-cone Dirac spinors (i.e., the quark spin states in the infinite momentum frame), that

$$\Delta q = \int_0^1 dx\, [q^\uparrow(x) - q^\downarrow(x)], \tag{3}$$

where $q^\uparrow(x)$ and $q^\downarrow(x)$ are the probabilities of finding, in the proton infinite momentum frame, a quark or antiquark of flavor q with fraction x of the proton longitudinal momentum and with polarization parallel or antiparallel to the proton spin, respectively. However, if one expresses the quark wave functions in terms of conventional instant form Dirac spinors (i.e., the quark spin state in the proton rest frame), it can be found, that

$$\Delta q = \int d^3\vec{p}\, M_q\, [q^\uparrow(p) - q^\downarrow(p)] = <M_q> \Delta q_L, \tag{4}$$

with

$$M_q = [(p_0 + p_3 + m)^2 - \vec{p}_\perp^2]/[2(p_0 + p_3)(m + p_0)] \tag{5}$$

being the contribution from the relativistic effect due to the quark transversal motions, $q^\uparrow(p)$ and $q^\downarrow(p)$ being the probabilities of finding, in the proton rest frame, a quark or antiquark of flavor q with rest mass m and momentum p_μ and with spin parallel or antiparallel to the proton spin respectively, and $\Delta q_L = \int d^3\vec{p}[q^\uparrow(p) - q^\downarrow(p)]$ being the net spin vector sum of quark flavor q parallel to the proton spin in the rest frame. Thus one sees that the quantity Δq should be interpreted as the net spin polarization in the infinite momentum frame if one properly considers the relativistic effect due to internal quark transversal motions.

Since $<M_q>$, the average contribution from the relativistic effect due to internal transversal motions of quark flavor q, ranges from 0 to 1 (or more properly, it should be around 0.75 for light flavor quarks and approaches 1 for heavy flavor quarks), and Δq_L, the net spin vector polarization of quark flavor q parallel to the proton spin in the proton rest frame, is related to the quantity Δq by the relation $\Delta q_L = \Delta q / <M_q>$, we have sufficient freedom to make the naive quark model spin sum rule, i.e., $\Delta u_L + \Delta d_L + \Delta s_L = 1$, satisfied while still preserving the values of Δu, Δd and Δs as parametrized from experimental data in appropriate explanations. Thereby we can understand the "spin crisis" simply because the quantity $\Delta\Sigma = \Delta u + \Delta d + \Delta s$ does not represent, in a strict sense, the vector sum of the spin carried by the quarks in the naive quark model. It is possible that the value of $\Delta\Sigma = \Delta u + \Delta d + \Delta s$ is small whereas the spin sum rule

$$\Delta u_L + \Delta d_L + \Delta s_L = 1 \tag{6}$$

for the naive quark model still holds, though the realistic situation may be complicated.

2 A Light-Cone Quark-Spectator-Diquark Model for Nucleons

From the impulse approximation picture of deep inelastic scattering, one can calculate the valence quark distributions in the quark-diquark model where the single valence quark is the scattered parton and the non-interacting diquark serves to provide the quantum number of the spectator[4]. The proton wave function in the quark-diquark model[9] is written as

$$\Psi_p^{\uparrow,\downarrow}(qD) = sin\theta \; \varphi_V \; |qV>^{\uparrow,\downarrow} + cos\theta \; \varphi_S \; |qS>^{\uparrow,\downarrow}, \tag{7}$$

with

$$|qV>^{\uparrow,\downarrow} = \pm \tfrac{1}{3}[V_0(ud)u^{\uparrow,\downarrow} - \sqrt{2}V_{\pm 1}(ud)u^{\downarrow,\uparrow} - \sqrt{2}V_0(uu)d^{\uparrow,\downarrow} + 2V_{\pm 1}(uu)d^{\downarrow,\uparrow}];$$
$$|qS>^{\uparrow,\downarrow} = S(ud)u^{\uparrow,\downarrow}, \tag{8}$$

where $V_{s_z}(q_1 q_2)$ stays for a $q_1 q_2$ vector diquark Fock state with third spin component s_z, $S(ud)$ stays for a ud scalar diquark Fock state, φ_D stays for the momentum space wave function of the quark-diquark with D representing the vector (V) or scalar (S) diquarks, and θ is a mixing angle that breaks SU(6) symmetry at $\theta \neq \pi/4$. We choose the bulk SU(6) symmetry case $\theta = \pi/4$. From Eq. (7) we get the unpolarized quark distributions

$$u_v(x) = \tfrac{1}{2}a_S(x) + \tfrac{1}{6}a_V(x);$$
$$d_v(x) = \tfrac{1}{3}a_V(x), \tag{9}$$

where $a_D(x)$ $(D = S$ or $V)$ is normalized such that $\int_0^1 dx a_D(x) = 3$ and denotes the amplitude for the quark q is scattered while the spectator is in the diquark state D. Therefore we can write, by assuming the isospin symmetry between the proton and the neutron, the unpolarized structure functions for nucleons,

$$F_2^p(x) = xs(x) + \tfrac{2}{9}xa_S(x) + \tfrac{1}{9}xa_V(x);$$
$$F_2^n(x) = xs(x) + \tfrac{1}{18}xa_S(x) + \tfrac{1}{6}xa_V(x), \tag{10}$$

where $s(x)$ denotes the contribution from the sea.

Exact SU(6) symmetry provides the relation $a_S(x) = a_V(x)$ which implies the valence flavor symmetry $u_v(x) = 2d_v(x)$. This gives the prediction[10] $F_2^n(x)/F_2^p(x) \geq 2/3$ for all x and is ruled out by the experimental observation $F_2^n(x)/F_2^p(x) < 1/2$ for $x \to 1$. It has been a well established fact that the valence flavor symmetry $u_v(x) = 2d_v(x)$ does not hold and the explicit $u_v(x)$ and $d_v(x)$ can be parameterized from the combined experimental data from deep inelastic scatterings of electron (muon) and neutrino (anti-neutrino) on the proton and the neutron *et al.*. In this sense, any theoretical calculation of quark distributions should reflect the flavor asymmetry between the valence u and d quarks in a reasonable picture. It has been shown[4] that the mass difference between the scalar and vector spectators can reproduce the up and down valence quark asymmetry that accounts for the observed ratio $F_2^n(x)/F_2^p(x)$ at large x.

The amplitude for the quark q is scattered while the spectator in the spin state D can be written as

$$a_D(x) \propto \int [\mathrm{d}^2\mathbf{k}_\perp] |\varphi_D(x, \mathbf{k}_\perp)|^2. \tag{11}$$

We adopt the Brodsky-Huang-Lepage prescription for the light-cone momentum space wave function[11] of the quark-spectator

$$\varphi_D(x, \mathbf{k}_\perp) = A_D exp\{-\frac{1}{8\beta_D^2}[\frac{m_q^2 + \mathbf{k}_\perp^2}{x} + \frac{m_D^2 + \mathbf{k}_\perp^2}{1 - x}]\}, \tag{12}$$

where $\mathbf{k}_\perp$ is the internal quark transversal momentum, m_q and m_D are the masses for the quark q and spectator D, and β_D is the harmonic oscillator scale parameter. The values of the parameters β_D, m_q and m_D can be adjusted by fitting the hadron properties such as the electromagnetic form factors, the mean charge radiuses, and the weak decay constants *et al.* in the relativistic light-cone quark model. We simply adopt $m_q = 330$ MeV and $\beta_D = 330$ MeV. The masses of the scalar and vector spectators should be different taking into account the spin force from color magnetism or alternatively from instantons[12]. We choose, e.g., $m_S = 600$ MeV and $m_V = 900$ MeV as estimated to explain the N-Δ mass difference. The mass difference between the scalar and vector spectators causes difference between $a_S(x)$ and $a_V(x)$ and thus the flavor asymmetry between the valence quark distribution functions $u_v(x)$ and $d_v(x)$. The calculated results[4] are in reasonable agreement with the experimental data and this supports the quark-spectator picture of deep inelastic scattering in which the difference between the scalar and vector spectators is important to reproduce the explicit SU(6) symmetry breaking while the bulk SU(6) symmetry of the quark model still holds.

For the polarized quark distributions, we take into account the contribution from the Wigner rotation[2,3]. In the light-cone or quark-parton descriptions, $\Delta q(x) = q^\uparrow(x) - q^\downarrow(x)$, where $q^\uparrow(x)$ and $q^\downarrow(x)$ are the probability of finding a quark or antiquark with longitudinal momentum fraction x and polarization parallel or antiparallel to the proton helicity in the infinite momentum frame. However, in the proton rest frame, one finds,

$$\Delta q(x) = \int [\mathrm{d}^2 \mathbf{k}_\perp] W_D(x, \mathbf{k}_\perp)[q_{s_z=\frac{1}{2}}(x, \mathbf{k}_\perp) - q_{s_z=-\frac{1}{2}}(x, \mathbf{k}_\perp)], \qquad (13)$$

with

$$W_D(x, \mathbf{k}_\perp) = \frac{(k^+ + m)^2 - \mathbf{k}_\perp^2}{(k^+ + m)^2 + \mathbf{k}_\perp^2} \qquad (14)$$

being the contribution from the relativistic effect due to the quark transversal motions, $q_{s_z=\frac{1}{2}}(x, \mathbf{k}_\perp)$ and $q_{s_z=-\frac{1}{2}}(x, \mathbf{k}_\perp)$ being the probability of finding a quark and antiquark with rest mass m and with spin parallel and anti-parallel to the rest proton spin, and $k^+ = x\mathcal{M}$ where $\mathcal{M} = \frac{m_q^2 + \mathbf{k}_\perp^2}{x} + \frac{m_D^2 + \mathbf{k}_\perp^2}{1-x}$. The Wigner rotation factor $W_D(x, \mathbf{k}_\perp)$ ranges from 0 to 1; thus Δq measured in polarized deep inelastic scattering cannot be identified with the spin carried by each quark flavor in the proton rest frame.

From Eq. (7) we get the spin distribution probabilities in the quark-diquark model

$$\begin{aligned}
u_V^\uparrow &= \tfrac{1}{18}; & u_V^\downarrow &= \tfrac{2}{18}; & d_V^\uparrow &= \tfrac{2}{18}; & d_V^\downarrow &= \tfrac{4}{18}; \\
u_S^\uparrow &= \tfrac{1}{2}; & u_S^\downarrow &= 0; & d_S^\uparrow &= 0; & d_S^\downarrow &= 0.
\end{aligned} \qquad (15)$$

Taking into account the Wigner rotation, we can write the quark helicity distributions for the u and d quarks

$$\begin{aligned}
\Delta u_v(x) &= u_v^\uparrow(x) - u_v^\downarrow(x) = -\tfrac{1}{18}a_V(x)W_V(x) + \tfrac{1}{2}a_S(x)W_S(x); \\
\Delta d_v(x) &= d_v^\uparrow(x) - d_v^\downarrow(x) = -\tfrac{1}{9}a_V(x)W_V(x),
\end{aligned} \qquad (16)$$

where $W_D(x)$ is the correction factor due the Wigner rotation. From Eq. (9) one gets

$$\begin{aligned}
a_S(x) &= 2u_v(x) - d_v(x); \\
a_V(x) &= 3d_v(x).
\end{aligned} \qquad (17)$$

204

Combining Eqs. (16) and (17) we have

$$\Delta u_v(x) = [u_v(x) - \tfrac{1}{2}d_v(x)]W_S(x) - \tfrac{1}{6}d_v(x)W_V(x);$$
$$\Delta d_v(x) = -\tfrac{1}{3}d_v(x)W_V(x). \tag{18}$$

Thus we arrive at simple relations between the polarized and unpolarized quark distributions for the valence u and d quarks. We can calculate the quark helicity distributions $\Delta u_v(x)$ and $\Delta d_v(x)$ from the unpolarized quark distributions $u_v(x)$ and $d_v(x)$ by relations (18), once the detailed x-dependent Wigner rotation factor $W_D(x)$ is known. On the other hand, we can also use relations (18) to study $W_S(x)$ and $W_V(x)$, once there are good quark distributions $u_v(x)$, $d_v(x)$, $\Delta u_v(x)$, and $\Delta d_v(x)$ from experiments. From another point of view, the relations (18) can be considered as the results of the conventional SU(6) quark model by explicitly taking into account the Wigner rotation effect and the flavor asymmetry introduced by the mass difference between the scalar and vector spectators, thus any evidence for the invalidity of Eq. (18) will be useful to reveal new physics beyond the SU(6) quark model.

We calculated the x-dependent Wigner rotation factor $W_D(x)$ in the light-cone SU(6) quark-spectator model[4] and noticed slight asymmetry between $W_S(x)$ and $W_V(x)$. Considering only the valence quark contributions, we can write the spin-dependent structure functions $g_1^p(x)$ and $g_1^n(x)$ for the proton and the neutron by

$$g_1^p(x) = \tfrac{1}{2}[\tfrac{4}{9}\Delta u_v(x) + \tfrac{1}{9}\Delta d_v(x)] = \tfrac{1}{18}[(4u_v(x) - 2d_v(x))W_S(x) - d_v(x)W_V(x)];$$
$$g_1^n(x) = \tfrac{1}{2}[\tfrac{1}{9}\Delta u_v(x) + \tfrac{4}{9}\Delta d_v(x)] = \tfrac{1}{36}[(2u_v(x) - d_v(x))W_S(x) - 3d_v(x)W_V(x)]. \tag{19}$$

We found[4] that the calculated A_1^N with Wigner rotation are in agreement with the experimental data, at least for $x \geq 0.1$. The large asymmetry between $W_S(x)$ and $W_V(x)$ has consequence for a better fit of the data.

As we consider only the valence quark contributions to $g_1^p(x)$ and $g_1^n(x)$, we should not expect to fit the Ellis-Jaffe sum data from experiments. This leaves room for additional contributions from sea quarks or other sources. We point out, however, it is possible to reproduce the observed Ellis-Jaffe sums $\Gamma_1^p = \int_0^1 g_1^p(x)\mathrm{d}x$ and $\Gamma_1^n = \int_0^1 g_1^n(x)\mathrm{d}x$ within the light-cone SU(6) quark-spectator model by introducing a large asymmetry between the Wigner rotation factors W_S and W_V for the scalar and vector spectators. For example, we need $< W_S >= 0.56$ and $< W_V >= 0.92$ to produce $\Gamma_1^p = 0.136$ and $\Gamma_1^n = -0.03$ as observed in experiments. This can be achieved by adopting a large difference between β_S and β_V which should be adjusted by fitting other nucleon properties in the model[12]. The calculated $A_1^p(x)$, $A_1^n(x)$, and $A_1^d(x)$ are in good agreement with the data[4]. This may suggest that the explicit SU(6) asymmetry could be also used to explain the EJSR violation (or partially) within a bulk SU(6) symmetry scheme of the quark model, or we take this as a hint for other SU(6) breaking source in additional to the SU(6) quark model.

We showed in the above that the u and d asymmetry in the lowest valence component of the nucleon and the Wigner rotation effect due to the internal quark transversal motions are important for re-producing the observed ratio F_2^n/F_2^p and the polarization asymmetries A_1^N for the proton, neutron, and deuteron. For a better understanding of the origin of polarized sea quarks implied by the violation of the Ellis-Jaffe sum rule, we still need to consider the higher Fock states implied by the non-perturbative meson-baryon fluctuations[5]. In the light-cone meson-baryon fluctuation model, the net d quark helicity of the intrinsic $q\bar{q}$ fluctuation is negative, whereas the net $\bar{d}$ antiquark helicity is zero. Therefore the quark/antiquark asymmetry of the $d\bar{d}$ pairs should be apparent in the d quark and antiquark helicity distributions. There are now explicit measurements of the helicity distributions for the individual u and d valence and sea quarks by SMC[13]. The helicity distributions for the u and d antiquarks are consistent with zero in agreement with the results of the light-cone meson-baryon fluctuation model of intrinsic $q\bar{q}$ pairs. The calculated quark helicity distributions $\Delta u_v(x)$ and $\Delta d_v(x)$ have been compared[4] with the recent SMC data. The data are still not precise enough for making detailed comparison, but the agreement with $\Delta u_v(x)$ seems to be good. It seems that the agreement with $\Delta d_v(x)$ is poor and there is somewhat evidence for additional source of negative helicity contribution to the valence d quark beyond the conventional quark model. This again supports the light-cone meson-baryon fluctuation model in which the helicity distribution of the intrinsic d sea quarks $\Delta d_s(x)$ is negative.

The standard SU(6) quark model gives the constraints $|\Delta u_v| \leq \frac{4}{3}$ and $|\Delta d_v| \leq \frac{1}{3}$. A global fit[14] of polarized deep inelastic scattering data together with constraints from nucleon and hyperon decay and the included higher-order perturbative QCD corrections leads to values for different quark helicity contributions in the proton: $\Delta u = 0.83 \pm 0.03$, $\Delta d = -0.43 \pm 0.03$, $\Delta s = -0.10 \pm 0.03$. In the light-cone meson-baryon fluctuation model, the antiquark helicity contributions are zero. We thus can consider the empirical values as the helicity contributions $\Delta q = \Delta q_v + \Delta q_s$ from both the valence q_v and sea q_s quarks. Thus the empirical result $|\Delta d| > \frac{1}{3}$ strongly implies an additional negative contribution Δd_s in the nucleon sea.

3 The Meson-Baryon Fluctuations of the Nucleon Sea

The light-cone meson-baryon fluctuation model of intrinsic $q\bar{q}$ pairs[5] leads to a consistent picture for understanding a number of empirical anomalies related to the composition of the nucleons in terms of their non-valence sea quarks. We have studied the sea quark/antiquark asymmetries in the nucleon wavefunction which are generated by a light-cone model of energetically-favored meson-baryon fluctuations. The model predicts striking quark/antiquark asymmetries in the momentum and helicity distributions for the down and strange contributions to the proton structure function: the intrinsic d and s quarks in the proton sea are negatively polarized, whereas the intrinsic $\bar{d}$ and $\bar{s}$ antiquarks give zero contributions to the proton spin. Such a picture is supported by experimental phenomena related to the proton spin problem: the recent SMC measurement of helicity distributions for the individual up and down valence quarks and sea antiquarks, the global fit of different quark helicity contributions from experimental data, and the negative strange quark helicity from the Λ polarization in $\bar{\nu}N$ experiments by the WA59 Collaboration. The light-cone meson-baryon fluctuation model also suggests a structured momentum distribution asymmetry for strange quarks and antiquarks which is related to an outstanding conflict between two different measures of strange quark sea in the nucleon[15]. The model predicts an excess of intrinsic $d\bar{d}$ pairs over $u\bar{u}$ pairs, as supported by the Gottfried sum rule violation[16]. We also predict that the intrinsic charm and anticharm helicity and momentum distributions are not identical.

The intrinsic sea model thus gives a clear picture of quark flavor and helicity distributions supported qualitatively by a number of experimental phenomena. It seems to be an important physical source for the problems of the Gottfried sum rule violation, the Ellis-Jaffe sum rule violation, and the conflict between two different measures of strange quark distributions.

4 Conclusion

The "intrinsic" $q\bar{q}$ pairs generated by the non-perturbative meson-baryon fluctuations in the nucleon sea are important source for a consistent picture of a number of empirical anomalies. The violations of the Gottfried sum rule and the Ellis-Jaffe sum rule are closely connected by the meson-baryon fluctuations in the nucleon, in contrary to most studies in which they are not related. The violation of the Ellis-Jaffe sum rule does not in conflict with the SU(6) quark model, provided that the relativistic effect from the Wigner rotation, the flavor asymmetry generated by the spin-spin interactions of the valence quarks, and the intrinsic $q\bar{q}$ pairs generated by non-perturbative meson-baryon fluctuations are taken into account. One important prediction of the model is the significant quark/antiquark asymmetries in the momentum and helicity distributions for the quarks and antiquarks of the nucleon sea, contrary to intuition based on perturbative gluon-splitting processes. It is important to test the meson-baryon fluctuation model by it's predictions in various physical processes, such as the forward energetic neutron events at HERA and the asymmetric quark/antiquark hadronization in e^+e^- annihilation[17].

Acknowledgments

I am very grateful to Professors Stan J. Brodsky, Tao Huang, and Qi-Ren Zhang for their valuable contribution, guidance and encouragement during the profitable and enjoyable collaborations from which the contents in this talk were developed.

References

1. For a recent review, see, e.g., M. Anselmino, A. Efremov, and E. Leader, Phys. Rep. **261** (1995) 1.
2. B.-Q. Ma, J. Phys. **G 17** (1991) L53; B.-Q. Ma and Q.-R. Zhang, Z. Phys. **C 58** (1993) 479.
3. S. J. Brodsky and F. Schlumpf, Phys. Lett. **B 329** (1994) 111.
4. B.-Q. Ma, Phys. Lett. **B 375** (1996) 320.
5. S. J. Brodsky and B.-Q. Ma, Phys. Lett. **B 381** (1996) 317.
6. B.-Q. Ma, Z. Phys. **A 345** (1993) (321).
7. T. Huang, B.-Q. Ma, and Q.-X. Shen, Phys. Rev. **D 49** (1994) 1490.
8. B.-Q. Ma and T. Huang, J. Phys. **G 21** (1995) 765.
9. M. I. Pavković, Phys. Rev. **D 13** (1976) 2128.
10. J. Kuti and V. F. Weisskopf, Phys. Rev. **4** (1971) 3419.
11. S. J. Brodsky, T. Huang, and G. P. Lepage, in: *Particles and Fields*, eds. A. Z. Capri and A. N. Kamal (Plenum, New York, 1983), p. 143.
12. H. J. Weber, Phys. Lett. **B 287** (1992) 14; Phys. Rev. **D 49** (1994) 3160.
13. SM Collab., B. Adeva *et al.*, Phys. Lett. **B 369** (1996) 93.
14. J. Ellis and M. Karliner, Phys. Lett. **B 341** (1995) 397.
15. B.-Q. Ma, Chin. Phys. Lett. **13** (1996) 648 (hep-ph/9604342).
16. See, e.g., B.-Q. Ma, A. Schäfer and W. Greiner, Phys. Rev. **D 47** (1993) 51, and references therein.
17. S.J. Brodsky and B.-Q. Ma, Phys. Lett. **B** in presss (hep-ph/9610304).

QUARK DISTRIBUTION FUNCTIONS IN A DIQUARK SPECTATOR MODEL

P.J. Mulders[1,2], J. Rodrigues[1,3]
[1] *National Institute for Nuclear Physics and High–Energy Physics (NIKHEF)*
P.O. Box 41882, NL-1009 DB Amsterdam, the Netherlands
[2] *Department of Physics and Astronomy, Free University*
De Boelelaan 1081, NL-1081 HV Amsterdam, the Netherlands
[3] *Instituto Superior Técnico*
Av. Rovisco Pais, 1100 Lisboa, Portugal

The representation of quark distribution functions in terms of nonlocal operators is combined with a simple diquark spectator model. This allows us to estimate these functions for the nucleon ensuring correct crossing and support properties.

1 Introduction

Quark distribution functions appear in the field-theoretical description of hard scattering processes as the parts that connect the quark and gluon lines to hadrons in the initial state. These parts are defined through (connected) matrix elements of nonlocal operators built from quark and gluon fields. The simplest, but most important ones, are the quark-quark correlation functions[1,2,3]. For each quark flavor one can write down

$$\Phi_{ij}(p, P, S) = \int \frac{d^4\xi}{(2\pi)^4} \, e^{-ip\cdot\xi} \, \langle P, S|\overline{\psi}_j(\xi)\psi_i(0)|P, S\rangle, \tag{1}$$

The hadron states are characterized by the momentum and spin vectors, $|P, S\rangle$. The quark momentum is denoted p. A summation over colors is understood in Φ.

In a particular hard process only certain Dirac projections of the correlation functions appear. For inclusive lepton-hadron scattering, where the hard scale Q is set by the spacelike momentum transfer $-q^2 = Q^2$, the correlation function Φ appears in the leading order result in an expansion in $1/Q$ and the structure functions can be expressed in quark distribution functions.

The hard momentum scale q and the hadron momentum P define the lightlike directions $n_\pm$. In inclusive lepton-hadron scattering they are related to the hadron momentum P and the hard momentum q as

$$P = \frac{M^2 x_b}{Q\sqrt{2}} \, n_- + \frac{Q}{x_b\sqrt{2}} \, n_+, \tag{2}$$

$$q = \frac{Q}{\sqrt{2}} \, n_- - \frac{Q}{\sqrt{2}} \, n_+. \tag{3}$$

The quark momentum p and the spin vector S are also expanded in the lightlike vectors,

$$p = \frac{xQ}{x_b\sqrt{2}} \, n_+ + \frac{x_b(p^2 + p_T^2)}{xQ\sqrt{2}} \, n_- + p_T, \tag{4}$$

$$S = \frac{\lambda Q}{M x_B \sqrt{2}} \, n_+ - \frac{\lambda M x_B}{Q\sqrt{2}} \, n_- + S_T. \tag{5}$$

Thus x represents the fraction of the momentum in the $+$ direction carried by the quark inside the hadron. The lightlike components $a^\pm \equiv a \cdot n_\mp$ are defined with the help of the vectors $n_\pm$.

It should be noted that there are other contributions in the leading cross section resulting from soft parts with A^+ gluon legs. These can be absorbed into the correlation function providing the link operator that renders the definition in Eq. 1 color gauge invariant. Choosing the $A^+ = 0$ gauge in the study of Φ the link operator reduces to unity.

Although hard cross sections can be expressed in terms of distribution and fragmentation functions, these objects cannot be calculated from QCD because they involve the hadronic bound states. We follow here a different route. We investigate the structure of lightcone correlation functions (including the effects of transverse separation of the quark fields), we illustrate the consequences of various constraints and we estimate the distribution functions in a simple model. Particularly suitable is a model in which the spectrum of intermediate states that can be inserted in Eq. 1 is replaced by one state, which if the hadron is a baryon will simply be referred to as a *diquark*.

2 Quark Correlation Functions

The most general expression for Φ is [4,5]:

$$\begin{aligned}
\Phi(p, P, S) &= M\,A_1 + A_2\,\slashed{P} + A_3\,\slashed{p} + M\,A_6\,\slashed{S}\,\gamma_5 + \frac{A_7}{M}\,p\cdot S\,\slashed{P}\,\gamma_5 \\
&\quad + \frac{A_8}{M}\,p\cdot S\,\slashed{p}\,\gamma_5 + i\,A_9\,\sigma^{\mu\nu}\,\gamma_5\,S_\mu P_\nu \\
&\quad + i\,A_{10}\,\sigma^{\mu\nu}\,\gamma_5\,S_\mu p_\nu + i\,\frac{A_{11}}{M^2}\,p\cdot S\,\sigma^{\mu\nu}\,\gamma_5\,p_\mu P_\nu,
\end{aligned} \tag{6}$$

where the amplitudes A_i depend on $\sigma \equiv 2p\cdot P$ and $\tau \equiv p^2$. Hermiticity requires all amplitudes $A_i(\sigma, \tau)$ to be real. Time reversal invariance has also been used.

210

As the variation along n_- up to order M^2/Q^2 is irrelevant in a hard process, one encounters the quantities

$$\Phi^{[\Gamma]}(x, \boldsymbol{p}_T) = \frac{1}{2} \int dp^- \, \text{Tr}\,(\Phi\Gamma)\Big|_{p^+ = xP^+}. \tag{7}$$

The projections of Φ on different Dirac structures define distribution functions. They are related to integrals over linear combinations of the amplitudes. The projections which contribute at leading order are

$$\Phi^{[\gamma^+]}(x, \boldsymbol{p}_T) \equiv f_1(x, \boldsymbol{p}_T^2),$$

$$\Phi^{[\gamma^+\gamma_5]}(x, \boldsymbol{p}_T) \equiv \lambda g_{1L}(x, \boldsymbol{p}_T^2) + \frac{\boldsymbol{p}_T \cdot \boldsymbol{S}_T}{M} g_{1T}(x, \boldsymbol{p}_T^2),$$

$$\Phi^{[i\sigma^{i+}\gamma_5]}(x, \boldsymbol{p}_T) \equiv S_T^i h_{1T}(x, \boldsymbol{p}_T^2) + \frac{p_T^i}{M}\left(\lambda h_{1L}^\perp(x, \boldsymbol{p}_T^2) + \frac{\boldsymbol{p}_T \cdot \boldsymbol{S}_T}{M} h_{1T}^\perp(x, \boldsymbol{p}_T^2)\right)$$

After integration over the quark transverse momentum, these leading twist functions have well known probabilistic interpretations. The function $f_1(x)$ gives the probability of finding a quark with light-cone momentum fraction x in the $+$ direction (and any transverse momentum). $g_1(x)$ is a chirality distribution: in a hadron that is in a positive helicity eigenstate ($\lambda = 1$), it measures the probability of finding a right-handed quark with light-cone momentum fraction x minus the the probability of finding a left-handed quark with the same light-cone momentum fraction (and any transverse momentum). $h_1(x)$ is a transverse spin distribution: in a transversely polarized hadron, it measures the probability of finding quarks with light-cone momentum fraction x polarized along the direction of the polarization of the hadron minus the probability of finding quarks with the same light-cone momentum fraction polarized along the direction opposite to the polarization of the hadron.

The subleading projections can also be expressed in distributions. The integrated distributions are $e(x)$, $g_T(x)$ and $h_L(x)$. Important relations are

$$g_T(x) = g_1(x) + \frac{d}{dx}\, g_{1T}^{(1)}(x), \tag{8}$$

$$h_L(x) = h_1(x) - \frac{d}{dx}\, h_{1L}^{\perp(1)}(x),, \tag{9}$$

where

$$F^{(n)}(x) \equiv \int d^2\boldsymbol{p}_T \left(\frac{\boldsymbol{p}_T^2}{2M^2}\right)^n F(x, \boldsymbol{p}_T^2). \tag{10}$$

The functions $g_2 = g_T - g_1$ and $h_2 = 2(h_L - h_1)$ thus satisfy the sum rules

$$\int_0^1 dx\, g_2(x) \;=\; -g_{1T}^{(1)}(0),\tag{11}$$

$$\int_0^1 dx\, h_2(x) \;=\; 2\,h_{1L}^{\perp(1)}(0),\tag{12}$$

If the functions $g_{1T}^{(1)}$ and $h_{1L}^{\perp(1)}$ vanish at the origin, we rediscover the Burkhardt-Cottingham sum rule[6] and the Burkardt sum rule[7].

3   The Diquark Spectator Model

The basic idea of the spectator model is to treat the intermediate states that can be inserted in Φ in Eq. 1 as a state with a definite mass M_R, called the spectator. In other words, we make a specific ansatz for the spectral decomposition of the correlation function. The quantum numbers of the intermediate state are those determined by the action of the quark field on the state $|P, S\rangle$, hence the name *diquark spectator*. The inclusion of antiquark and gluon distributions requires a more complex spectral decomposition of intermediate states. We restrict ourselves to the simplest case. Φ is then given by

$$\Phi_{ij}^R = \frac{1}{(2\pi)^3}\, \langle P, S|\overline{\psi}_j(0)|X^{(\lambda)}\rangle\, \theta(P_R^+)\, \delta\left[(p-P)^2 - M_R^2\right]\, \langle X^{(\lambda)}|\psi_i(0)|P, S\rangle\tag{13}$$

where $P_R = P - p$ and $X^{(\lambda)}$ represents the spectator, whose possible spin states are indicated with λ.

The spin of the diquark state can be 0 (scalar diquark s) or 1 (axial vector diquark a). The matrix element appearing in the RHS of (13) is given by

$$\langle X_s|\psi_i(0)|P, S\rangle \;=\; \left(\frac{i}{\not{p} - m}\right)_{ik} \Upsilon_{kl}^s\, U_l(P, S),\tag{14}$$

$$\langle X_a^{(\lambda)}|\psi_i(0)|P, S\rangle \;=\; \epsilon_\mu^{*(\lambda)}\left(\frac{i}{\not{p} - m}\right)_{ik} \Upsilon_{kl}^{a\mu}\, U_l(P, S),\tag{15}$$

for a scalar and a vector diquark, respectively. The matrix elements consist of a nucleon-quark-diquark vertex Υ yet to be specified, the Dirac spinor for the nucleon $U_l(P, S)$, a quark propagator for the untruncated quark line (m is the mass of the quark) and a polarization vector $\epsilon_\mu^{*(\lambda)}$ in the case of an axial vector diquark. The next step is to fix the Dirac structure of the vertex Υ. We

212

assume that

$$\Upsilon^s = \mathbb{1}\, g_s(p^2), \tag{16}$$

$$\Upsilon^a_\mu = \frac{g_a(p^2)}{\sqrt{3}}\, \gamma_5 \left(\gamma^\mu + \frac{P^\mu}{M}\right), \tag{17}$$

where $g_R(p^2)$ (with R being s or a) are form factors that take into account the composite structure of the nucleon and the diquark. In the choice of vertices, the factors and projection operators are chosen to assure that in the target restframe the vector diquark states are pure spatial and the nucleon spinors have only upper components, in which case e.g. the axial vector diquark vertex reduces to $\chi_N^\dagger \boldsymbol{\sigma} \cdot \boldsymbol{\epsilon} \chi_q$. With our choices we obtain

$$\Phi^R = \frac{|g^R(p^2)|^2}{2(2\pi)^3} \frac{\delta\left(\tau - \sigma + M^2 - M_R^2\right)}{(p^2 - m^2)^2} (\not{p} + m)\,(\not{P} + M)\,(1 + a_R \gamma_5 \not{S})\,(\not{p} + m), \tag{18}$$

where $a_s = 1$ and $a_a = -1/3$. In obtainig this result we used as the polarization sum for the axial vector diquark $\sum_\lambda \epsilon_\mu^{*(\lambda)} \epsilon_\nu^{(\lambda)} = -g_{\mu\nu} + P_\mu P_\nu / M^2$, which is consistent with the choice that the axial vector diquark spin states are purely spatial in the nucleon restframe. We will use the same form factors for scalar and axial vector diquark being of the form

$$g(\tau) = N\, \frac{\tau - m^2}{|\tau - \Lambda^2|^\alpha}. \tag{19}$$

Λ is another parameter of the model and N is a normalization constant. This form factor has the advantage of killing the pole of the quark propagator [8].

4 Results and Discussion

Using Eq. 18 we can compute the amplitudes A_i shown in Eq. 6. Taking out some common factors by defining

$$A_i = \frac{N^2}{2(2\pi)^3} \frac{\delta\left(\tau - \sigma + M^2 - M_R^2\right)}{|\tau - \Lambda^2|^{2\alpha}} \tilde{A}_i \tag{20}$$

we obtain $\tilde{A}_4 = \tilde{A}_5 = \tilde{A}_{12} = 0$ and

$$\tilde{A}_1 \;=\; -\tilde{A}_6/a_R = \frac{m}{M}\left((M + m)^2 - M_R^2\right) + (\tau - m^2)\left(1 + \frac{m}{M}\right),$$

$$\tilde{A}_2 \;=\; -\tilde{A}_9/a_R = -\left(\tau - m^2\right),$$

$$\tilde{A}_3 \;=\; -\tilde{A}_{10}/a_R = (M+m)^2 - M_R^2 + (\tau - m^2),$$
$$\tilde{A}_7 \;=\; 2\,a_R\,mM,$$
$$\tilde{A}_8 \;=\; -\tilde{A}_{11}/a_R = 2\,a_R\,M^2.$$

Introducing the function

$$\lambda_R^2(x) = \Lambda^2(1-x) + xM_R^2 - x(1-x)M^2, \tag{21}$$

one gets, e.g., for $f_1(x,\boldsymbol{p}_T^2)$ the result

$$f_1(x,\boldsymbol{p}_T^2) = \frac{N^2\,(1-x)^{2\alpha-1}}{16\pi^3}\,\frac{(xM+m)^2 + \boldsymbol{p}_T^2}{\left(\boldsymbol{p}_T^2 + \lambda_R^2\right)^{2\alpha}}. \tag{22}$$

Although there is a certain freedom in the choice of the parameters, one immediately sees that the occurence of singularities in the integration region will cause problems that can be avoided if we require $\lambda_R^2(x)$ to be positive, what imposes the condition $M_R > M - \Lambda$. Provided this condition is fulfilled one obtains the integrated distributions:

$$f_1(x) \;=\; \frac{N^2\,(1-x)^{2\alpha-1}}{32\pi^2\,(\alpha-1)(2\alpha-1)}\left[\frac{2(\alpha-1)\,(xM+m)^2 + \lambda_R^2(x)}{\left(\lambda_R^2(x)\right)^{2\alpha-1}}\right],$$

$$g_1(x) \;=\; \frac{N^2 a_R\,(1-x)^{2\alpha-1}}{32\pi^2\,(\alpha-1)(2\alpha-1)}\left[\frac{2(\alpha-1)\,(xM+m)^2 - \lambda_R^2(x)}{\left(\lambda_R^2(x)\right)^{2\alpha-1}}\right],$$

$$h_1(x) \;=\; \frac{N^2 a_R\,(1-x)^{2\alpha-1}}{16\pi^2(2\alpha-1)}\,\frac{(Mx+m)^2}{\left(\lambda_R^2(x)\right)^{2\alpha-1}}.$$

Explicit expressions for the twist three functions can also be written [9].

An example of a $\boldsymbol{p}_T^2/2M^2$-weighted distribution is

$$g_{1T}^{(1)}(x) = \frac{N^2 a_R\,(1-x)^{2\alpha-1}}{32\pi^2\,(\alpha-1)(2\alpha-1)}\,\frac{\left(x + \frac{m}{M}\right)}{\left(\lambda_R^2(x)\right)^{2\alpha-2}}. \tag{23}$$

Up to now, we have not specified flavor in the distributions. For the nucleon we distinguish two types of distributions, f_1^s and f_1^a, etc. This corresponds to the spectator having spin 0 or spin 1, respectively. In that case spin 0 diquarks are in a flavor singlet state (S) and spin 1 diquarks are in a flavor triplet state (T) in order to combine to a symmetric spin-flavour wave function. Since the coupling of the spin has already been included in the vertices, we need the flavor coupling

$$|p\rangle = \frac{1}{\sqrt{2}}\,|u\,S_0\rangle + \frac{1}{\sqrt{6}}|u\,T_0\rangle - \frac{1}{\sqrt{3}}\,|d\,T_1\rangle, \tag{24}$$

to find that the flavor distributions are $f_1^u = \frac{3}{2} f_1^s + \frac{1}{2} f_1^a$, $f_1^d = f_1^a$ and similarly for g_1, h_1, e, g_T and h_L. The proportionality of the numbers is obtained from Eq. 24, while the overall factor is chosen to reproduce the sum rules for the number of up and down quarks if f_1^s and f_1^a are normalized to unity upon integration over p_T and x. This will fix the normalization N in the form factor. The factors $a_s = 1$ and $a_a = -1/3$ in the distribution functions will produce different u and d weighting for unpolarized and polarized distributions. Further differences between u and d distributions can also be induced by different choices of M_R, Λ or α. We take for the nucleon $\alpha = 2$ to reproduce the right large x behavior of f_1^u, i.e. $(1-x)^3$, as predicted by the Drell-Yan-West relation and reasonably well confirmed by data. We refrain from tuning the large x behaviour of f_1^d to match the $(1-x)^4$ form indicated by data. Since f_1^d is only affected by vector diquarks, this could be easily obtained by choosing a different form factor for the latter. We feel that this kind of fine-tuning would take things too far with the simple model we use. Similarly, we will only consider one value of Λ. We will, however, allow for different masses for scalar and vector diquark spectators. The color magnetic hyperfine interaction, responsible for the nucleon-delta mass difference, will also produce a mass difference between singlet and triplet diquark states. Neglecting dynamical effects, group-theoretical factors lead to a difference $M_a - M_s = 200$ MeV.

Another important constraint comes from the axial charge of the nucleon

$$g_A = \int_0^1 dx \left[g_1^u(x) - g_1^d(x) \right] = \int_0^1 dx \left[\frac{3}{2} g_1^s(x) + \frac{1}{2} g_1^a(x) \right]. \qquad (25)$$

The sensitivity to the parameters M_R and Λ is best illustrated by considering some characteristic values. We take a quark mass of 0.36 GeV, two different values for M_R (0.6 and 0.8 GeV) and three values for Λ (0.4, 0.5 and 0.6 GeV.) In Table 1 the values of some moments are given.

Table 1: The second moment of f_1, $\langle x \rangle = \int dx\, x f_1(x)$ and the first moments $g_1 = \int dx\, g_1(x)$ and $h_1 = \int dx\, h_1(x)$ are given for two diquark masses and for three values of Λ.

Λ (GeV)	$M_R = 0.6$ GeV			$M_R = 0.8$ GeV		
	$\langle x \rangle^R$	g_1^R	h_1^R	$\langle x \rangle^R$	g_1^R	h_1^R
0.4	0.367	$0.922\,a_R$	$0.965\,a_R$	0.232	$0.643\,a_R$	$0.821\,a_R$
0.5	0.373	$0.792\,a_R$	$0.896\,a_R$	0.255	$0.524\,a_R$	$0.760\,a_R$
0.6	0.383	$0.664\,a_R$	$0.832\,a_R$	0.278	$0.412\,a_R$	$0.707\,a_R$

Fig. 1 shows the twist two distributions for different values of the mass of the spectator and of the parameter Λ. We can see that an increase of M_R

induces a shift on the peak of the valence distribution $f_1(x)$ towards lower values of x and a decrease in its second moment. This is due to the fact that a spectator with higher mass corresponds to more massive intermediate states which can contain sea quarks. In these circumstances the momentum carried by valence quarks decreases. In the non relativistic limit the equality $f_1(x) = g_1(x) = h_1(x)$ holds, a result that could be anticipated.

We now use the axial charge of the nucleon to find the most suitable values for Λ and the diquark masses. The values $\Lambda = 0.5$ GeV, $M_s = 0.6$ GeV and $M_a = 0.8$ GeV give $g_A = 1.25$, close to the experimental value. Fig. 2 shows the distribution $f_1(x)$ multiplied by x for these values of the parameters. We find a satisfactory qualitative agreement with the valence distributions of Glück, Reya and Vogt [10].

Finally, Fig. 3 shows the distributions $g_2^u(x)$ and $g_2^d(x)$, for which we find a small violation of the Burkhardt-Cottingham sum rule, in agreement with Eq. 11.

Acknowledgments

This work was suported by the Foundation for Fundamental Research on Matter (FOM), the National Organization for Scientific Research and the Junta Nacional de Investigacao Cientifica (JNICT, PRAXIS XXI).

References

1. D.E. Soper, Phys. Rev. D **15** (1977) 1141; Phys. Rev. Lett. **43** (1979) 1847
2. J.C. Collins and D.E. Soper, Nucl. Phys. **B194** (1982) 445
3. R.L. Jaffe, Nucl. Phys. **B229** (1983) 205
4. J. P. Ralston and D. E. Soper, Nucl. Phys. B **152** (1979) 109.
5. R. D. Tangerman and P. J. Mulders, Phys. Rev. D **51** (1995) 3357.
6. H. Burkhardt and W. N. Cottingham, Ann. Phys. (N.Y.) **56** (1970) 453
7. M. Burkardt, in Proceedings of 10th International Symposium on High Energy Spin Physics, Nagoya, Japan, 1992, ed. Hasegawa et al. (Universal Academy Press, Tokyo, 1993)
8. W. Melnitchouk, A. W. Schreiber and A. W. Thomas, Phys. Rev D **49** (1994) 1183.
9. R. Jakob, P.J. Mulders and J. Rodrigues; in preparation.
10. M. Glück, E. Reya and A. Vogt; Z. Phys. C 67, 433 (1995).

216

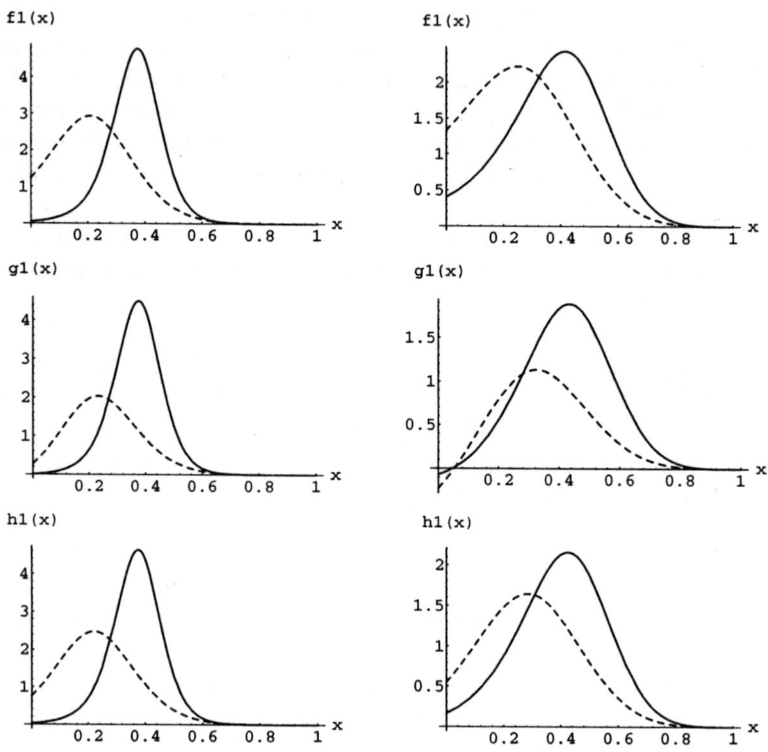

Figure 1: Twist two distributions for the nucleon. The plots on the top represent $f_1(x)$, the ones on the middle show $g_1(x)/a_R$ and at the bottom we have $h_1(x)/a_R$. The plots on the left correspond to $\Lambda = 0.4$ GeV and the ones on the right to $\Lambda = 0.6$ GeV. The full line corresponds to $M_R = 0.6$ GeV and the dashed line to $M_R = 0.8$ GeV.

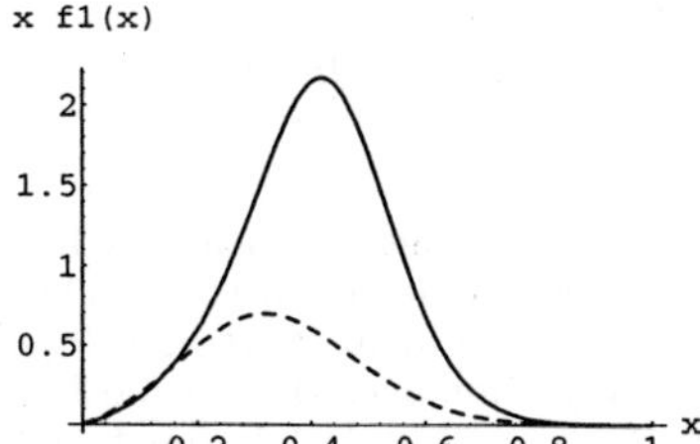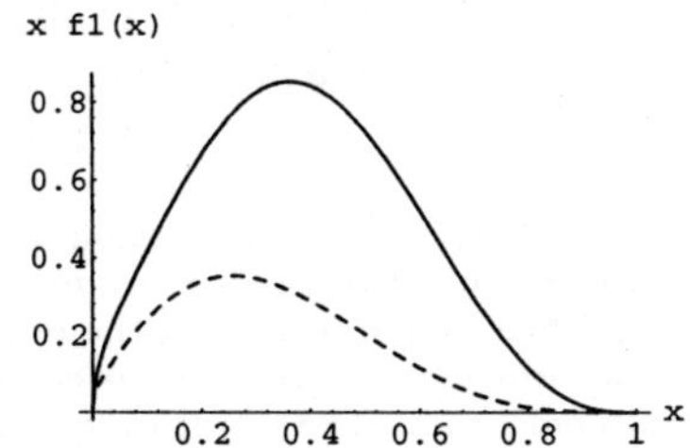

Figure 2: Twist two distributions for the nucleon. The plot on the left shows $x f_1^s(x)$ (full line) and $x f_1^d(x)$ (dashed line) for $M_s = 0.6$ GeV, $M_a = 0.8$ GeV and $\Lambda = 0.5$ GeV. The plot on the right shows the "low scale" valence distributions of GRV.

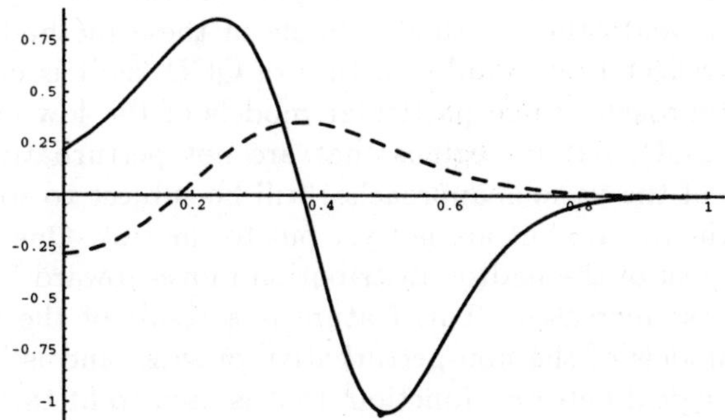

Figure 3: Distributions $g_2^u(x)$ (solid line) and $g_2^d(x)$ (dashed line) for $M_s = 0.6$ GeV, $M_a = 0.8$ GeV and $\Lambda = 0.5$ GeV.

SPIN-DEPENDENT FRAGMENTATION FUNCTIONS FOR BARYONS IN A DIQUARK MODEL

ANATOLY D. ADAMOV, GARY R. GOLDSTEIN

*Department of Physics, Tufts University, Medford,
MA 02155, USA*

The perturbative QCD calculation of heavy quark fragmentation into a heavy meson is extended to predict fragmentation into heavy flavor baryons. This is accomplished by implementing the quark-diquark model of the baryons. Several diquark form factors are used to enable the integration over the virtual heavy quark momentum. The resulting spin independent functions for charmed quarks to fragment into charmed baryons with spin 1/2 and 3/2 are compared with recent data. Predictions are made for the spin dependent fragmentation functions as well, particularly for the functions $\hat{g}_1$ in the case of spin 1/2 baryons.

1 Introduction

Fragmentation functions have received considerable attention in recent years. While experimental information beyond the pion distribution (presumably from light quarks) has been slow in accumulating, theoretical interest has been growing. The particular functional form for heavy flavored quarks to fragment into heavy mesons has been studied using Operator Product Expansion techniques, light cone quantization, QCD perturbation theory, Heavy Quark Effective Theory, and other methods. Some of these methods yield general properties that reflect the overall structure of QCD, as it is currently understood. Other approaches take particular models of the low energy behavior expected from QCD, but in regions that are not perturbatively calculable. The particulars of the various approaches will be subject to some experimental scrutiny in the future, but are not yet put to the test. One general feature is known - the peak of the hadron distribution moves toward higher momenta as the quark mass increases. This feature is a result of the kinematics implicit in most models of the non-perturbative process, and is incorporated in the phenomenological Peterson function[1] that is used to fit the sparse data on heavy quark fragmentation[2].

Even more difficult to test experimentally are the spin dependences of the fragmentation processes. Yet these dependences are important to know. They reflect the details of the primarily non-perturbative mechanism by which parton polarization is passed on to the hadrons. In this sense, the spin-dependent fragmentation involves the reverse of the process by which the nucleon spin is shared by its partons (the "spin crisis"), and may reveal a similarly mysterious decoupling of valence quark spin and hadron spin for some regions of kinematics.

Over the last several years a number of theorists have noticed that for fragmentation of a heavy flavor quark into "doubly heavy" mesons, like the c-quark into the J/Ψ or the b-quark into the B_c, perturbative QCD may be applicable[3,4]. If this is the case, the fragmentation functions are calculable, at an appropriate scale, and QCD radiative corrections can be obtained from the Renormalization Group or the Altarelli-Parisi equations. Such calculations have been performed and scrutinized. It has been shown that in the heavy quark limit (i.e. the mass goes to infinity) the functions have the form expected from more general considerations[5]. This corresponds to the heavy meson taking all of the heavy quark's momentum; the distribution becomes a delta function at $z = 1$. The $1/m_Q$ corrections are calculated also[a]. In any case, this approach can predict the spin-dependent fragmentation functions along with their momentum and mass dependences. In the heavy mass limit, of course, the spin of the heavy quark is conserved, so the spin dependence is simple. What is of phenomenological interest is the next order correction, at least, since that has non-trivial spin dependence.

In some circumstances the spin dependence of fragmentation is most readily studied experimentally by observing baryons rather than mesons. This is true for the production of hyperons or heavy hyperons (Λ_c, Λ_b, etc.), wherein the weak, parity violating decays provide polarization analyses[6]. To consider fragmentation into baryons in this perturbative scheme, the three quark system has to be confronted. A simple alternative is to consider the baryons as quark-diquark bound states, and to use the same perturbative method as for the mesons. In order for the perturbative calculation to be useful the creation of a heavy pair of quarks or diquarks must be an intermediate step. Hence, doubly heavy baryon fragmentation is an appropriate testing ground for these ideas. It is not expected that sufficient data to study this process will be available in the near future, however.

To begin to see the structure it will be worthwhile to stretch the region of applicability to the "singly" heavy baryons. We have been carrying out this program to see the expected spin and kinematic dependences, with the hope of providing an experimentally testable model. Of immediate interest is the question of whether the baryon fragmentation functions have the same kinematic dependence as the meson case. In general the answer is no in this model, but the details have to be studied. Furthermore, the spin dependent

---

[a]It is not clear to the present authors that all sources for such terms have been considered, particularly corrections coming from the treatment of the Bethe-Salpeter bound state wavefunction for the meson

fragmentation is interestingly distinct from the naive heavy quark limit. The perturbative calculation and its results will be presented below, along with a comparison with some recent data.

2 Perturbative calculation

The first calculations of the fragmentation functions in the perturbative scheme were applied to some of the inclusive heavy flavor meson decays of the Z^0, as produced at LEP. The partial width for this inclusive process can be written in general for a hadron H as

$$d\Gamma(Z^0 \to H(E) + X) = \sum_i \int_0^1 dz\, d\hat\Gamma(Z^0 \to i(E/z) + X, \mu)\, D_{i \to H}(z, \mu), \quad (1)$$

where H is the hadron of energy E and longitudinal momentum fraction z relative to the parton i, while μ is the arbitrary scale whose value will be chosen to avoid large logarithms. The fragmentation function enters here in a factorized form (that can be maintained through the evolution equations).

Now, consider the final state with one heavy flavor meson, say the B_c for definiteness. To leading order the B_c meson arises from the production of a pair of b-quarks, in which one of the quarks fragments into the meson. As Fig.1 illustrates, with $Q = b, Q' = c, \bar{Q}' = \bar{c}$, the perturbative contribution involves the virtual b-quark radiating a hard gluon (in axial gauge, so that there is no contribution from the opposite quark). The hard gluon produces a heavy flavor pair of c-quarks. The gluon must have energy at least twice the charm mass, so that the coupling $\alpha_s(Q^2)$ is small.

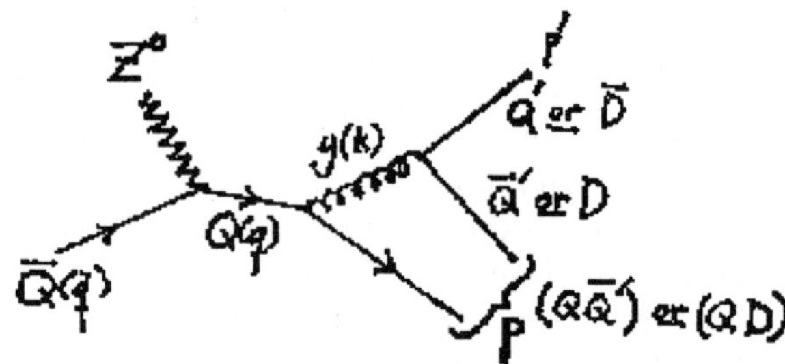

Figure 1: The amplitude for $Z^0 \to Meson(Q\bar{Q}') + X$ or Baryon(QD)+X.

For matching momenta (or relative 3-velocity zero) the b and $\bar{c}$ form the B_c meson with probability given by the square of the Bethe-Salpeter wave function. Since the doubly heavy mesons are weakly bound objects, the wave function at the origin is known from non-relativistic quark models for the heavy-heavy system.

The amplitude for Fig.1, A_1, can be evaluated explicitly from perturbation theory. The decay rate for unpolarized $Z^0 \to B_c + \bar{c} + b$ can be written

$$\Gamma_1 = \frac{1}{2M_Z} \int [d\bar{q}][dp][dp']\, (2\pi)^4 \delta^4(Z - \bar{q} - p - p')\frac{1}{3}\sum |A_1|^2, \qquad (2)$$

where $\bar{q}$, p, and p' are the 4-momenta of the $\bar{b}$, B_c, and c, respectively.

To obtain the full inclusive width the unobserved quarks must be integrated over. The phase space integration can be simplified considerably when the limiting case of $M_Z \to \infty$ is approached by taking leading order in m_b/M_Z (with $m_c < m_b$). The two body phase space for p and p' can be written as an integration over z and s, with transverse momentum fixed for each such pair. In the large M_Z approximation the transverse momentum of the hadron is small and $p = zq$. Once the square of the amplitude A_1 is summed over spins and simplified by dropping non-leading contributions, the width for $Z^0 \to \bar{c}c$ can be factored out of the expression Eq. 1 leaving an integral over the fragmentation function. Then

$$\int_0^1 dz\, D_{c \to H}(z) = \frac{8\alpha_s^2 |R(0)|^2}{27\pi m_c} \int_0^\infty ds \int_0^1 dz\, \Theta(s - \frac{4m_c^2}{z} - \frac{m_b^2}{1-z}) F(s, z), \quad (3)$$

where R(0) is the Bethe-Salpeter wavefunction at the origin, and F is the remaining integrand, which depends on $s = q^2$, z and the quark masses. The upper limit on the s integration appears as the $M_Z \to \infty$ limit. So the partial width for $Z^0 \to H + X$ is given by an integral over the virtuality of the heavy quark and the phase space of the unobserved degrees of freedom.

The same procedure can be applied directly to the baryons, if the quark-diquark model of the baryons is used. The hard gluon in the process must produce a diquark–anti-diquark pair, and the diquark (color anti-triplet) combines with the quark to form the baryon. Note that an alternative scenario has the heavy quark fragment into a diquark first, and then the diquark dresses itself to form the baryon[9]. This leads to very different results, as pointed out in Ref.[10], and is not justifiable herein, where the diquark is not necessarily heavy flavor. These latter authors [10] have performed a calculation that is similar in spirit to part of the spin independent procedure we follow below.

We now proceed with the calculation of fragmentation functions for (singly) heavy flavor baryons. The basic covariant coupling of diquarks to gluons was written long ago[7]. There is one coupling constant for the scalar diquark color octet vector current coupling to the gluon field—a color charge strength, along with a possible form factor F_s. The momentum space color octet current (which couples to the gluon field vector) is

$$J_\mu^{A(S)} = g_s F_s(k^2)(p + p')_\mu S^{\alpha\dagger} \lambda_{\alpha\beta}^A S^\beta, \qquad (4)$$

222

where p and p' are the scalar diquark 4-momenta and $k = p' - p$. For the vector diquark there are three constants - color charge, anomalous chromomagnetic dipole moment κ, and chromoelectric quadrupole moment λ, along with the corresponding form factors, F_E, F_M, and F_Q.

$$
\begin{aligned}
J_\mu^{A(V)} \;=\; & g_s(\lambda^A)_{\beta\alpha} \left\{ F_E(k^2)[\epsilon^\alpha(p) \cdot \epsilon^{\beta\dagger}(p')](p+p')_\mu \right. \\
& +(1+\kappa)F_M(k^2)[\epsilon_\mu^\alpha(p)p \cdot \epsilon^{\beta\dagger}(p') + \epsilon_\mu^{\beta\dagger}(p')p' \cdot \epsilon^\alpha(p)] \\
& \left. +\frac{\lambda}{m_D^2}F_Q(k^2)[\epsilon_\rho^\alpha(p)\epsilon_\nu^{\beta\dagger}(p') + \tfrac{1}{2}g_{\rho\nu}\epsilon^\alpha(p) \cdot \epsilon^{\beta\dagger}(p')]k^\rho k^\nu(p+p')_\mu \right\} ,
\end{aligned}
$$
$$(5)$$

where A is the color octet index, α, β, ..., are color anti-triplet indices, the ϵ's are polarization 4-vectors for the diquarks.

In the perturbative diagrams involved here, the virtual heavy quark emits a time-like off-shell gluon, that produces a diquark-antidiquark pair, while attaining nearly on-shell 4-momentum. The diquark combines with the heavy quark to form a heavy flavor baryon, whose amplitude for formation is related to the Bethe-Salpeter wavefunction for the diquark-quark system. As in the meson production calculations, it is assumed that the constituents are heavy enough so that the binding is relatively weak, i.e. the quark and diquark are both on-shell and the binding energy is negligibly small. This is expected to be true for constituents with masses well above Λ_{QCD}, and even the light flavor diquarks almost satisfy this constraint. The basic perturbative amplitude is shown in Fig.1 with the Q'-quark line replaced by an (anti-)diquark D line.

It should be realized that the integration (over s, the square of the virtual heavy quark mass) involved in the calculation would diverge for point-like vector diquarks, since the gluon coupling to a pair, Eq. 5 carries momentum factors. The virtual mass in the integration, $\sqrt{s}$, is passed on to the gluon and, subsequently, to the gluon-diquark vertex. Hence it is essential to regulate the integrand by some means. This is best accomplished via the chromoelectromagnetic form factors for the gluon coupling to the diquark. The form factor approach makes physical sense - it is a result of the compositeness of the diquarks. And for consistency, once the vector has form factors, the scalar diquark must have one also.

There is no direct information about the chromoelectromagnetic form factors. We may expect that the ordinary electromagnetic form factors will have the same functional form as their QCD counterparts—the source of both sets of form factors is the matrix element of a conserved vector current operator. In the relevent case here, though, the vector operator is the gluon field — a color octet. Also, what is of concern here is the time-like region of the form factor. For diquarks, of course, there is not any direct empirical evidence about their electromagnetic form factors, but diquark-quark models of the nucleon have constrained the parameterization of the form factors. Dimensional counting rules require that asymptotically, $F_S \sim 1/|q|^2$, $F_{E\,or\,M} \sim 1/|q|^4$, and $F_Q \sim 1/|q|^6$. Jacob, *et al.*[11], have obtained electromagnetic form factors for the diquarks (as has a recent study of higher twist contributions to the nucleon structure functions[12]). The diquark form factors are assumed to have simple pole or dipole forms, and the resulting pole positions appear near 1 GeV. If we make the assumption that the color form factors have the same functional form as these empirical electromagnetic form factors, we can proceed. In the integration that will be done here, the time-like q^2 region begins at $4m^2_{Diquark}$ (below the $N\bar{N}$ threshold) for the value $z = 1/(1 + m_D/m_B)$, and at higher values for other choices of z. This implies that the integration region either overlaps or comes near to overlapping the pole positions. We treat the poles as real resonance positions, by including a sizeable imaginary part (of 1 GeV). This is sensible physically, since the color octet form factor would be dominated by color octet vector mesons, and the latter are not expected to be strongly bound or narrow. Hence we have the Breit–Wigner forms and their squares, with pole positions as given by Jacob, *et al.*. This choice hides our ignorance and provides an interpolation between the space-like and time-like asymptotic regions.

The amplitudes for the baryon production can now be calculated. The spin 1/2 ground state baryons are composed of a scalar diquark and a heavy quark in an s-state. There is only one coupling, and it involves the F_S. The amplitude is

$$A_{S\,1/2} = -\frac{\psi(0)}{\sqrt{2m_d}} F_S(k^2)\bar{U}_B g_s [k_\lambda - 2m_d v_\lambda] P^\lambda, \tag{6}$$

where

$$P^\lambda = \triangle^\lambda_\nu g_s \gamma^\nu \frac{m_Q(1+\mathbf{v}) + \mathbf{k}}{(s - m_Q^2)}\Gamma. \tag{7}$$

For the vector diquark baryons, there are two form factors (we take the quadrupole to be zero – it falls as $1/|q|^6$ asymptotically). The chromomagnetic coupling involves a parameter κ, the "anomalous chromomagnetic moment". This is taken as 1.39. The s-state baryons are spin 3/2 and 1/2, which we will refer to as 1/2′. The 1/2′ lies between the 3/2 and the ground state 1/2 baryon. The amplitude for vector diquarks to be produced, along with the heavy quark, contributes to both 3/2 and 1/2′ states. The amplitude is conveniently divided into a chromoelectric and chromomagnetic part, involving the two distinct form factors. The chromoelectric part contributing to the spin

$1/2'$ baryon is

$$A_{E\,1/2} = -\frac{\psi(0)}{\sqrt{3}m_d}F_E(k^2)\bar{U}_B\gamma_5\gamma^\mu\frac{1+\mathbf{v}}{2}g_s\bar{\epsilon}^*_\mu[k_\lambda - 2m_d v_\lambda]P^\lambda. \tag{8}$$

The chromomagnetic contribution to the spin $1/2'$ baryon is

$$A_{M\,1/2} = \frac{\psi(0)}{\sqrt{3}m_d}F_M(k^2)(1+\kappa)\bar{U}_B\gamma_5\gamma^\mu\frac{1+\mathbf{v}}{2}g_s[g_{\mu\lambda}(\bar{\epsilon}^*v)m_d - \bar{\epsilon}^*_\lambda k_\mu]P^\lambda. \tag{9}$$

For the spin $3/2$ baryon the corresponding amplitudes are

$$A_{E\,3/2} = -\frac{\psi(0)}{\sqrt{2}m_d}F_E(k^2)\bar{\Psi}^\mu_B g_s\bar{\epsilon}^*_\mu[k_\lambda - 2m_d v_\lambda]P^\lambda, \tag{10}$$

and

$$A_{M\,3/2} = \frac{\psi(0)}{\sqrt{2}m_d}F_M(k^2)(1+\kappa)\bar{\Psi}^\mu_B g_s[g_{\mu\lambda}(\bar{\epsilon}^*v)m_d - \bar{\epsilon}^*_\lambda k_\mu]P^\lambda. \tag{11}$$

In these amplitudes, $\psi(0)$ is the Bethe-Salpeter wavefunction at the origin (for the s-state Q-diquark system), m_d is the appropriate diquark mass, U_B is a spin $1/2$ Dirac spinor for the baryon, Ψ^μ_B is the Rarita-Schwinger spinor for the spin $3/2$ baryon, $v = p/M$ for the heavy baryon of mass M, k is the 4-momentum of the gluon and $\triangle^\lambda_\nu$ is the corresponding propagator in axial gauge, Γ is the production vertex for the heavy quark–antiquark pair. Each amplitude should be multiplied by the color factor $4/3\sqrt{3}$.

Considerable simplification of these amplitudes follows. There are 3 cases to consider— 3 final state baryons. For the two states resulting from the vector diquark, the electric and magnetic amplitudes must be added together. Then for each baryon, the amplitude is squared and a trace

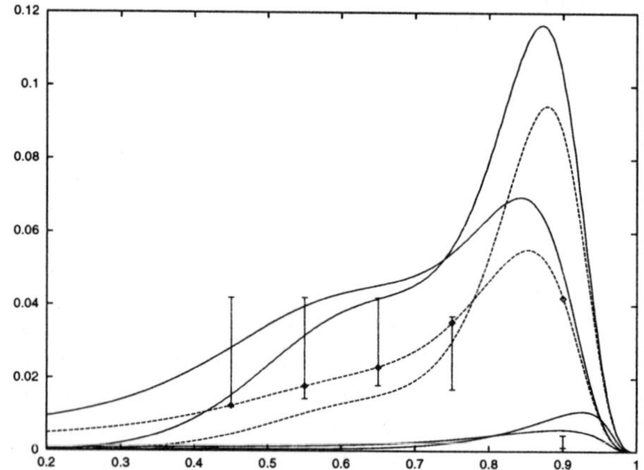

Figure 2: $\hat{f}_1(z,Q^2)$ for $1/2'$ (upper pair), $3/2$ (middle), $1/2$ baryons, each at $Q = \mu$ and 5.5 GeV.

is taken to sum over spins (including spin projection operators for the spin dependent cases). The analog of Eq. 2 is obtained for each baryon. By carefully organizing the terms in the integrand, the width for the inclusive production of the virtual heavy quark can be divided out to yield the analog of Eq. 3 for each baryon. Finally the integration over $s = q^2$ can be performed numerically—the form factors make it difficult to write an analytic expression for each case. The resulting z dependent fragmentation functions are the "boundary" functions, obtained at a scale μ^2 at the threshold $4m_D^2$. To consider higher momentum scales, the Altarelli-Parisi evolution equations are used.

We have taken some particular cases to illustrate the results. For the c(su) or c(sd) baryons, the Ξ_c states, the diquark is given a mass of 0.95 GeV/c^2 and the ratio of diquark to hadron mass is $r = 0.33$. The resulting function, $\hat{f}_1(z, Q^2)$ is shown in Fig.2 for the boundary value at $\mu = 2m_d$ and for $Q = 5.5$ GeV. The three s-states lead to different behavior and overall probability. It is particularly noteworthy that the 1/2 ground state is produced far less frequently than the 3/2 state or the 1/2$'$ state from the vector diquark. As we will see, the observed 1/2 ground states are primarily from the decays of these vector diquark states.

In Fig.3 the corresponding longitudinal fragmentation function $\hat{g}_1(z, Q^2)$ is plotted for the two spin 1/2 s-tates. Note that it is very similar in shape to the spin averaged case. Lastly, the behavior of the transversity function $\hat{h}_1(z, Q^2)$ remains to be determined.

The spin dependent fragmentation functions for the spin 3/2 baryons are even richer

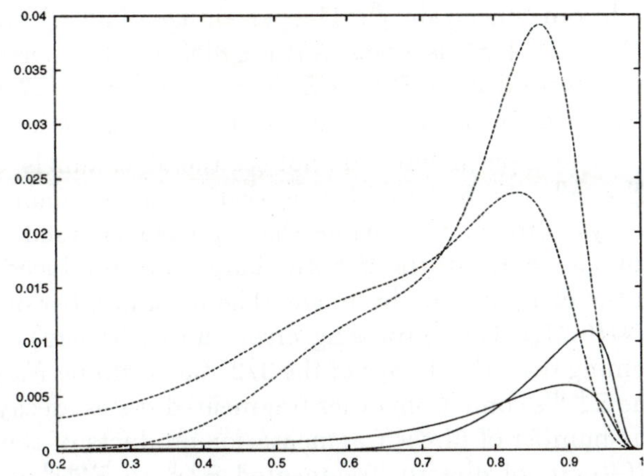

Figure 3: $\hat{g}_1(z, Q^2)$ for 1/2$'$ (upper pair) and 1/2 baryons, each at $Q = \mu$, 5.5 GeV.

in complexity. There are seven such functions at leading twist, many of which will be accessible from the decay distributions of these states into the 1/2 state plus a pion.

3   Comparison with some data and summary

The production of charmed baryons is becoming sizeable at CESR and data now exist from the CLEO collaboration[2] for some of the Ξ_c states. In particular, spin independent fragmentation functions have been determined for the lowest mass spin 1/2 and 3/2 states. The lowest $1/2^+$ states are the $c + [u, s]$ and $c + [d, s]$ states, Ξ_c^+ and Ξ_c^0, involving the antisymmetric, spin 0 diquarks. The spin $3/2^+$ states are $c + \{u, s\}$ and $c + \{d, s\}$ baryons, Ξ_c^{*+} and Ξ_c^{*0}, involving the symmetric, spin 1 diquarks. The spin $1/2^+$ partners, Ξ_c', of $3/2^+$ states have not been seen yet. They are presumed to have a mass below the $\Xi_c + \pi$ threshold, so must be seen in radiative decay channels. Note that these latter Ξ_c' $1/2^+$ states have the same isospin as the lower lying ground states $1/2^+$ Ξ_c and could mix with them, in principle. In any case, the measured fragmentation functions provide a crude test of the model. The data are fit by the experimenters with a common parameterization of the Peterson function[1]. It is easy to see in Fig.2 that the data fall nicely on $\hat{f}_1$ of our model(with arbitrary normalization), evolved to $Q = 5.5$ GeV, but are not sufficiently accurate to be a crucial test of the model. Note that the experimental variable $x_p{}^2$ does not correspond exactly to our z, the light cone variable.

The ratio of the 3/2 to 1/2 production can be extracted from the data with some uncertainty[8]. The percentage of all Ξ_c^+ states that arose from decays $\Xi_c^{*0} \to \Xi_c^+ + \pi^-$ is given as $(27 \pm 8)\%$ and the percentage of all Ξ_c^0 states that arose from decays $\Xi_c^{*+} \to \Xi_c^0 + \pi^+$ is given as $(17 \pm 6)\%$. (Note that we have combined the statisitical and systematic errors here.)

The experimenters do not see the π^0 channels, $\Xi_c^{*+} \to \Xi_c^+ + \pi^0$ and $\Xi_c^{*0} \to \Xi_c^0 + \pi^0$. From isospin conservation these channels account for 1/3 of the decays into $\Xi_c + \pi$, while the reported charged π channels constitute 2/3. Suppose N Ξ_c^* states of both charges are produced. Then 2/3 N will be seen in the charged π decay mode. The total number of $\Xi_c^{+,0}$'s seen will be $N_{+,0} = \frac{2}{3}N/(0.27, 0.17)$ (supressing errors until the end). The number of $\Xi_c^{+,0}$'s not coming from the decays of the 3/2 states will be $N_{+,0} - N$. Assume that $n_{+,0}$ of the $\Xi_c^{+,0}$'s come from other fragmented states' decays. Then $N_{+,0} - N - n_{+,0}$ is the number of direct fragmentation products of the charmed quark. The ratio $R(+ \text{ or } 0)$ of directly fragmented $\Xi_c^{+,0}$ to $\Xi_c^{*+,0}$ is given thereby as $R(+) = 1.5 \pm 0.7 - n_+/N : 1$ and $R(0) = 2.9 \pm 1.4 - n_0/N : 1$.

The numbers $n_{+,0}$ will come from the radiative decays of the heavier 1/2 states, as well as higher Ξ_c states (radial and orbital excitations of the $c + (su)$ and $c + (sd)$ systems). We have calculated the fragmentation functions for the spin 1/2 quark–vector-diquark states and hence the number of Ξ_c' spin $1/2'$ states vs. Ξ_c^* spin 3/2 states. That is 1.7:1 for the parameterization used in Fig.2. Assuming $n_{+,0}$ is due entirely to these $1/2'$ states decaying 100% into the ground state $\Xi_c^{+,0}$, we have for the different charge states $R(+) = -0.2 \pm 0.7$ and $R(0) = 1.2 \pm 1.4$, both of which are consistent with the small ratio of 1/9 predicted by the same model calculation.

Hence an interesting feature of our model is that the directly fragmented gound state baryon is very rare compared to the vector diquark states. In the calculations for heavy-heavy baryons by Marteynenko and Saleev [10], a ratio closer to unity was obtained. It will be interesting to see whether the small ratio, as suggested by our model, persists as more data are obtained.

Acknowledgments

This work was supported, in part by a grant from the US Department of Energy. G.R.G. appreciates the hospitality of the organizers of Diquark III. We appreciate helpful communications with J. Yelton regarding CLEO results.

References

1. C. Peterson, *et al.*, *Phys. Rev.* D **27**, 105 (1983).
2. L. Gibbons, *et al.*, *Phys. Rev. Lett.* **77**, 810 (1996); P. Avery, *et al.*, *Phys. Rev. Lett.* **75**, 4364 (1995); K.W. Edwards, *et al.*, *Phys. Lett.* B **373**, 261 (1996).
3. C.-H. Chang and Y.-Q. Chen, *Phys. Lett.* B **284**, 127 (1992).
4. E. Braaten, K. Cheung, S. Fleming, T.C. Yuan, *Phys. Rev.* D **51**, 4819 (1995); and references contained therein.
5. R. L. Jaffe and L. Randall, *Nucl. Phys.* B **412**, 79 (1994).
6. K. Chen, G. R. Goldstein, R. L. Jaffe, X. Ji, *Nucl. Phys.* B **445**, 380 (1995)
7. G. R. Goldstein, J. Maharana, *Nuovo Cimento* **59**, 393 (1980).
8. J. Yelton, private communication.
9. A.F. Falk, *et al.*, *Phys. Rev.* D **49**, 555 (1994).
10. A.P. Marteynenko and V.A. Saleev, hep-ph/9604259 (1996).
11. R. Jacob, P. Kroll, M. Schurmann, W. Schweiger, *Z. Phys.* A **347**, 109 (1993).
12. M. Anselmino, *et al.*, *Z. Phys.* C **71**, 625 (1996).

FAST ANTIBARYON PRODUCTION IN pp COLLISIONS AND STRING FUSION

N. ARMESTO, E. G. FERREIRO and C. PAJARES

*Departamento de Física de Partículas, Universidade de Santiago de Compostela,
15706-Santiago de Compostela, Spain*

Yu. M. SHABELSKI

*Petersburg Nuclear Physics Institute, Gatchina,
Sanct-Petersburg 188350, USSR*

The inclusion of string fusion in Dual Parton Model results in the appearance of diquark-antidiquark pairs in the sea with comparatively large Feynman-x. Such antidiquarks can fragment into fast antibaryons, thus increasing several times the corresponding yields in high energy pp collisions for $x_F \sim 0.8 \div 0.9$. String fusion also results in an unusual dependence of inclusive spectra on the multiplicity of secondaries. Some numerical estimations are presented.

1 Introduction

Models based on pomeron exchange are very popular for the description of multiparticle production on nucleon and nuclear targets at high energies (see for example [1,2]). In the simplest approach pomerons are assumed not to interact one with each other. However the string fusion effects (i.e. pomeron interactions)[3] should exist with a small probability in the case of high energy pp collisions and result in antibaryon production with very high Feynman-x. Model estimations predict a yield of such antibaryons several times larger than in the case of a variant without string fusion. For numerical estimations we use the Quark-Gluon String Model (QGSM)[1] which describes quite successfully the spectra of different secondaries produced in high energy hadron-nucleon and hadron-nucleus collisions. This model is very close to the Dual Parton Model [2]. In particular the spectrum of Λ_s, to which the fragmentation of a valence diquark gives the main contribution[a] at large x_F, is described well enough so we can hope that the parameters of antibaryon production by sea-antidiquark fragmentation are more or less fixed.

String fusion also changes the dependence of inclusive spectra on the multiplicity. Usually the spectrum of any particle for events with multiplicity higher than the mean one becomes more and more narrow, due to the division of the initial energy between many cut pomerons. However for a particle whose

---

[a]In the case of high x_F proton production in pp collisions triple-reggeon diagrams give the main contribution.

spectrum in the large x_F region is determined by string fusion the behaviour becomes the opposite, i.e., the spectrum gets constant or even broader.

2 Inclusive spectra of secondary hadrons in QGSM

In the QGSM high energy hadron-nucleon interactions are considered as proceeding via the exchange of one or several pomerons. Each pomeron corresponds to a cylindrical diagram, so in the case of one cut pomeron two showers of secondaries are produced, see Fig. 1. The inclusive spectrum of secondaries is determined by the convolution of the diquark and valence and sea quark distributions $u(x, n)$ in the incident particles with the fragmentation functions of quarks and diquarks into secondary hadrons $G(z)$. The diquark and quark distribution functions depend on the number n of cut pomerons in the considered diagram. For a nucleon target the inclusive spectrum of a secondary hadron h has the form [1]

$$\frac{x_E}{\sigma_{inel}} \frac{d\sigma}{dx_F} = \sum_{n=1}^{\infty} w_n \phi_n^h(x) \, , \tag{1}$$

where the function $\phi_n^h(x)$ determines the contribution of the diagram with n cut pomerons and w_n is the probability of this diagram. We neglect diffractive dissociation, as its effect is known to be comparatively small in the case of antibaryon production.

For pp collisions:

$$\begin{aligned}
\phi_n^h(x) &= f_{qq}^h(x_+, n) f_q^h(x_-, n) + f_q^h(x_+, n) f_{qq}^h(x_-, n) + \tag{2} \\
&+ 2(n-1) f_s^h(x_+, n) f_s^h(x_-, n) \, ,
\end{aligned}$$

$$x_\pm = \frac{1}{2}[\sqrt{4m_T^2/s + x^2} \pm x] \, , \tag{3}$$

where f_{qq}, f_q and f_s correspond to the contribution to the inclusive spectrum of diquarks, valence and sea quarks respectively. They are determined by the convolution of diquark and quark distributions with fragmentation functions,

$$f_q^h(x_+, n) = \int_{x_+}^1 u_q(x_1, n) G_q^h(x_+/x_1) \, dx_1 \, . \tag{4}$$

Both diquark and quark distributions and fragmentation functions are expressed via their Regge-asymptotics, taking into account the conservation laws [1,4].

In the calculations we use quark and diquark distributions in the proton of the form [1]:

$$u_{uu}(x, n) = C_{uu} \, x^{2.5}(1-x)^{n-1.5} \, , \tag{5}$$

$$u_{ud}(x,n) = C_{ud}\, x^{1.5}(1-x)^{n-1.5} \ , \tag{6}$$

$$u_u(x,n) = C_u\, x^{-0.5}(1-x)^{n+0.5} \ , \tag{7}$$

$$u_d(x,n) = C_d\, x^{-0.5}(1-x)^{n+1.5} \ , \tag{8}$$

$$u_{\overline{u}}(x,n) = u_{\overline{d}}(x,n) = C_{\overline{u}}\, x^{-0.5}[(1+\delta/2)(1-x)^{n+0.5}(1-x/3)-\delta\,(1-x)^{n+1}/2] \ , \tag{9}$$

$$n > 1 \ ,$$

$$u_s(x,n) = C_s\, x^{-0.5}(1-x)^{n+1} \ , \ n > 1 \ . \tag{10}$$

$\delta = 0.2$ is the relative probability to find a strange quark in the sea and the factors C_i are determined from the normalization condition

$$\int_0^1 u_i(x,n)dx = 1 \ . \tag{11}$$

Quark and diquark fragmentation functions into secondary hadrons are taken from [1,4].

In the case of baryon B production in pp collisions there are two different contributions [5]. The first one corresponds to the central production of a $B\overline{B}$ pair and can be described by the formulas written above. The second contribution is connected with the direct fragmentation of the initial proton into B with conservation of the string junction. To account for this possibility we input into Eq. (2) two additional items $f_{qq2}(x_+,n)$ for $x_F > 0$ and $f_{qq2}(x_-,n)$ for $x_F < 0$ which are not multiplied by $f_q(x_-,n)$ and $f_q(x_+,n)$ respectively. The form of $f_{qq2}(x_+,n)$ and $f_{qq2}(x_-,n)$ is determined by the corresponding fragmentation functions. For example, in the case of Λ_s production they are

$$G_{uu2}^{\Lambda_c} = a_{02}\, z^2(1-z)^{1+\lambda-\alpha_\varphi(0)} \tag{12}$$

and

$$G_{ud2}^{\Lambda_c} = a_{02}\, z^2(1-z)^{\lambda-\alpha_\varphi(0)} \ , \tag{13}$$

with $\lambda = 0.5$ and $\alpha_\varphi(0) = 0$.

The probability of a process with n cut pomerons is calculated in the quasieikonal approximation [1,6].

3   String fusion contribution in QGSM

A possible mechanism of string fusion can be inferred if we consider the intermediate states of a triple-reggeon diagram. Let us consider such a diagram in which a triple-reggeon exchange connects sea quarks of two interacting nucleons and let all reggeons be cut, Fig. 2a[b]. One possible inelastic intermediate state is shown in Fig. 2b (we present only one of the two produced chains, compare to Fig. 1c). If in the upper part of the diagrams in Fig. 2 two pomerons are connected with two sea quarks, in the lower part a reggeon will interact with two sea antiquarks because every chain as a whole has to be white. So the ends of the strings (two antiquarks in this case) can fuse and produce some new object which will fragment into secondary hadrons as a whole. We will call this object a sea antidiquark.

In the QGSM an incident fast quark is assumed to fragment into a hadron (say, a meson) producing a slower quark which will fragment again and so on, until it fuses with a target diquark or antiquark, forming respectively a baryon or a meson. In the case shown in the Fig. 2b the same process will occur in the upper part. However a sea diquark will appear at a rapidity value approximately equal to the rapidity of the triple-reggeon vertex. Then two possibilities appear: it can fragment into a baryon B as shown in Fig. 2b or into mesons until its annihilation with the sea antidiquark. It is in the first possibility that we can obtain a comparatively fast antibaryon in the backward hemisphere. Let us note that the lower reggeon in Fig. 2a corresponds to the pomeron at high energies. This is clear if we compare the sea diquark-antidiquark interaction in Fig. 2b to the valence diquark-antidiquark interaction in the case of high energy $p\bar{p}$ interactions.

The process of fusion of two sea quarks (antiquarks) into one sea diquark (antidiquark) becomes possible for $n \geq 3$ in Eq. (1). It seems that the best place to see the effects of string fusion is in the spectra of fast antibaryons produced in pp collisions because the contribution of all other processes is comparatively small. The enhancement of these spectra in the large x_F region in the case of string fusion arises from two facts: i) the x-distribution of sea antidiquarks is harder than that of sea antiquarks and ii) the fragmentation functions are harder.

For the numerical estimations which are, of course, model dependent, we take the probability of two sea quarks fusing into a diquark equal to $v_0 = 0.02$. So the probability to find two fused quarks in a diagram with a n-pomeron exchange is:

$$v_n = v_0 \, (n - 1) \, (2n - 3) \tag{14}$$

(the probability that these quarks are both antiquarks is separately accounted for). As a result, the total probability for string fusion in inelastic pp collisions

---

is

$$w_{fus} = \sum_{n=3}^{\infty} v_n \, w_n \, , \tag{15}$$

which is about $1/20$ at $\sqrt{s} = 39$ GeV. This is close to the standard contribution of triple-reggeon diagrams at this energy.

For the x-distributions of sea antidiquarks we use the simplest assumption: it is a convolution of two sea quark distributions,

$$u_{\overline{qq}}(x,n) = \int u_{\overline{q}}(x_1,n) u_{\overline{q}}(x_2,n) \, \delta(x - x_1 - x_2) \, dx_1 dx_2 \tag{16}$$

and we use the standard functions [1,4] for their fragmentation into antibaryons.

The results for the x_F-spectra of $\overline{p}$ and $\overline{\Lambda}_s$ produced in pp collisions at $\sqrt{s} = 39$ GeV are shown in Fig. 3. Model predictions without string fusion are shown by solid lines and with string fusion by dashed lines. In all cases the difference in the small x_F region is very small and decreasing with energy. The reason is that sea antidiquarks have an average x_F value not so small – about $0.1 \div 0.15$, so they are far (in rapidity) from the central region. In the large x_F region the difference becomes larger and it is about one order of magnitude at $x_F = 0.85 \div 0.9$ (almost independently on the initial energy).

4 The shape of x_F-spectra in events with different multiplicity

Let us take a sample of multiparticle production data without low multiplicity events, say with $n_{ch} \geq\, < n_{ch} >$. Usually the spectra of secondaries in such a sample become more narrow than that for standard events. The reason is that events corresponding to one-pomeron cut have on the average a smaller multiplicity of secondaries in comparison with multipomeron events, so one-pomeron events are more frequently suppressed from our high multiplicity sample. Then such a sample is rich in multipomeron events. But it is in these events where the x-distributions of quarks and diquarks (ends of strings), Eqs. (5)-(10), are more narrow (simply because the energy has to be divided between more partons), so these quarks and diquarks will generate a more narrow spectrum of secondaries.

However if string fusion is taken into account the behaviour of the spectrum is more complicated. For secondary mesons or baryons produced in pp collisions we can only expect some small quantitative difference because the string fusion contribution is small practically at all x_F. But in the case of the antibaryon spectrum string fusion dominates the region of large x_F (Fig. 3). After substracting the low multiplicity events the inclusive cross section in this region will change only slightly[c]. On the other hand it will decrease more or less significantly in the small x_F region where one-pomeron processes give about one half of the inclusive cross section. So we can expect that the spectrum of antibaryons becomes even wider in the set of high multiplicity events.

For numerical estimations let us define the variable

$$z = \frac{n_{ch}}{<n_{ch}>} \qquad (17)$$

and consider the spectra of secondary antibaryons for events with $z \geq z_0$. $z_0 = 0$ means that we take all events (minimal bias sample), for $z_0 = 1$ we take only events with multiplicity of charged secondaries larger than the mean one and so on.

As the differences in the shapes of the are not very large it is better to consider the ratio of the inclusive cross section in two different points, x_1 and x_2,

$$R\left(\frac{x_F = x_1}{x_F = x_2}\right) = \frac{[x_E/\sigma_{inel}\ d\sigma/dx_F]|_{x_F=x_1}}{[x_E/\sigma_{inel}\ d\sigma/dx_F]|_{x_F=x_2}} , \qquad (18)$$

as a function of z_0.

The model estimations of these ratios for $x_1 = 0.6$ (where the cross section is not very small) and $x_2 = 0$ are presented in Fig. 4 for $\bar{p}$ and $\overline{\Lambda}_s$ produced in pp collisions at energy $\sqrt{s} = 39$ GeV. It can be seen that in both cases there exists a large difference in the behaviour of the ratios without (solid curves) and with (dashed curves) string fusion. The experimental investigation of such ratios could give evidence on the existence of string fusion.

5 Conclusion

The existence of triple-reggeon interaction has been firmly confirmed by experiment. The string fusion mechanism is a simple possibility to incorporate the contribution of such diagrams into a model of multiparticle production. The difference in the content of secondaries between high energy nucleon-nucleon and nucleus-nucleus collisions can be explained [7] if one takes into accountthe contribution coming from string fusion. However some numerically small effect should also exists in the case of high energy pp (or $\bar{p}p$) interactions. It can be observed as an enhancement of secondary antibaryon production and an unusual dependence of the shapes of their inclusive spectra on the multiplicity of secondaries.

---

[c]The x-distribution of sea antidiquarks is wider than those of sea antiquarks, as it results from adding the x's of the two parent sea antiquarks, Eq. (18).

Acknowledgments

We thank the Dirección General de Política Científica and the CICYT of Spain for financial support. E.G.F also thanks the Xunta de Galicia for for financial support The paper was supported in part by INTAS grant 93-0079.

References

1. A.B.Kaidalov and K.A.Ter-Martirosyan. Yad.Fiz. 39 (1984) 1545; 40 (1984) 211.
2. A.Capella, U.P.Sukhatme, C.-I.Tan and J.Tran Thanh Van. Phys.Rep. 236 (1994) 225.
   K.Werner. Phys.Rep. 232 (1993) 87.
3. M.A.Braun and C.Pajares. Phys.Lett. B287 (1992) 154; Nucl.Phys. B390 (1993) 542; 559.
4. Yu.M.Shabelski. Yad.Fiz. 44 (1986) 186.
5. A.B.Kaidalov and O.I.Piskunova. Yad.Fiz. 43 (1986) 1545; Z.Phys. C 30 (1986) 145.
6. K.A.Ter-Martirosyan. Phys.Lett. 44B (1973) 377.
7. N.S.Amelin, M.A.Braun and C.Pajares. Phys.Lett. B306 (1993)312; Z.Phys. C63 (1994) 507.
   N.Armesto, M.A.Braun, E.G.Ferreiro and C.Pajares. Phys.Lett. B344 (1995) 301.
   H.Sorge, M.Berenguer, H.Stöcker and W.Greiner. Phys.Lett. B289 (1992) 6.

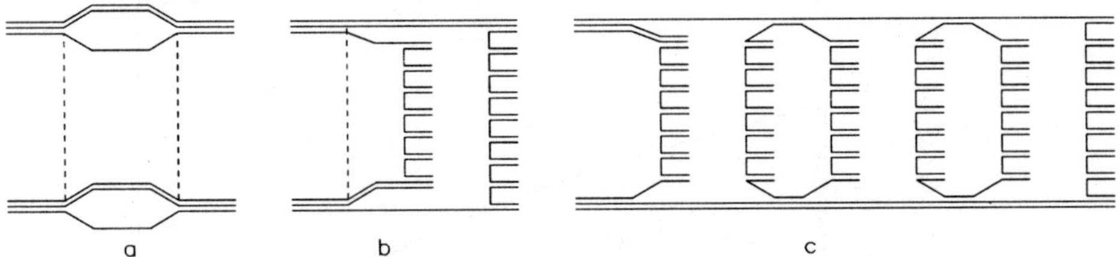

Figure 1: Cylindrical diagram which corresponds to the one-pomeron exchange contribution to elastic pp scattering (a). Its cut which determines the contribution to inelastic pp cross section (b). The diagram which corresponds to the cut of three pomerons (c).

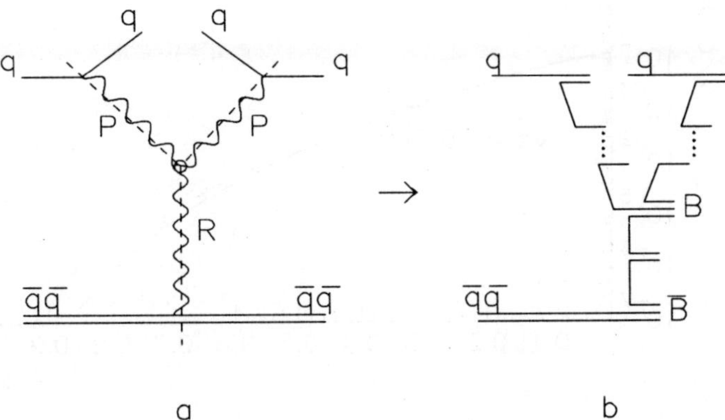

Figure 2: Cut triple-reggeon diagram (a) and the corresponding inelastic intermediate state (b) with baryon-antibaryon production via string fusion.

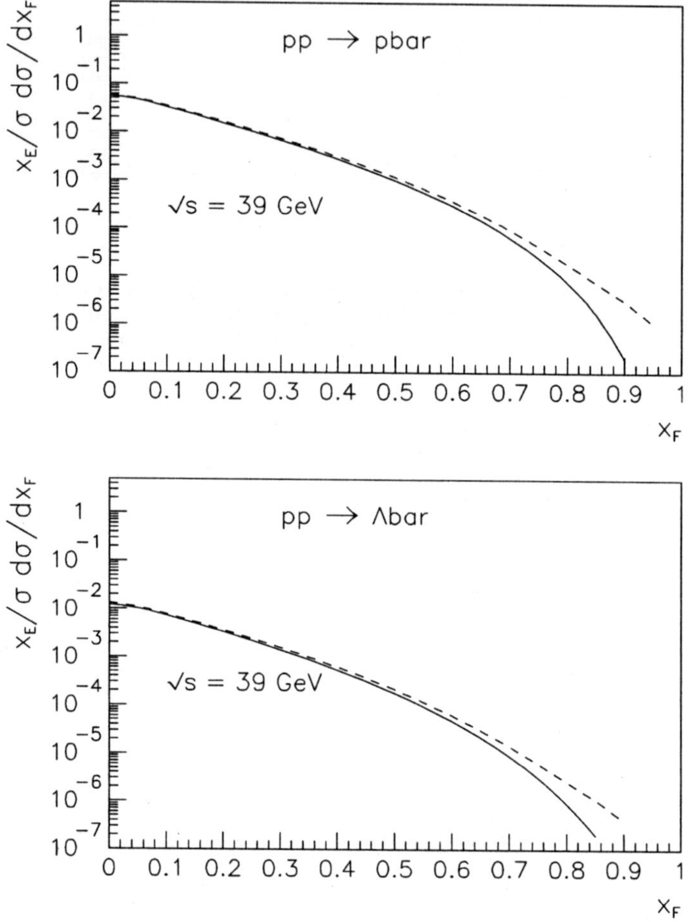

Figure 3: The predicted inclusive spectra of antiprotons and $\overline{\Lambda}_s$ in *pp* collisions at c.m. energy 39 GeV without (solid curves) and with (dashed curves) string fusion.

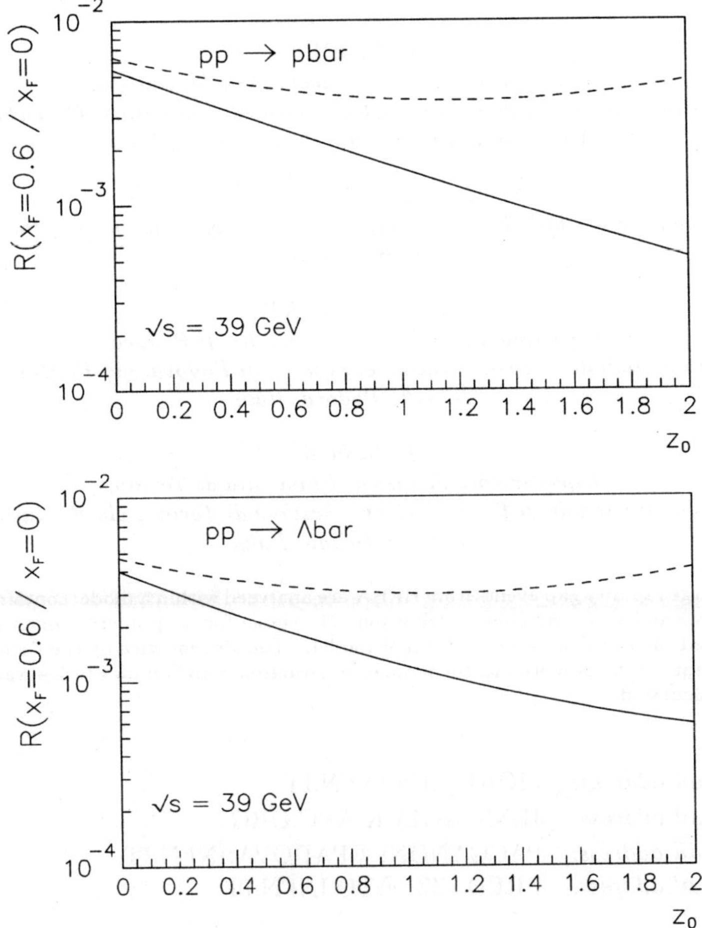

Figure 4: The predicted ratios of $\bar{p}$ and $\overline{\Lambda}_s$ inclusive spectra in pp collisions at c.m. energy 39 GeV without (solid curves) and with (dashed curves) string fusion.

NON-FACTORIZABLE CONTRIBUTIONS TO THE LARGE RAPIDITY GAP AT HERA

R. FIORE[†]

Dipartimento di Fisica, Università della Calabria,
Istituto Nazionale di Fisica Nucleare, Gruppo collegato di Cosenza,
Arcavacata di Rende, I-87030 Cosenza, Italy

L. L. JENKOVSZKY[‡]

Bogoliubov Institute for Theoretical Physics, Academy of Sciences of the Ukrain,
Kiev, Ukrain

F.PACCANONI[*]

Dipartimento di Fisica, Università di Padova,
Istituto Nazionale di Fisica Nucleare, Sezione di Padova, via F. Marzolo 8,
I-35131 Padova, Italy

E.Predazzi[°]

Dipartimento di Fisica, Universitá di Torino,
Istituto Mazionale di fisica Nucleare, Sezione di Torino, via P.Giuria 1,
I-10125 Torino, Italy

The large rapidity gap events from HERA are analyzed within a model containing a pomeron and an $f-$ reggeon contribution. The choice for the pomeron contribution is based on the Donnachie-Landshoff model. The dependence of the "effective intercept" of the pomeron on the momentum fraction β and on its Bjorken variable ξ is calculated.

[†] *email address:*   FIORE @CS.INFN.IT
[‡] *email address:*   JENK @GLUK.APC.ORG
[*] *email address:*   PACCANONI @PADOVA.INFN.IT
[°] *email address:*   PREDAZZI @TO.INFN.IT

The increasing precision of the HERA-measurements and the extension of the kinematical domain where diffractive deep inelastic scattering (DIS) was measured, necessitates the perfection of the relevant theoretical calculations beyond the simple, single, factorizable pomeron exchange, as it was originally introduced in [1] and was used in subsequent papers (for a partial list of recent papers on the subject see [2]). Although the role of other than a single pomeron contribution to the rapidity gap has been realized by the theorists long ago [3], it was only recently [4] that the effect was measured experimentally.

Studies of diffractive DIS beyond the simple pole exchange approximation have two, albeit interrelated aspects: one is rather technical and has to deal with various contributions that break factorization; the other one is conceptual, dealing with the nature and in particular the internal structure of the pomeron(s).

The pomeron itself appears to be a complicated object; according to the perturbative QCD calculations [5], it corresponds to an infinite number of singularities in the complex angular momentum plane accumulating at the rightmost point, $J = 1 + \delta$, $\delta \geq 0.3$.

Below we consider a simple model for the pomeron: that of Donnachie and Landshoff [3], corresponding to a single "supercritical" Regge pole exchange. Other options are possible, for example the dipole pomeron (DP) model (see [6]) with two terms - one constant and the other one rising with energy logarithmically. This DP model is close to that of D-L as to the numerical fits - both producing a moderate ("soft") energy dependence, but the latter one may be used also as a polygon for studying the effects coming from the pomeron non-factorizability. Here we will limit ourselves to consider the first choice.

Other possible contributions are those allowed in elastic hadron (e.g. pp) scattering: the odderon, f, ω, etc reggeons, daughter trajectories, π exchange etc. We confine our analysis by considering two major contributions, namely the pomeron (single and double poles) plus an effective reggeon, essentially dominated by f. Other contributions are either negligibly small or they may be absorbed by the above ones (subtle details - like π-exchange - have not been settled in the literature even in the case of the much better known case of elastic scattering).

Explicitly, the spin-averaged differential cross-section for a diffractive DIS, ignoring spin and the proton mass, is (see Fig. 1 for kinematics and notations):

$$d\sigma = \frac{(2\pi)^{-5}}{2pl} \frac{d^3\vec{l'}}{2l'^0} \frac{d^3\vec{p}\,'}{2p'^0} \sum_n \left(\frac{d^3\vec{k}_n}{2k_n^0} \delta(p+l-l'-k_n-p') |T(p+l \rightarrow p'+l'+k_n)|^2 \right).$$

We consider the scattering amplitude corresponding to two Regge exchanges R_i $(R_i = P, f)$:

$$T \equiv T(p+l \rightarrow p'+l'+k_n) = \sum_i T^{(i)} = \sum_i F(R_i+l \rightarrow l'+k_n)\Phi_i(\xi,t),$$

where

$$\Phi_i(\xi,t) = \left(e^{i\pi/2}\xi\right)^{-\alpha_i(t)}\beta_i(t),$$

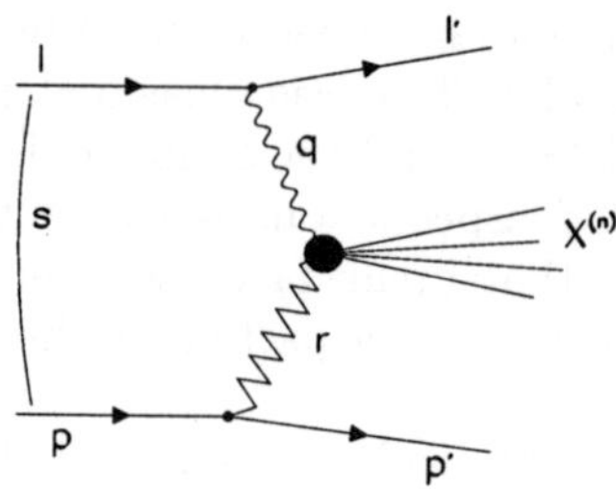

Figure 1: Diagram for the diffractive deep-inelastic scattering.

and the slowly varying function $\sin(\pi\alpha/2)$ has been absorbed by the residue.
By denoting

$$F(R_i + l \to l' + k_n) = \frac{e}{q^2} F(R_i + q \to k_n) = F_{R_i+l}$$

we get in the case of two Regge exchanges, P and f,

$$|T|^2 = |F_{P+l}\Phi_P(\xi,t)|^2 + |F_{f+l}\Phi_f(\xi,t)|^2 + 2\Re[F_{f+l}F_{P+l}^*\Phi(\xi,t)\Phi_P^*(\xi,t)]. \quad (1)$$

The first two terms in Eq. (1) assume an immediate physical interpretation.
Really, the cross section for $R_i + l \to l' + X^n$ is

$$d\sigma_{R_i+l} = \frac{1}{2rl}\frac{d^3\vec{l'}}{2l'^0}\sum_n \left(\frac{d^3\vec{k_n}}{2k_n^0}\delta^{(4)}(r+l-l'-k_n)|F_{R_i+l}|^2\right).$$

Thus, for the $p+l \to p'+l'+X^{(n)}$ differential cross section with the exchange
of a single reggeon R_i one has

$$d\sigma^{(i)} = \frac{rl}{pl}\frac{d^3p'}{2p'^0}|\Phi(\xi,t)|^2 d\sigma_{R_i+l} \simeq \frac{\pi}{2}d\xi dt\xi|\Phi(\xi,t)|^2 d\sigma_{R_i+l},$$

where the relations $(rl)/(pl) \simeq (rq)/(pq) = \xi$ and $(d^3\vec{p}\,')/(2p'^0) \simeq 1/4d\phi_p' d\xi dt$, valid for small ξ, were used.

The last term in the r.h.s. of Eq. (1) is related to the imaginary part of the $f + l \to P + l'$ transition amplitude, and consequently, in $d\sigma$ a term proportional to

$$\frac{d^3\vec{l}'}{2l'^0} \sum_n \frac{d^3\vec{k}_n}{2k_n^0} \delta^4(p + l - p' - l' - k_n) F_{f+l} F_{P+l}^* \tag{2}$$

emerges, corresponding to an f meson dissociating into a pomeron. Following an "educated guess" by N.N.Nikolaev, W.Schäfer and B.G.Zakharov [7], we ignore this contribution as being small compared to the elastic case.

As recently noticed by J.Ellis and G.G.Ross [8], the rapidity cuts commonly employed in diffractive DIS measurements require the struck parton in the pomeron be far off shell in a sizable region of parameter space. The H1 cuts correspond to very small k_{min}^2 that justify the treatment of the pomeron structure function.

Once the above-mentioned approximation has been accepted, one may write

$$d\sigma_{R_i+l} = d\beta G_{q/R_i}(\beta) dq^2 d\phi_{l'} \frac{d\hat{\sigma}}{dq^2 d\phi_l'}(k + l \to k' + l'), \tag{3}$$

in terms of the parton-lepton cross section $d\hat{\sigma}$, where $k = \beta r$ and G_{q/R_i} is the structure function of the Reggeon R_i.

The above detailed presentation of the formalism was intended to fix the basic notions of diffractive DIS, to avoid further confusion and misunderstanding.

Now we proceed towards a quantitative analysis of the phenomenon by using explicit models for the reggeon fluxes and their structure functions, based on our experience in treating both elastic and inelastic scattering.

Let us write our basic formula (1) in terms of the commonly used notation for diffractive DIS structure functions:

$$F_2^{D(4)}(x, t; \beta, Q^2) = A[\Phi_{q/P}(\xi, t) G_P(\beta, Q^2) + a\Phi_{q/f}(\xi, t) G_f(\beta, Q^2)], \tag{4}$$

where, apart from the overall normalization factor A, a is the only free parameter of the model.

The f-Reggeon flux is determined uniquely from standard fits to elastic hadron scattering (see e.g. [9]):

$$\Phi_f(\xi, t) = g_f^2(t)\xi \exp\left[(1 - 2\alpha_f(t))L\right],$$

where $g_f(t) = \exp\left[b_f \alpha_f(t)\right]$, $\alpha_f(t) = \alpha_f(0) + \alpha_f' t$ and $L \equiv \ln(1/\xi)$.

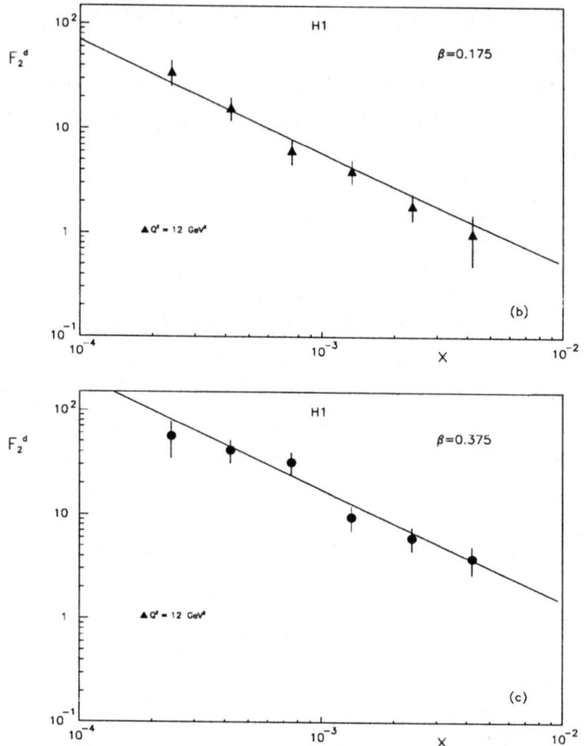

Figure 2: A representative fit of $F_2^{D(3)}(x,\beta,Q^2)$ to the published H1 data [12] for fixed $Q^2 = 12\,GeV^2$ and $\beta = 0.175, 0.375$.

We use "world average" values for the above parameters[a] : $b_f = 5$, $\alpha_f(0) = 0.65$ and $\alpha'_f = 0.9\ GeV^{-2}$.

The pomeron flux

$$\Phi_{q/P}^{D-L}(\xi,t) = [\exp b(\alpha - 1)\xi^{-\alpha)}]^2\xi, \tag{5}$$

where $\alpha \equiv \alpha_P(t) = 1 + \delta + 0.25t$, $\delta = 0.08$, is integrated in t. (For an explicit treatment of the $t-$dependence see [2]c).)

We make standard choices for the structure functions. For the f-reggeon

---

[a]Note that the f-reggeon intercept as extracted from various fits to the data has a trend to increase progressively with the time. The "old-fashioned" value of $\alpha_f(0) = 0.5$ is incompatible with the data, while another extreme, 0.8 is claimed by some authors [10] to be compatible with the data. We (see e.g. [9]) consider $\alpha_f(0) = 0.65$ to be a reasonable value both in the scattering ($t < 0$) and resonance ($t > 0$) region. Due to the strong $P - f$ mixing, the pomeron and f-reggeons intercepts are strongly correlated.

we choose the following parametrization by Glück, Reya and Vogt [11,4]:

$$\beta G_f(\beta, Q^2) = N\beta^a(1 + A\sqrt{\beta})(1 - \beta)^D \tag{6}$$

(originally intended for pions). The values of the parameters $N, a,$ A and D, together with their explicit Q^2-dependence may be found in Ref. [11]. The parameter N will be rescaled by our fitting procedure. Actually, we are not too much concerned with any Q^2 dependence since it is known [4] to be weak anyway.

A representative fit to the diffractive structure function $F_2^{D(3)}$ is shown in Fig. 2 for fixed $Q^2 = 12\,GeV^2$ and $\beta = 0.175, 0.375$.

The non-trivial dependence of the effective power $n(\beta, \xi) = \partial \ln F / \partial \ln \xi$, calculated from our model, is shown in Fig. 3 as a function of β and ξ. Factorization breaking, that manifests itself in the β dependence of the effective power n, is evident at larger ξ values, where the f contribution is no more negligible with respect to the pomeron one. Since we use in this analysis the published 1993 H1 data [12], differences in the β-dependence of $n(\beta)$ from the result of [4] are expected. A maximum of $n(\beta)$, around $\beta \approx 0.6$ at large ξ, seems to be a feature present also in the 1994 preliminary H1 data [4] where different selection cuts, e.g. in the t-dependence, are applied.

To conclude, we have presented an explicit model for factorization-violating diffractive DIS. Some details of the model may still vary but two essential points remain invariant, namely: 1) the overwhelming contribution in the kinematical region of diffractive DIS as measured at HERA comes from the pomeron and the f-reggeon. The separation and identification of these objects is a complicated but at the same time interesting problem in doing phenomenology; 2) the pomeron emitted from the lower vertex in Fig.1 is the same as it appears in elastic hadron scattering.[b] Therefore, one can rely on pomeron models fitted to high energy pp and $\bar{p}p$ data. These fits unambiguously fix [9] the energy dependence of the pomeron, which is "soft". In fitting Eq. (4) to the data, the rest of the parameters may be let free.

The pomeron, which is the central object of the present analysis, itself may be parametrized in various ways. Here, we have presented a typical model of the pomeron, other possibilities will be considered elsewhere. Notice that a different type of the pomeron (see [7] and earlier references therein) gives rise to a strong rise of $n(\beta)$ at small β, that however may be compensated [7] by various subleading contributions.

---

[b]For the sake of definiteness, we did not consider the case of double diffraction dissociation. Our results can be generalized to the case when the incident proton dissociates as well.

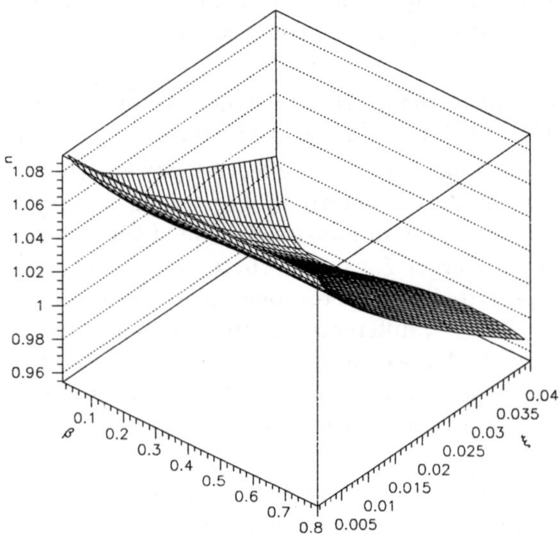

Figure 3: ξ and β dependence of $n(\xi,\beta) = \partial \ln F/\partial \ln \xi$ with the parameters fitted to the data of Ref. [12]. The departure from factorization, manifest in the β-dependence of n, tends to increase with increasing ξ.

References

1. G.Inglemann and P.Schlein, Phys.Lett. **B 152**, 256 (1985).
2. a) A.Donnachie and P.V.Landshoff, Phys. Lett. **B 191**, 309 (1987); b) K.Goulianos, Phys.Lett. **B 358** 379 (1995); c) R.Fiore, L.Jenkovszky and F.Paccanoni, Phys. Rev. **D 54** (1996); d) G.Ingelmann and K.Prytz, Z.Phys. **C 58**, 285 (1993).
3. A.Donnachie and P.V.Landshoff, Phys. Lett. **191**, 309 (1987); A.Donnachie and P.V.Landshoff, Nucl. Phys. **B 303**, 634 (1988).
4. H1 Collaboration, A Measurement and QCD Analysis of Diffractive Structure Function $F_2^{D(3)}$, Paper pa02-061, Submitted to the 28-th International Conference on HEP, Warsaw, Poland, July 1996.
5. E.A.Kuraev, L.N.Lipatov and V.S.Fadin, Sov. Phys. JETP, **45**, 199 (1977).
6. A.N.Vall, L.L.Jenkovszky and B.V.Struminsky, Sov.J.Particles and Nuclei, **19**, 77 (1988).
7. N.N.Nikolaev, W.Schäfer and B.G.Zakharov, Diffractive DIS: back to triple-Regge phenomenology? KFA-IKP(Th)-1996-06 preprint.
8. J.Ellis, Graham G.Ross, Phys.Lett. **B 384** 293 (1996).
9. P.Desgrolard, M.Giffon and E.Predazzi, Z.Phys. **C63** 241 (1994).
10. P.Desgrolard, M.Giffon, A.Lengyel, E.Martynov, Nuovo Cim. **107A** 637 (1994).
11. M.Glück, E.Reya and A.Vogt, Z.Phys. **C53** 651 (1992).
12. H1 Collab. Phys.Lett. **B 348** 681 (1995).

Medium effects in the production of hadronic resonances on nuclei (mass shift and collision broadening)

L.A. Kondratyuk

Institute of Theoretical and Experimental Physics, 117259 Moscow, Russia

Ye.S. Golubeva

Institute of Nuclear Research, 117312 Moscow, Russia

The influence of nuclear matter on the properties of hadronic resonance, formed or produced on nuclear target, is discussed. We consider collision broadening of baryon resonances in total photoabsorption cross sections on nuclei and medium effects in the production of ω and ρ resonances in pion-nucleus collisions.

1 Introduction

The in-medium properties of hadronic resonances has been widely investigated during the last years both experimentally and theoretically. Recent measurements of the total photoabsorption cross section on different nuclei in the resonance region, performed with tagged photon beams at Frascati and Mainz [1,2] have revealed an interesting result: the $D_{13}(1520)$ and $F_{15}(1680)$ resonances, clearly seen in the previous photoabsorption experiments on hydrogen and deuterium, desappear on nuclei. To explain this result (see ref. [3]) it was necessary to take into account the effect of collisional broadening, which is proportional to nuclear density and total resonance-nucleon interaction cross section. The measurements of dilepton spectra at the SPS in proton-nucleus and nucleus-nucleus collisions [4,5,6] have found the enhancement in $S + Au$ reactions compared to $p + Au$ collisions in the invariant mass range $0.3 \leq M \leq 0.7$ GeV which might be due to a shift of the ρ meson mass [7]. Actually, effective Lagrangian models [8,9] or approaches based on QCD sum rules [10,11] predicted that the masses of the vector mesons ρ, ω and ϕ should decrease with the nuclear density.

Therefore both experimental and theoretical results suggest that in the medium the mass and width of hadronic resonance should be different from their vacuum values. We discuss here the collisional broadening of baryon resonances in the total photoabsorption cross sections on nuclei and two-component structure of the dilepton ω and ρ peaks in pion-nucleus collisions.

2 Collisional broadening of baryon resonances in total photoabsorption cross section

The Frascati data on the total photoabsorption cross section on uranium (see ref. [1]) were analysed in ref. [3]. Proton and neutron cross sections have been considered separately as a sum of a smooth background plus eight Breit-Wigner resonances ($P_{33}(1232)$, $P_{11}(1440)$, $D_{13}(1520)$, $S_{11}(1535)$, $D_{15}(1675)$, $F_{15}(1680)$, $D_{33}(1700)$, $F_{37}(1950)$), which contribute to the total γN cross section below 1.2 GeV. The total nuclear photoabsorption cross section can be expressed through the imaginary part of the forward Compton scattering amplitude $\overline{\mathcal{I}}(\omega, \theta = 0)$ as

$$\frac{\sigma_{\gamma A}}{A} = -\frac{1}{2m\omega}\,\mathrm{Im}\,\overline{\mathcal{I}}(\omega, \theta = 0), \tag{1}$$

where ω is the photon energy, m is the nucleon mass and $\overline{\mathcal{I}} = \sum_i\left[(Z/A)\,\overline{\mathcal{I}}_{i,\gamma p} + (N/A)\,\overline{\mathcal{I}}_{i,\gamma n}\right]$. The elementary amplitude is given by a convolution over the nucleon momentum distribution $|\phi(\vec{p})|^2$ as

$$\overline{\mathcal{I}}_{\gamma N}(\omega, \theta = 0) = \int \frac{d\vec{p}}{(2\pi)^3}\,\frac{|\phi(\vec{p})|^2 g_{\gamma NN^*}^2}{[(\vec{p} + \vec{k})^2 - M_R^{*2} + iM_R^*\Gamma_R^*]}, \tag{2}$$

where $k = (\omega, \vec{k})$ and $p = (E, \vec{p})$ are the photon and nucleon four-momenta, and $g_{\gamma NN^*}$ is the vertex function for the $\gamma N \to N^*$ process. The effects of the medium are contained in the mass shift δM_R ($M_R^* = M_R + \delta M_R$) and in the width Γ_R^*,

$$\Gamma_R^* = \Gamma_R B_P + \delta\Gamma_R, \tag{3}$$

where B_P is the Pauli blocking factor and $\delta\Gamma_R$ the broadening due to collisions and interactions. The Pauli blocking factor takes into account the increased lifetime of the resonance due to the other nucleons occupying the momentum space below the Fermi momentum p_F. $\delta\Gamma_R$ and δM measure the broadening of the resonance due to the interaction with surrounding nucleons and can be written in the following form:

$$\delta M = -\frac{2\pi}{M_R}\,Re f_{NR}(0)\rho \tag{4}$$

and

$$\delta\Gamma_R = \frac{4\pi}{M_R}\,Im f_{NR}(0)\rho, \tag{5}$$

where $f_{NR}(0)$ is the forward NR scattering amplitude and ρ is the nuclear density.

Using the optical theorem, the broadening $\delta\Gamma_R$ can be written as:

$$\delta\Gamma_R = \rho\sigma_{(NR)}v\gamma, \tag{6}$$

where $\sigma_{(NR)}$ is the total cross section for the NR interaction, γ is the Lorentz factor and v the propagation velocity.

A fit to the photofission cross section for ^{238}U in ref. [3] was done assuming $\delta M = 0$ for all resonances except for the Δ and a fixed broadening $\delta\Gamma_R$ of $P_{11}(1440)$, $S_{11}(1535)$, $D_{15}(1675)$ and $F_{37}(1950)$, which corresponds to $\sigma_{(RN)}v=40$ mb. In fact these parameters do not contribute significantly to the χ^2.

The data were described with broadening 75, 315, 320 and 335 MeV for $P_{33}(1232)$, $D_{13}(1520)$, $D_{15}(1675)$ and $D_{33}(1700)$ respectively. Those values of $\delta\Gamma_R$ give the following estimates for the total RN cross sections $v\sigma_{tot}=21$ mb for the $P_{33}(1232)$ at $p_{lab}=0.34$ GeV/c, $v\sigma_{tot}=82$ mb for the $D_{13}(1520)$ at $p_{lab}=0.76$ GeV/c, $v\sigma_{tot}=80$ mb for the $P_{15}(1680)$ at $p_{lab}=1.03$ GeV/c, $v\sigma_{tot}=83$ mb for the $D_{33}(1700)$ at $p_{lab}=1.07$ GeV/c.
These values of RN cross sections do not violate the unitary limits and the magnitude of the cross section for ΔN interaction is consistent with the experimental data on the time reversed reaction $NN \rightarrow N\Delta$.

A density dependence of collisional broadening in $\sigma_{(tot)}(\gamma A)$ was investigated in paper [12] where the Frascati data [13] on ^{12}C, ^{64}Cu and ^{208}Pb were analysed.. The values of $\sigma_{(RN)}v$ for $P_{33}(1232)$, $D_{13}(1520)$, $F_{15}(1680)$ and $D_{33}(1700)$ were taken from ref. [3]. A constant $\rho = \rho_0$ has been assumed, where ρ_0 is the normalization constant of the usual Woods-Saxon form for $\rho(r)$. The Fermi smearing and the Pauli blocking factor also depend on the nucleus through p_F, which is 0.221 GeV for ^{12}C, 0.240 GeV for ^{64}Cu and 0.265 GeV for ^{208}Pb. Finally, some dependence of the mass shift δM of the Δ on the number of nucleons has to be expected, as already found in electron scattering [14].

The results are shown in Fig. 1. While heavier nuclei keep no trace of the second and third peak of the free-nucleon cross section, in the case of ^{12}C it is possible to see a little broad bump corresponding to the second free-nucleon peak in both the theoretical curve and the data. On the other hand, the Δ peak is well pronounced mainly due to the rather strong Pauli blocking effect. Nevertheless, it is clear that also the Δ peak is damped for heavier nuclei, where the Fermi smearing and collision broadening are more important. In agreement with data, the area under the resonance curve is unchanged, but the width increases with A. For ^{12}C we have $\delta\Gamma_R = 51$ MeV for the Δ, while it is between 218 MeV and 235 MeV for the other resonances ($D_{13}(1520)$, $F_{15}(1680)$, $D_{33}(1700)$). Similarly, $\delta\Gamma_R = 67$ MeV for the Δ and between 280 MeV and 305 MeV for the other resonances in the case of ^{63}Cu and $\delta\Gamma_R = 74$ MeV for the Δ and between 315 MeV and 340 MeV for the others in the case of ^{208}Pb. Moreover, the Δ-mass shift δM_R shows slight dependence on A. Data on ^{12}C are well described by $\delta M_R = 12$ MeV, while $\delta M_R = 15$ MeV for ^{63}Cu and $\delta M_R = 18$ MeV for ^{208}Pb. These values are comparable with those extracted in ref. [13] from experimental data, which can be interpolated by a curve linearly dependent on ρ.

3 Medium effects in the production of ω- and ρ-resonances in pion-nucleus interactions

3.1 Two-component formula for the invariant mass distribution in ω- and ρ decays

In previous section we considered the formation of baryon resonances by photons on nuclei, where the form of the resonance peak can be investigated looking for the energy dependence of the total cross section. Different situation happens when we consider the resonance production where the form of the resonance peak can be investigated by the measuring of the invariant mass spectrum of resonance decay products. An important point is that the resonance production takes place on finite nuclei and its decay can occur inside or outside nucleus. Therefore the invariant mass distribution of resonance decay will contain two components : a broad one which corresponds to its decay inside a nucleus and a narrow one which corresponds to its decay outside. In previous case, considered in section 2, the *in* component was only important, because the probability for baryon resonance to decay outside nucleus was negligibly small. Another situation takes place when we consider the production of ω resonance on nuclei. Essential fraction of ω mesons will decay outside nucleus. In this case both components should be taken into account.

Two-component formula for the resonance invariant mass distribution was derived in papers[15,16] for the case of coherent photoproducton of vector mesons on nuclei. In refs. [17,18] it was generalised to the incoherent production of hadronic resonances on nuclei.

Let us consider the production of the vector mesons ρ and ω in the nuclear target and their decay into the e^+e^- pair:

$$\pi^- A \to RX \to e^+e^- X. \tag{7}$$

The vector mesons ρ, ω are produced in the first hard pion-nucleon collision in the target and propagate through the rest of the nucleus. If the wave length of the produced meson is much smaller than the nuclear radius, it is possible to use the eikonal approximation for the Green function describing its propagation from the point $\vec{r} = (\vec{b}, z)$ to $\vec{r}' = (\vec{b}', z')$ [15,16],

$$G_k(\vec{b}', z'; \vec{b}, z) = \frac{1}{2ik} exp\{i \int_z^{z'} [k + \frac{1}{2k}(\Delta + 4\pi f(0)\rho(\vec{b}, \zeta)]d\zeta\} \tag{8}$$

$$\times \delta(\vec{b} - \vec{b}')\theta(z' - z),$$

where the z-axis is directed along the resonance momentum $\vec{k}$, $\vec{b}$ is the impact parameter, $f(0)$ is the resonance–nucleon forward scattering amplitude, ρ is the nuclear density and Δ is the inverse resonance propagator

$$\Delta = P^2 - M_R^2 + iM_R\Gamma_R \tag{9}$$

with M_R, Γ_R and P being the mass, width and four momentum of the resonance. The latter can be defined through the four momenta of its decay products

$$P = p_1 + p_2 + ... \tag{10}$$

If the resonance R is created at $(\vec{b}, z)$ and decays in $(\vec{b}', z')$, the exponent of the Green function (1) can be separated into the contributions from two regions

1) $z \leq z' \leq z_s = \sqrt{R^2 - \vec{b}^2}$ and
2) $z' \geq z_s = \sqrt{R^2 - \vec{b}^2}$.

When the resonance decays inside the nucleus of radius R, only the first part contributes and the inverse resonance propagator has the form

$$\Delta^* = \Delta + 4\pi f(0)\rho_0 = P^2 - M_R^{*2} + iM_R^*\Gamma_R^* \tag{11}$$

where M_R^* and Γ_R^* are defined in previous section.

If the resonance decays outside the nucleus, both regions contribute and the probability for lepton pair production with total momentum $\vec{P}$ and invariant mass squared P^2 can be written as

$$|M(\vec{P}, P^2)|^2 = N|f_{\pi-N\to RX}|^2 \int d^2\vec{b}dz\, \rho(\vec{b}, z)|A_{in}(\vec{P}, P^2; \vec{b}, z) + A_{out}(\vec{P}, P^2; \vec{b}, z)|^2. \tag{12}$$

In Eq. (12) the contributions from the first and second part can be written as

$$A_{in}(\vec{P}, P^2; \vec{b}, z) = \frac{1 - exp[i(\Delta^*/2k)(z_s - z)]}{\Delta^*}, \tag{13}$$

and

$$A_{out} = \frac{exp[i(\Delta^*/2k)(z_s - z)]}{\Delta}, \tag{14}$$

where $f_{\pi - N \to RX}$ is the amplitude for the resonance production on a nucleon in the channel RX and N is the normalization factor which contains the branching ratio $R \to e^+ e^-$.

When the resonance decays inside the nucleus, its form is described by the Breit–Wigner formula with Δ^* from (11), that contains the effects of collisional broadening and a shift of the meson mass.

The sign of the mass shift depends on the sign of the real part of the forward RN scattering amplitude which, in principle, also depends on the momentum of the resonance. For example, at low energy various authors [8,9,10,11] predict a decreasing mass of the vector mesons ρ, ω and ϕ with the nucleon density, whereas Eletsky and Ioffe have argued recently [19] that the ρ might become heavier in nuclear matter at energies of 2-7 GeV.

3.2 Dilepton spectra from ω- and ρ-peaks in pion-nucleus interactions

Medium effects in the production of ω- and ρ-resonances in pion-nucleus interaction were analysed in ref.[20], where the reaction $\pi^- A \to VX \to e^+ e^- X$ for Pb-targets at a pion momentum of 1.3 GeV/c was considered. The yields of ω – and ρ– mesons were calculated within the framework of the intranuclear cascade model developed in Ref. [21]. The nucleus was devided into series of concentric zones which helped to follow the propagation of each produced particle from one zone to another. The decay of the resonances to $e^+ e^-$ with their actual spectral shape was performed as in Refs. [17,18]. The calculation proceeded as follows: when the resonance decayd into dileptons inside the nucleus its mass was generated according to the Breit–Wigner distribution with $M_R^* = M_R + \delta M_R$ and $\Gamma_R^* = \Gamma_R + \delta \Gamma_R$, where the collisional broadening and the mass shift were calculated according to the local nuclear density. Its decay to dileptons was recorded as a function of the corresponding invariant mass bin and the local density. If the resonance left the nucleus, its spectral function automatically coincided with the free distribution because δM_R and $\delta \Gamma_R$ were zero in this case.

In Fig. 2a, taken from ref.[20], we present the $e^+ e^-$ mass spectrum from ω-decays, when collisional broadening was neglected, but the medium dependent mass shift as suggested by Hatsuda and Lee [10]

$$M_R^* = M_R(1 - 0.18\rho_B(r)/\rho_0), \tag{15}$$

was taken into account. Here $\rho_B(r)$ is the nuclear density at the resonance decay and $\rho_0 = 0.16 fm^{-3}$. Fig. 2a describes the full mass distribution whereas

Fig. 2b includes a cut in the ω momentum for $p_{lab} \leq 0.4 GeV/c$. The shaded histograms describe the contributions of the 'in' components, while the solid histograms are the sum of the 'in' and 'out' components. Due to sizeable surface effects (with lower nucleon density) the mass distribution of the 'in' components is spread out. Nevertheless, we still observe two sharp peaks corresponding to the in-medium and vacuum masses, respectively. As expected from Fig. 1, a cut in the momentum distribution for $p_{lab} \leq 0.4$ GeV/c enhances the contribution from the 'in' component relative to the 'out' component (Fig. 2b).

In Fig. 2c we show the dilepton spectrum from ω decays when a mass shift and collisional broadening are taken into account.Again the relative contribution of the 'in' component (shaded histograms) in the e^+e^- invariant mass spectrum considerably increases when ω-mesons with momenta less than 0.4 GeV/c are selected (Fig. 2d). Due to collisional broadening the width of the 'in' component increases substantially relative to the free ω width such that a pronounced peak for the 'in' component can no longer be observed.

In Fig. 3 and 4, taken from ref.[20], we show the dilepton spectra from ρ and ω decays for $\pi^- + $ Pb at 1.3 GeV/c compared with the different sources of background. The dominant background processes are seen to result from pion-nucleon (πN) bremsstrahlung, the η Dalitz decay (denoted by η; solid line) and the ω Dalitz decay ($\omega \rightarrow \pi^0 e^+ e^-$). The Δ Dalitz decay (Δ) as well as proton-neutron (pn) bremsstrahlung are of minor importance. In Fig.3 no mass shifts of the ω and ρ mesons in the medium were taken into account, however collisional broadening was included. In Fig.4 both these effects were taken into account. The upper solid curves in Fig. 3 and 4 represent the sum of all contributions. We note that the width of the ρ–peak is about 220 MeV, which is essentially larger than its vacuum width $\Gamma_\rho \approx 150$ MeV due to collisional broadening as described above. Nevertheless, above about 0.65 GeV of invariant mass M the spectrum is dominated by the vector meson decays with a low background.

The background processes dominate the dilepton spectrum for $M \leq 0.6$ GeV. When comparing the total spectrum from Fig. 3 with that from Fig. 4 we find an enhancement by a factor of 2 for $0.65 \leq M \leq 0.75$ GeV for the dropping vector meson masses as well a decrease of the spectrum for $M \geq 0.85$ GeV due to the shifted ρ mass because the ρ almost completely decays inside the Pb-nucleus.

A mass shift of the ρ and ω mesons can be seen experimentally at GSI (Darmstadt)[22] by an enhanced yield in the mass region $0.65 \leq M \leq 0.75$ GeV and a mass shift of the ρ meson especially for $M \geq 0.85$ GeV because the ρ almost completely decays inside a Pb-nucleus. The in-medium modifications of the ω mesons are most pronounced for small momentum cuts on the e^+e^- pair in the lab. system.

We can conclude that :
i) recent data on the total photoabsorption cross sections on nuclei in resonance region demonstrate clearly the importance of collision broadening of baryon resonances in nuclear medium and give the evidence of the Δ mass shift depending on nuclear density;
ii)medium effects are quite pronounced in the production of vector mesons (ρ and ω) in pion-nucleus collisions at intermediate energies, which gives a good chance to detect them at GSI.

The authors acknowledge many helpful discussions with S.Boffi, W.Cassing and A.Iljinov.

References

1. N.Bianchi et al., Phys. Lett. **B 299** (1993) 219; M.Anghinolfi et al., Phys. Rev.**C 47** (1993) R922; N.Bianchi et al., Phys. Lett. **B 309** (1993) 5; N.Bianchi et al., Phys. Lett. **B 325** (1994) 333

2. Th.Frommohold et al. Phys. Lett. **B 295** (1992) 28

3. L.A. Kondratyuk, M.I. Krivorushenko, N. Bianchi et al, , Nucl.Phys.**A579** (1994) 453

4. G. Agakichiev et al., Phys. Rev. Lett. 75 (1995) 1272.

5. M.A. Mazzoni, Nucl. Phys. A566 (1994) 95c.

6. T. Akesson et al., Z. Phys. C68 (1995) 47.

7. G.Q. Li, C.M. Ko, and G.E. Brown, Phys. Rev. Lett. 75 (1995) 4007.

8. G. Brown and M. Rho, Phys. Rev. Lett. 66 (1991) 2720.

9. C.M. Shakin and W.-D. Sun, Phys. Rev. C 49 (1994) 1185.

10. T. Hatsuda and S. Lee, Phys. Rev. C 46 (1992) R34.

11. M. Asakawa and C.M. Ko, Phys. Rev. C 48 (1993) R526.

12. S.Boffi, Ye.Golubeva, L.Kondratyuk et al. Yad.Fiz. 60 (1997) No.4.

13. N. Bianchi et al., Phys. Rev. C54 (1996) 1688

14. R. M. Sealock et al., Phys. Rev. Lett. 62 (1989) 1350.

15. K.G. Boreskov, J. Koch, L.A. Kondratyuk and M.I. Krivoruchenko, Proc. of the 3rd International Conference on Nucleon-Antinucleon Physics (NAN'95) Yad. Fiz. (Phys. of Atomic Nucl.) 59 (1996) 1908.

16. K. Boreskov, L. A. Kondratyuk, M.Krivoruchenko and J.Koch., nucl-th/97010331; Nucl.Phys. A (1997) to be published

17. Ye.S. Golubeva, A.S. Iljinov and L.A. Kondratyuk, Proc. of the 3rd International Conference on Nucleon-Antinucleon Physics (NAN'95),

Yad.Fiz.(Phys. of Atomic Nuclei) 59 (1996) 1894.

18. Ye.S. Golubeva, A.S. Iljinov and L.A. Kondratyuk, Proc. of the 3rd International Conference "Mesons-96", Cracow, May 1996, to be published by World Scientific, Singapore, 1997.
19. V. Eletsky and B.L. Ioffe, Preprint KFA-IKP(TH)-1996-10.
20. W. Cassing, Ye.S. Golubeva, A.S. Iljinov and L.A. Kondratyuk, nucl-th/9612026; Phys.Lett. B (1997) to be published
21. Ye.S. Golubeva, A.S. Iljinov, B.V. Krippa and I.A. Pshenichnov, Nucl. Phys. A 537 (1992) 393.
22. W.Schoen, H.Bokemeyer, W.Koenig and V.Metag, Acta Physica Polonica B27 (1996) 2959.

Figure captions

Fig. 1: The total photoabsorption cross section for ^{12}C, ^{63}Cu and ^{208}Pb. Data from [13].

Fig. 2: The e^+e^- mass spectrum from ω-decay when including a medium dependent ω-mass shift according to Hatsuda and Lee ([10]): $a)$ and $b)$ - without collisional broadening; $c)$ and $d)$ - with collisional broadening. Figs. $a)$ and $c)$ describe the full mass distributions, $b)$ and $d)$ events with a ω momentum cut $p_{lab} \leq 0.4 GeV/c$. The hatched histograms describe the contribution of the 'in' component, while the solid histograms display the total contribution from the 'in' and 'out' components, respectively.

Fig. 3: The dilepton spectra for $\pi^- +$ Pb reactions at 1.3 GeV/c in logarithmic representation when a mass shift of the vector mesons is neglected and collisional broadening according to Eq. (6) is taken into account. The upper solid curve is the sum of all contributions; the individual channels correspond to the η Dalitz decay (η), pion-nucleon bremsstrahlung (πN), the ω Dalitz decay ($\omega \to \pi^0 e^+e^-$), the Δ Dalitz decay (Δ), proton-neutron bremsstrahlung (pn), and the direct decays of the vector mesons ρ and ω. The mass resolution adopted is $\Delta M = 10$ MeV.

Fig. 4: The dilepton spectra for $\pi^- +$ Pb reactions at 1.3 GeV/c in logarithmic representation when including a mass shift of the vector mesons according to Eq. (15) and collisional broadening according to Eq. (6). The notations are the same as in Fig. 3.

cross section (μb)
400
300
200
12C
400
300
200
63Cu
400
300
200
208Pb
0.2
0.4
0.6
0.8
1
1.2
E (GeV)

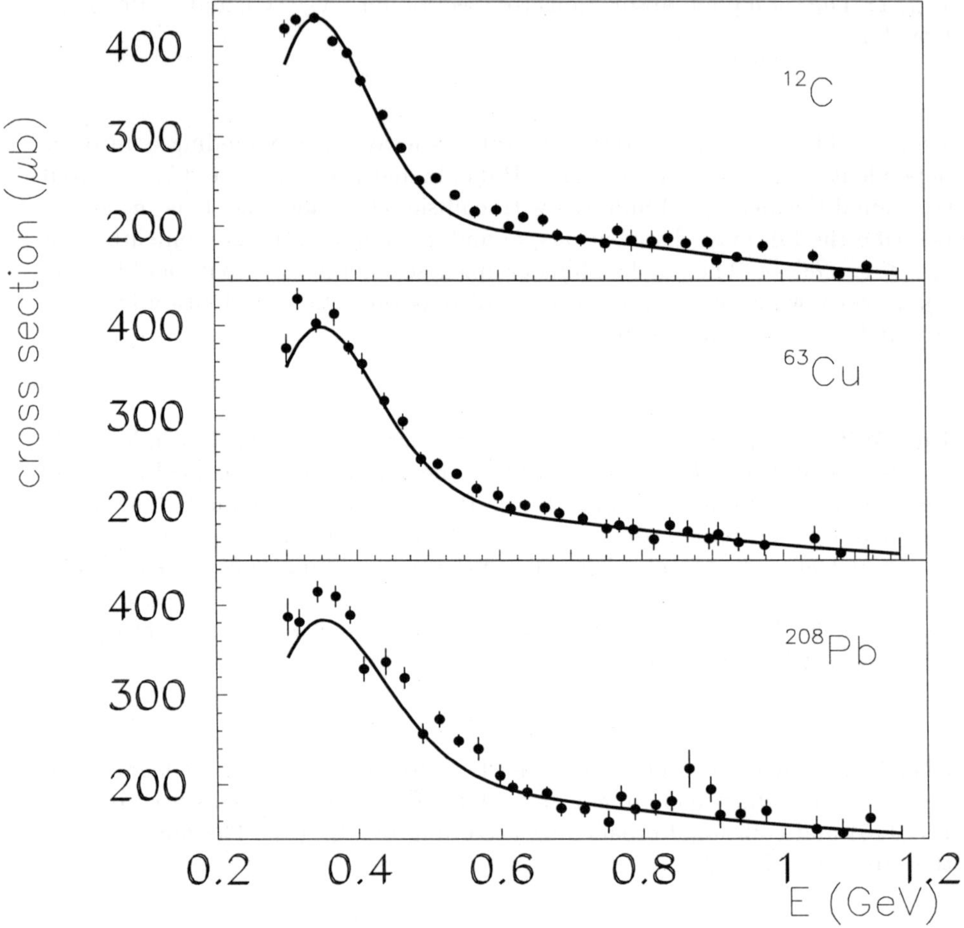

Fig. 1

π⁻ + Pb, 1.3 GeV/c
a)
300
250
200
150
100
50
0
events
0.6
0.7
0.8
M (GeV)
b)
p_lab < 0.4 GeV/c
40
30
20
10
0
events
0.6
0.7
0.8
M (GeV)
c)
300
250
200
150
100
50
0
events
0.6
0.7
0.8
M (GeV)
d)
p_lab < 0.4 GeV/c
50
40
30
20
10
0
events
0.6
0.7
0.8
M (GeV)

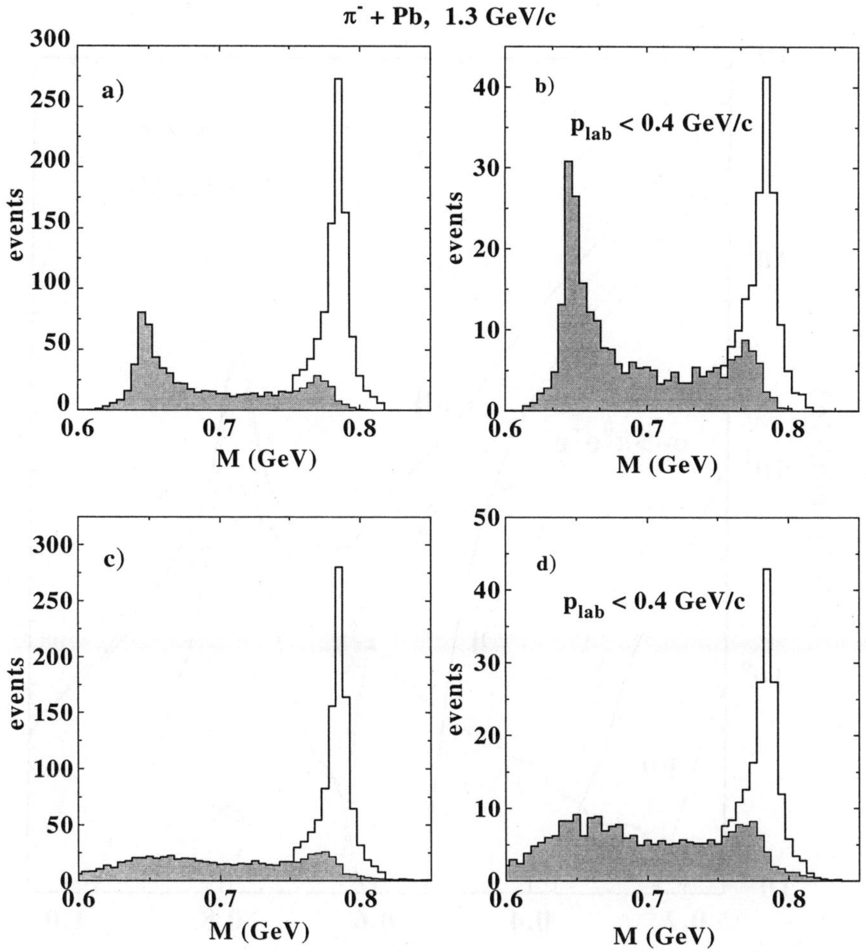

Fig. 2

10^3
10^2
10^1
10^0
10^{-1}
dσ/dM (μb/GeV)
$\pi^- + $ Pb
1.3 GeV/c
all
$\omega \to \pi^0 e^+ e^-$
ω
ρ
Δ
η
pn
πN
0.2
0.4
0.6
0.8
1.0
M (GeV)

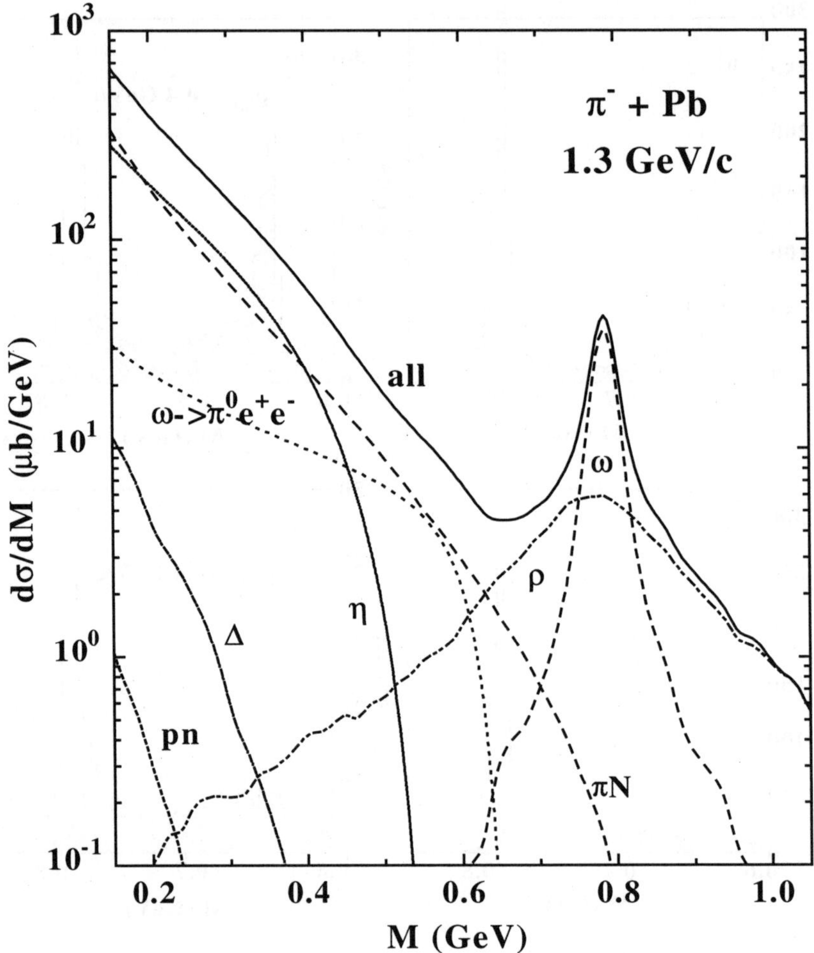

Fig. 3

π⁻ + Pb
1.3 GeV/c
dσ/dM (μb/GeV)
all
ω->π^0e$^+$e$^-$
η
ρ
ω
Δ
pn
πN
M (GeV)

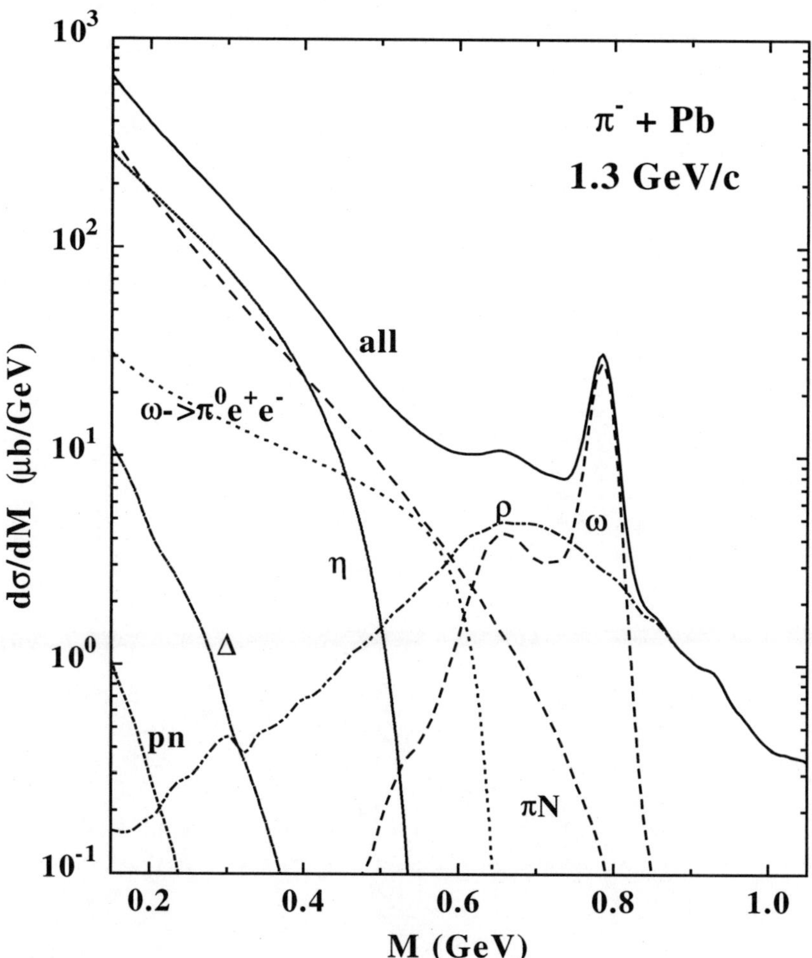

Fig. 4

Fig. 4